AF553632

ENCYCLOPAEDIA OF FISH AND FISHERIES

Vol. I

FISH BREEDING

By

Dr. Arvind N. Shukla

School of Studies of Zoology & Biotechnology

Vikram University

Ujjain

DISCOVERY PUBLISHING HOUSE PVT. LTD.

NEW DELHI-110 002

First Published-2008

ISBN 978-81-8356-303-1 (Set)

Published by:

DISCOVERY PUBLISHING HOUSE PVT. LTD.
4831/24, Ansari Road, Prahlad Street,
Darya Ganj, New Delhi-110002 (India)
Phone: 23279245 • Fax: 91-11-23253475
E-mail: dphbooks@rediffmail.com
dphtemp@indiatimes.com

Printed at:

Sachin Printers, Delhi

Preface

The present title *"Encyclopaedia of Fish and Fisheries"* has been designed for undergraduate and postgraduate students of all Indian Universities. The text of the present title is organized on some major areas like fishing techniques, fish behaviour, and fish physiology, fisheries of important Indian forms. Fishes inhabit every kind of aquatic environment, and their wide distribution resulted in many different designs for their special mode of life. The present title, with a skilful mix of breadth combined with detail, reviews what is known about fishes. Biological inter-relationship are fully discussed, and fishes with features of special interest are highlighted, whether they are economically significant or not. The present text brings together many scattered observations as well as the results of recent investigations. It is hoped that the general biologist, the zoologist and the Icthyologist will find wealth of exciting information in it.

The present title gives a comprehensive overview of the fishery science and also offers diverse information on the subject of fisheries to make the readers up-to-date with the latest in this field where proliferation of scientific information has not only been fast but enormous too in the last decade. Written in scholarly yet easy to understand language. It combines the usefulness of a reference work with the readability of a browsing book, perfect for any one attuned to the splendor of our natural world. It is hoped that the book will prove indispensable to fisheries organizations, lecturers and the young zoologists engaged in teaching and research.

In the preparation of this book large number of books and research papers have been consulted. So no authenticity is claimed.

The author expresses his gratitude to Mr. Wasan and staff of M/s Discovery Publishing House for their whole hearted co-operation in the publication of this book.

The author tried hard to be accurate and upto date in statement and realises the impossibility of completely avoiding errors therefore, the author will greatly appreciate having his attention called to any questionable statement.

Author

Contents

1

INTRODUCTION

The fishes, like all the other vertebrates, reproduce sexually. In the vast majority of species, eggs and spermatozoa are formed in separate individuals (dioecious), and the gametes are expelled into the surrounding water where fertilisation takes place immediately and is promptly followed by the development of a new generation. Within this broad pattern there is an amazing array of curious modifications so that the fishes as a group exemplify almost every device known among sexually reproducing animals; indeed, they display some variations which may be unique in the animal kingdom.

Although most fishes are dioecious, hermaphroditism does occurparticularly among the cyclostomes and teleosts; in some species it is the normal way of life. Parthenogenesis has not been observed among fishes in nature but has been produced experimentally. Gynogenesis occurs in *Poecilia formosa* and perhaps also in some populations of *Carassius auratus*. These modes of reproduction are, however, unusual; and most of the 20,000 or more known species of fish have male and female organs in separate individuals.

At one extreme, the two sexes are externally indistinguishable. They swim together during the breeding season to discharge their genital products into the water without specialised mating behaviour. At the other extreme, marked differences in morphology and colouration are characteristic of the males and females and these play distinctive roles in elaborate presexual behaviour, courtship, mating, and parental care. A synchronised spawning without copulation is usual but copulation occurs at all evels in fish phylogeny; it sometimes merely insures close proximity of eggs and sperm discharged into the water but more frequently involves insemination of the females. Although internal fertilisation may be followed by the laying of newly fertilised eggs or the release of primitive larvae, viviparity is highly specialised in some elasmobranchs and teleosts with young born in advanced stages of development; in *Cymatogaster aggregata* the males may even be sexually mature at birth. Sexual activities are normally followed by the prompt fertilisation of eggs but sometimes sperm are stored in the female for prolonged periods prior to fertilisation; occasionally the fertilised eggs develop only after a period of diapause; sometimes development occurs but hatching is delayed during adverse environmental conditions.

Breder and Rosen have summarised these and many other curious specialisations in an extended chart; this tabulates information on secondary sex characters, mating, breeding, and parental behaviour for each of the many families of fishes.

There are many specialisations associated with reproduction in fishes but only a few of them are considered in this chapter. Breeding behaviour, fertilisation, and early development are considered in other chapters. The present account is confined to a discussion of the gonads and their associated ducts, the special physiology of viviparity, and the endocrinology of reproduction. The latter topic has been repeatedly reviewed, but the other two aspects of fish reproduction have not been recently summarised.

THE GONADS AND THEIR DUCTS

Embryology and Phylogeny

A knowledge of early embryology, as well as adult morphology, is essential to an understanding of several features of the comparative physiology of the reproductive system of fishes. In particular, the two very different patterns of gonadal ontogeny may account for the variable frequency of intersexuality in different groups while comparative studies of spawning, fertilisation, and gestation must be based on an understanding of the diversified anatomy of the gonads and their ducts.

The gonads of all vertebrates arise in the dorsolateral lining of the peritoneal cavity—one gonad on each side of the dorsal mesentery. Their development is intimately associated with that of the nephric system. In most of the vertebrates each gonad has a double origin, developing from two distinct but closely associated cellular proliferations. The more laterally located *cortex* or cortical portion arises as an elongated ridge of peritoneal wall and is destined to become an ovary. The *medulla* or medullary portion which is destined to form the testis arises from a more medial cellular proliferation which also produces

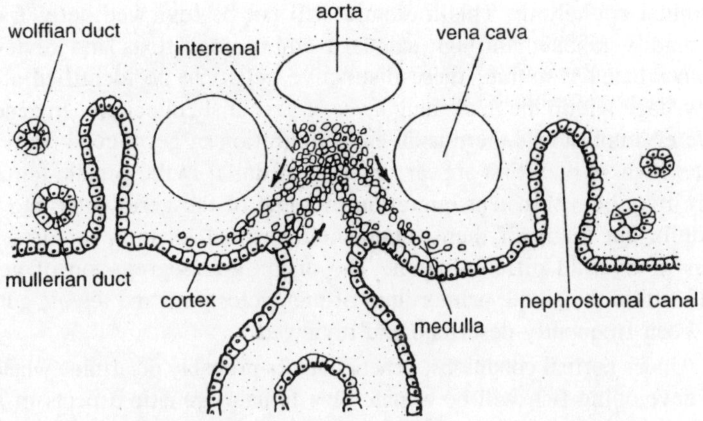

Figure 1.1 : Cross section through a 22-mm embryo dogfish, Scyliorhinus caniculus, to show the origins of the cortex and medulla of the gonad.

the adrenocortical tissue (interrenal or mesonephric blastema). Usually, one of these portions grows rapidly while the other fails to develop and the sex of the individual is thus determined at a very early stage. A section through the undifferentiated gonad of a 22-mm dogfish embryo.

This pattern of gonad differentiation from two different components is characteristic of the elasmobranchs and all of the tetrapods. In contrast, the gonads of cyclostomes and telcosts develop from single primordia.

In cyclostomes and teleosts the entire gonad develops directly in the peritoneal epithelium and corresponds to only the cortex of other vertebrates. There is evidently no contribution from the interrenal (mesonephric) blastema. Atz has summarised the literature and provided an extensive bibliography; D'Ancona did much of the early work on teleosts. It has been suggested that these differences in embryology may account for the more widespread occurrence of intersexuality among the cyclostomes and teleosts.

Whether the ontogeny of the gonad is by way of a single or a double primordium, distinctive cells (the primordial germ cells) which are destined to form gametes first appear within or migrate into the cortical portion of the gonad. They can usually be identified as conspicuously larger cells within this modified layer of proliferating mesothelium (germinal epithelium). There is a voluminous literature on the origin of the germ cells with a considerable body of evidence for a widespread development of these cells and a subsequent migration of them into the germinal epithelium. This literature will not be reviewed here; it can be readily reached through standard embryology texts and reviews. From whatever source, these distinctive cells can be identified at an early stage within the thickening layer of germinal epithelium. In genetic male elasmobranchs there is an early migration of germ cells from the cortex, where they first appear, into the medulla; in the genetic females there may be a transient migration of some of the germ cells into the medulla but many of them retain their cortical location to form the basis of ovarian differentiation. The origin and segregation of germ cells within the single primordium of the cyclostome and teleost gonad has been frequently described and reviewed.

Under normal conditions, genetic factors probably determine whether the developing fish will be a male or a female; genetic aspects of sex determination are considered in the chapter by Yamamoto, this volume. Earlier workers postulated male and female inductor substances or hormones which controlled the course of development. Witschi who carried out the pioneer work on amphibians referred to them as "medullarin" and "corticin," for male and female, respectively; D'Ancona working with the teleosts called these theoretical substances "androgenin" and "gynogenin." Reinboth does not consider evidence for the existence of these factors to be at all convincing; Atz agrees with Reinboth's conclusions. Whether or not there are special embryonic hormones concerned with the determination of sex, it is well established that the differentiation of the gonadal primordium can be readily modified with gonadal steroids similar to those found in adults. Androgens stimulate the development of testes, and estrogens promote ovarian differentiation. Findings are consistent and the literature has been frequently reviewed.

The gonads of vertebrates always originate from bilateral primordia, but many species as adults possess only one reproductive gland. In some of the fishes there is a fusion of the two primordia during development (as in the ovaries of lampreys), while in other cases one of the gonads fails to develop (as in the myxinoid ovary). The literature reveals a range of specialisations in all groups of fishes from complete fusions to partial fusions involving only the posterior portion of the gonads or just the gonoducts; sometimes one of the gonads is rudimentary or merely smaller but still present. Franchi *et al.* summarise the pertinent literature with specific examples.

The comparative anatomist has found some of his most interesting problems in the phylogeny of the gonoducts and their relationships to the mesonephric tubules and ducts. Gonoducts are absent in the

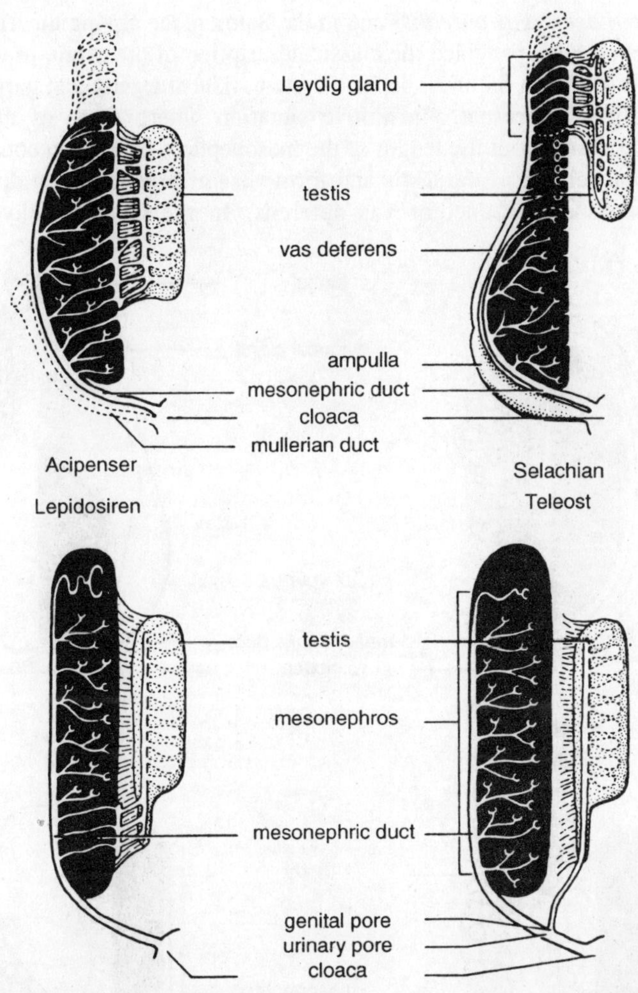

Figure 1.2 : Representative types of urinogenital systems in male fishes.

cyclostomes. Spermatozoa or ova are discharged from the surface of the gonad directly into the body cavity and then pass through pores into the urinary sinus or urinary duct; they are expelled through the cloaca or urinogenital papilla-depending on the anatomy of the urinogenital opening in the particular sex and species.

Gonoducts are present in all groups of the gnathostomes although they may be secondarily lost in some fishes (for example, in the Greenland shark, *Laemargus borealis*, and in the Salmonidae among the Teleostei, Balfour). Kerr provided the classic description of the origin of the male gonoducts from the mesonephric system. The sturgeon and garpike are thought to represent a primitive situation where some of the renal tubules throughout the length of the mesonephros have been conscripted into the service of the testis and form vasa efferentia which drain into the mesonephric duct or vas deferens. In the Chondrichthyes (and

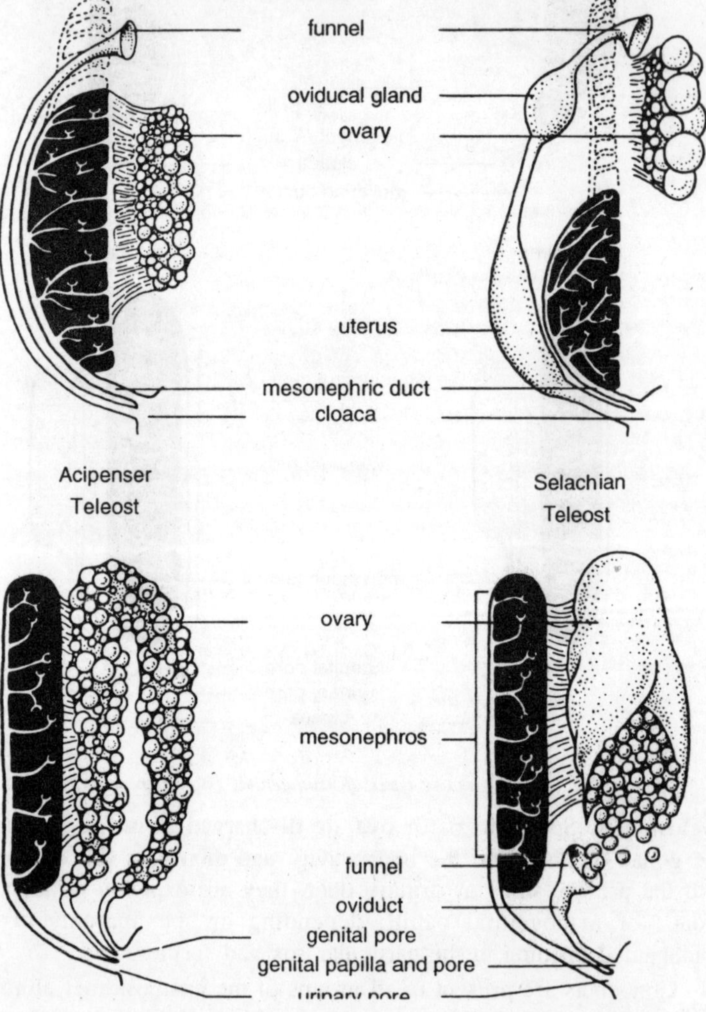

Figure 1.3 : Representative types of urinogenital systems in female fishes.

Amphibia), the' testis is thought to have taken over a group of anterior mesonephric tubules which cease to have any relationship with the excretory system; in *Lepidosiren*, the vasa efferentia are formed from some of the posterior mesonephric tubules. In *Polypterus* and the Teleostei there is no connection between the mesonephros and the gonad at maturity; and the vas deferens is quite separate from the ureter or mesonephric duct. It is generally assumed, however, that the main gonoduct has been derived from the mesonephric duct during phylogeny. More complete descriptions of this system will be found in Goodrich, Gerard and van den Broek.

In all of the vertebrates, except some of the more specialised fishes, the ova are discharged into the peritoneal cavity and find their way to the outside through oviducts (Miillerian ducts), which pass from open anterior funnels to the cloaca. In these groups with naked ovaries (*gymnovarian* condition) and open ovarian funnels, the genital ducts are derived as in the male from the mesonephric ducts although the evidence for this origin is completely lost in the land vertebrates.

In the Teleostei, the oviducts are posterior continuations of the ovarian tunic. The embryology of the ovary and its duct varies so that the ovary in some teleosts has a central ovarian cavity continuous with the oviduct while in others the oviducts are para-ovarian; in any case the oviducts are formed by the backward growth of the same peritoneal folds which enclose the ovary during its development (*cystovarian* condition). The gonoducts are also continuous with the ovaries in the holostean *Lepidosteus*, thus providing the exception to the rule that only teleosts fail to discharge their ova into the peritoneal cavity. Some of the teleosts are themselves exceptional in that they do release their eggs directly into the body cavity. In the Salmonidae, Galaxiidae, Hyodontidae, Notopteridae, Osteoglassidae, and the cyprinid *Misgurnus*, the oviducts degenerate in whole or in part so that the ova pass into the peritoneal cavity and thence through pores or funnels, depending on the degree of degeneration, to the exterior. In the Anguillidae the loss of gonoducts occurs in both males and females.

The Male

The Testis and Spermatogenesis

Spermatozoa are formed from the sperm mother cells or spermatogonia through a series of cytological stages collectively referred to as "spermatogenesis," This process involves a proliferation of spermatogonia through repeated mitotic divisions and growth to form primary spermatocytes; these then undergo reduction division to form secondary spermatocytes; the division of the secondary spermatocytes produces the spermatids which then metamorphose into the motile and potentially functional gametes-spermatozoa, spermia or sperm. This process of spermatid metamorphosis is often called "spermiogenesis." Details of the cytological changes are similar in all vertebrates as described in standard textbooks of histology and embryology. Physiologists are interested in the factorsboth environmental and hormonal-which trigger waves of spermatogenesis at different seasons and control the essential steps of meiosis (division of primary to secondary spermatocytes) and the metamorphosis of the spermatid with eventual release of mature sperm. In some species particularly the elasmobranchs and viviparous teleosts-sperm production involves the packaging of sperm into sperm balls or spermatophores which are transferred to the female.

Spermatogenesis occurs within testicular units which may take the form of small sacs, ampullae, lobules, or tubules; in many groups of fishes these differ radically from the familiar seminiferous tubules of the mammalian testis. In the cyclostome, spermatogenesis occurs within small bladders, follicles, or ampullae. These are separated by a delicate connective tissue; a number of units may be grouped together and bounded by somewhat thicker connective tissue to form lobules. Spermatogenesis is almost synchronous throughout the many follicles and just prior to spawning the follicles filled with mature sperm rupture to release their contents into the body cavity. Walvig summarises the cytological details of spermatogenesis in *Myxine*.

In elasmobranchs, spermatogenesis occurs within a mass of ampullae arranged in a manner which seems to be unique among the vertebrates. The testis of the basking shark, *Cetorhinus maximus*, carefully described by L. H. Matthews, *is* divided by connective tissue trabeculae into many lobules each of which corresponds to the entire testis of the dogfish, *Scylliorhinus canicula*, as described by Fratini and Mellinger. The spermatogenetic units, usually called "ampullae," are proliferated from a mesoventral area of the testis referred to as the "tubulogenic zone." Within this zone, nests of cells-somewhat like primary ovarian follicles-arise and proliferate to form small tubules or ampullae which gradually shift toward the dorsal side of the organ while spermatogenesis occurs within them. By the time the ampullae reach the dorsal surface of the testis, the sperm (a constant number in each ampulla) are ready for discharge into the efferent ducts which emerge from the testis at this point. At this stage, the Sertoli cells surround the ampullae and are clearly associated with packets of sperm. According to L. H. Matthews, the ampullae shrink after sperm are discharged into these collecting tubules and the Sertoli cells are resorbed. It is of interest that all of the gonocytes within any one ampulla are in the same stage of spermatogenesis and that within the testis, distinct zones are evident from ventral to dorsal surface with all the tubules of a particular zone in a similar stage of development. Thus, in studies of the pituitary regulation of spermatogenesis, Dodd and his colleagues were readily able to spot a distinct zone of degeneration in the primary spermatocytes when it appeared following hypophysectomy. The cytology of the elasmobranch testis, including spermatogenesis, the development of the Sertoli cells.

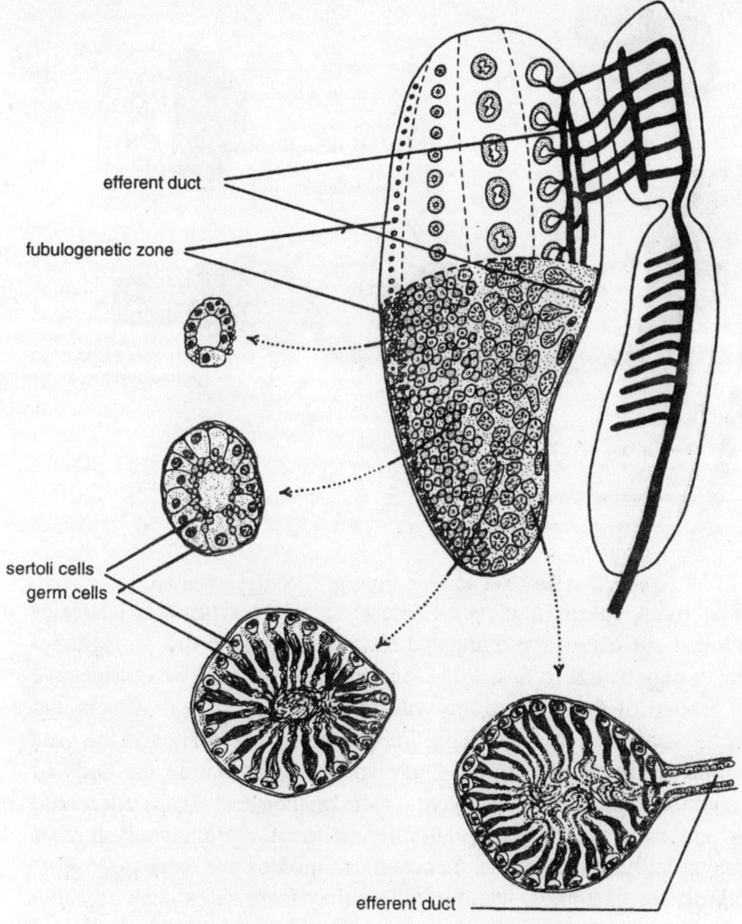

Figure 1.4 : Structure of the chondrichthyan testis.

There are now good descriptions of the testicular histology of several species of teleosts. Among the early papers, the following are particularly helpful: C. L. Turner's description of the spermary of the perch, Craig-Bennett's account of the stickleback, S. A. Matthews' report on *Fundulus*, and Cooper's study of the crappies. Many other investigations are cited in the bibliographies of these papers and in the reviews by Hoar and Dodd. More recent descriptions are available for the minnow *Couesius*, the rockfish *Sebastodes*, the sea perch *Cymatogaster*, and the guppy *Poecilia*. Testes of different species vary in complexity; the brief description which follows is a generalised one.

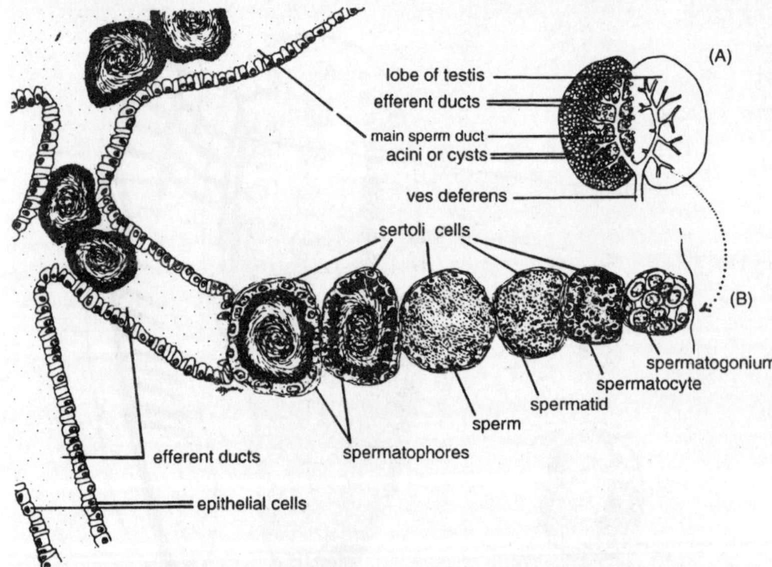

Figure 1.5 : Structure of testis of guppy, Poecilia reticulata.

The main sperm duct (vas deferens) arises from the posterior mesodorsal surface of the elongated testis and leads to the urinogenital papilla. It may be traced anteriorly for a variable distance in a connective tissue groove of the testis along with the spermatic blood vessels and nerves. In many teleosts, the paired testes fuse posteriorly and the vasa deferentia are combined into a single sperm duct. Within the body of the testis, the main sperm ducts give rise to smaller ducts (vasa efferentia) which penetrate ventrally and laterally to form a drainage system of variable complexity. In some species these tubules are extremely short (poeciliids, for example), while in others they form an extensive system of seminiferous tubules which can be followed almost to the periphery of the organ (*Fundulus*, the rockfishes, and the cottids). Testes of the poeciliid type are sometimes referred to as "acinar" while those with the extensive duct systems are called "tubular." This difference is one of degree rather than kind.

It is to be noted that the seminiferous tubules of the teleost-in contrast to those of the higher vertebrates-lack a permanent germinal epithelium. Whether the testis is acinar or tubular, nests of spermatogonia proliferate from the resting germ cells near the margin of the organ. In the acinar type, these nests of cells or cysts undergo the various stages of maturation as they are displaced toward the sperm ducts into which they eventually discharge their contents. In the tubular testis, the resting germ cells are particularly evident and packed together at the blind ends of the tubules near the periphery, but many of them migrate or are displaced along the walls of the tubules. In active spermatogenesis, nests of spermatogonia proliferate both from the ends of the tubules and from the resting germ cells along their walls. Thus, at the end of spermiogenesis, the seminiferous tubules are packed with sperm as the masses of gametes from a multitude of matured cysts combine within the tubules. During maturation all of the cells within one of the cysts are in approximately the same stage of development; the degree of synchrony among the many cysts varies in different species.

The Endocrine and Supporting Tissues of the Testis

Current histological and histochemical techniques have now resolved a long-standing uncertainty about the tissues responsible for the production of the testicular androgens. Well-vascularised clusters of cells similar to those described by Leydig many years ago in the mammalian testis have been identified between the seminiferous alveoli and tubules of many different fishes. In routine H and E sections, these large cells with spherical or oval nuclei often appear vacuolated because of the removal of lipoidal substances. Their endocrine nature was first postulated from the marked seasonal proliferation which occurs just prior to the breeding season in the stickleback *Gasterosteus*; in this same species Gottfried and van Mullem have recently shown a convincing correlation between the histological development of the tissue and its biochemical content of androgen. Nests of typical Leydig cells or interstitial tissue have now been identified in the cyclostomes, in all groups of elasmobranchs, in the lungfishes and the coelacanth *Latimeria* as well as the testes of many teleosts (*Gasterosteus Tilapia*, *Tinca*, *Solea*, *Lebistes*, *Cymatogaster*, *Oncorhynchus*, and others). Staining of the cytoplasmic droplets with sudanophilic colouring agents and the demonstration of steroid dehydrogenases leaves little doubt of their role in steroidogenesis.

In some species of teleosts, however, the hormone-producing cells are located in the walls (basement membrane) of the seminiferous tubules. Marshall and Lofts who first noted this difference in their studies on the pike, *Esox lucius*, and the char, *Salvelinus willughbii*, referred to this tissue as the "lobule boundary cells." The difference between interstitial and lobule boundary cells is largely one of distribution since both tissues arise from the same source and are similar histochemically. It may be of interest to note that the Urodela also possess lobule boundary cells, in contrast to the Anura and all other groups of tetrapods.

Sertoli cells are also prominent in the testes of all groups of fishes. The spermatogenetic units-whether cysts, ampullae, or tubules-are bounded by a thin layer of connective tissue (the basement membrane) and contain two types of cells; one of these is the gonocyte giving rise to the several generations of spermatogenetic cells, while the other is the Sertoli or supporting cell believed to play a nutritive role during spermiogenesis. The process whereby spermatids become embedded in the centripetal end of the Sertoli cells to undergo metamorphosis is well described in the higher vertebrates. Cytological details probably vary in the different species, but in all cases the spermatids become intimately associated with these nurse cells and presumably draw nourishment from them during transformation. The fully developed sperm are attached to the Sertoli cells prior to spermiation; this association can be very nicely seen in fishes such as the sharks, the poeciliids, and the embiotocids which form spermatophores. In these groups, where all of the sperm in a unit mature at the same time, the Sertoli cells form an almost complete layer just inside the basement membrane; in many other fishes the situation is similar to that of the higher vertebrates where

Sertoli cells are interspersed at intervals along the basement membrane with the groups of spermatogonia between them. L. H. Matthews and Fratini have detailed the cytogenesis of the Sertoli cell system in the basking shark and the dogfish. Pandey describes Sertoli cell development in the guppy.

In addition to their nutritive role, the Sertoli cells may be concerned with the phagocytosis of unused sperm and involved in hormone production. The presence of hydroxysteroid dehydrogenases has been demonstrated in the Sertoli cells of dogfishes and in the surfperch. In summary, the cytology of the Sertoli cell suggests three functions: nutritive, phagocytic, and hormonal; in this and in their embryology they seem very similar to the granulosa cells of the ovarian follicle which are described later.

Secretions of the Sperm Ducts and Male Secondary Sex Characters

Following spermatogenesis, the mature sperm are made ready for discharge or spermiation. The glandular epithelial lining of the sperm duct probably always contributes to the seminal discharge but there seems to be no systematic study of this. However, a wide variation is recognised-at one extreme, the formation of specialised packets of sperm (spermatophores) and, at the other, a mere thinning of the semen through the hydration of the testis and accumulation of fluid in the testicular passages.

L. H. Matthews has detailed the formation of the complex spermatophores of the basking shark, *Cetorhinus maximus*, where these sperm packets range in size from a few millimeters up to 25 or 30 mm and have a cortex of translucent hyaline material surrounding a central mass of opaque white sperm; the sperm mass may be 10 mm in diameter. The general process, as described by L. H. Matthews, seems to be representative of the elasmobranchs although the structure of the spermatophores varies in different species from simple sperm aggregations to the hyaline packets of the basking shark. In the elasmobranchs, sperm released from the ampullae into the efferent canals pass through a mass of coiled glandular tubules (gland of Leydig) which are derived from the anterior nonurinary portion of the mesonephros. Sperm contained in the secretions of the Leydig gland then pass into an expansion of the vas deferens known as the ampulla. As they move through the complex system of septae in this structure they are consolidated and receive additional secretions such as the hyaline cortex of the basking shark spermatophore. Borcea's monograph should be consulted for anatomical details of the gonoducts of elasmobranchs. Spermatophores are also regularly formed in the viviparous teleosts. They have been described in the embiotocids and in the poeciliids. In these groups, the aggregations of sperm formed within the seminiferous acini or tubules become arranged with their tails directed centrally and the heads oriented peripherally to form the sperm balls. As these pass through the efferent ducts, they seem to receive a gelatinous secretion which binds them together so that they remain intact during transfer to the female. More complex spermatophores have also been described in teleosts, but there are very few studies of either the cytogenesis or the physiology of fish spermatophores.

Spermiation in the goldfish is typical of the much simpler process involving only a thinning of the semen. Clemens and his associates have described the weight changes in the testis and established a pituitary regulation of the spermiation process. Yamazaki and Donaldson have used the spermiation of the goldfish in the bioassay of salmon pituitary gonadotropin.

In some teleosts (Ariidae, Gobeiidae, and Blennoidae), large structures often referred to as "seminal vesicles" are found as glandular developments from the sperm ducts-occasionally from the testis as in *Tachicorystes*. These "seminal vesicles" do not store sperm and are not comparable to the structures of the same name in the higher vertebrates. They provide secretions which are of importance in sperm transfer or other breeding activities. Descriptions of these glands and the many other interesting secondary sex characters of fishes are beyond the scope of this review. Good general accounts with bibliographies have been given by Berlin and Breder and Rosen. Secretory activities of the sperm ducts, accessory glands, and secondary sex characters show a marked seasonal development and are under the control of the androgenic secretions of the testis.

The Female

The History of the Ovarian Follicle

In structure, the fish ovary ranges from an expanded mesentery (mesovarium) which dehisces mature ova from its ventral margin in the hagfish *Myxine* to a complex hollow organ in the viviparous teleosts where the gonad produces eggs, stores sperm, serves as a site for fertilisation, and provides nourishment for the development of young to an advanced stage. Comprehensive reviews of the earlier literature are available. The description which follows is a generalised one.

Ovarian follicles develop from or in association with the germinal epithelium which covers the surface of the ovary as an extension of the peritoneum (mesovarium). As described earlier, this germinal epithelium also lines the cavity of the hollow teleost (cystovarian) ovary. The ovary of the basking shark, *Cetorhinus* maximus-single organ-is evidently exceptional among the elasmobranchs in that the germinal epithelium invaginates to form a series of tubular ramifications within a gonad which is superficially similar to that of the teleosts. It differs, however, both in its embryology and in its gonoduct. The cavities of the basking shark ovary open into a pocket on the right side of the organ and ova discharged into this pocket pass via the peritoneum into the open end of the Mullerian duct. These hollow ovaries of the teleosts and the

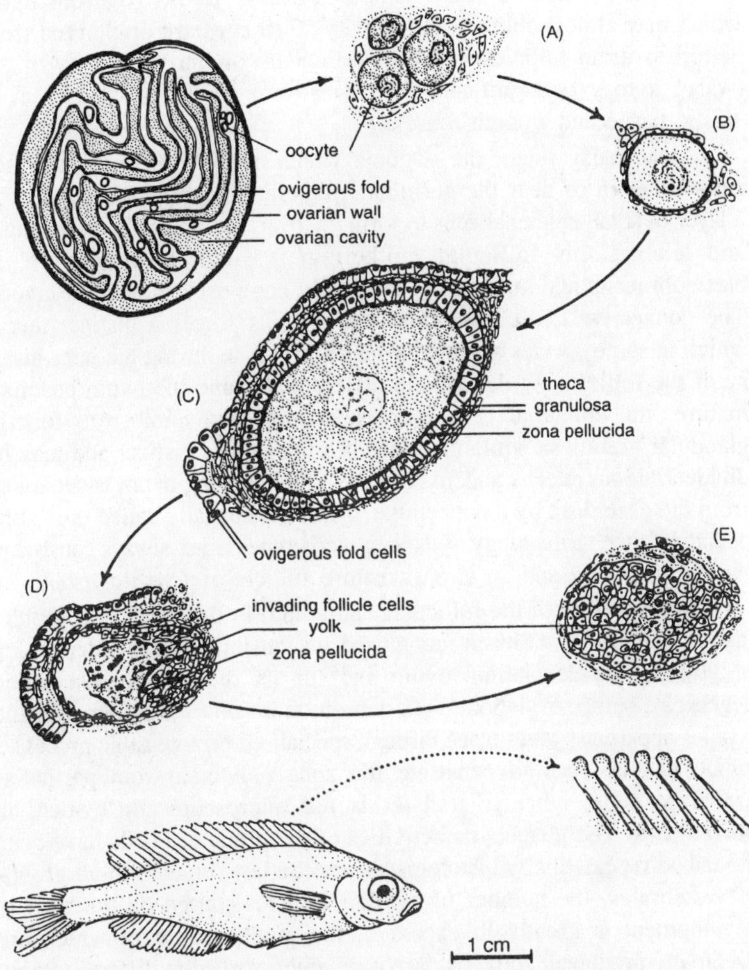

Figure 1.6 : Structure of the ovary in Cymatogaster.

basking shark are unique among vertebrates and quite different from the hollow ovaries of some other elasmobranchs and the amphibians, where the lining is not germinal epithelium and where the cavities develop as large lymph spaces within the stroma (medulla) of the gland.

The numerous ovarian follicles of the fish ovary are supported by a richly vascular connective tissue stroma which extends into the gland from the somewhat denser connective tissue layer (tunica albuginea) just under the germinal epithelium. The internal lining of the hollow teleost ovary is thrown into a complex series of folds (ovigerous folds) which may almost obliterate the cavity. Fish eggs are discharged from mature ovarian follicles into the peritoneum or into the cavity of the ovary; it may be significant that the stroma of the ovary is rich in elastic tissue and smooth muscle.

At an early stage, the oogonia which arise from primordial sex cells either in or near the germinal epithelium become surrounded by a layer of small epithelial cells to form the ovarian follicle. In cyclostomes and teleosts this follicular epithelium is single-layered while in elasmobranchs and amniotes it is usually composed of several layers. The connective tissue near this nest of cells forms a distinct theca, which in some species assumes a very active role during the later history of the follicle. As the follicle differentiates and the ovum becomes mature, the epithelial cells increase in size and number to form a glandular granulosa while the theca becomes more distinct and may be divided into an interna and an externa. The maturing ovum is separated from the granulosa by a noncellular membrane usually called the "zona pellucida"; the terminology of the egg membranes is not always consistent. An early follicle and an almost mature follicle of *Cymatogaster*.

The functions of the follicular epithelium in fishes are still problematic. The granulosa has a recognised responsibility for the deposition of yolk in the developing ovum and for its removal in ova which degenerate before ovulation. Yolk deposition in some elasmobranchs and reptiles apparently takes place through specialised protoplasmic processes which can be seen to penetrate the zona pellucida from particular granulosa cells; nutritive transfer is not microscopically evident in other fishes. The phagocytic activities of the granulosa cells have been described repeatedly by histologists since the last century. In all groups of vertebrates, the number of ovarian follicles started on the road to development is greatly in excess of the number of eggs which are eventually produced; some ova are normally resorbed in different stages of development.

In addition to its nutritive and phagocytic functions, the granulosa may also be concerned with the elaboration of the ovarian hormones. The fish ovary does not contain interstitial tissue comparable in development and histochemistry to the Leydig cells of the testis; the theca of the follicle, which in some mammals participates in the formation of the corpus luteum and probably secretes progesterone-perhaps also estrogen-never shows a sudanophilia in fishes and is evidently a simple fibroelastic connective tissue. It is evident that the ovarian hormones must be synthesised by the ovum or the granulosa or the corpus luteum (corpus atreticum) which develops from the granulosa. Major interest has centered around the corpora, which in this review are termed "corpora lutea" even though they may not be physiologically equivalent to the corpora lutea of mammals which are known to produce progesterone and are under the control of the pituitary gland. Hisaw and Hisaw and Chieffi agree with this terminology.

Corpora lutea have now been observed in all groups of fishes. Their histogenesis has been described repeatedly since the beginning of the century and a rich literature is cited in several reviews of fish endocrinology. Although there is a remarkable variation in the extent of these proliferations, they regularly appear when yolky ova become atretic (preovulatory corpora lutea or corpora atretica) or during the postovulatory history of the follicle (postovulatory or corpora lutea of ovulation). The two opposing views concerning their physiology have been ably maintained from the studies of elasmobranchs by Hisaw who argues that they are concerned with yolk phagocytosis in the preovulatory follicles or the removal of tissue fragments and blood cells in the postovulatory ones, while Chieffi finds substantial histochemical evidence for steroidogenesis and attributes an endocrine function to them. Their function is just as problematic in other groups of fishes. The controversy was detailed recently and will not be repeated here; the conclusion reached at that time still seems valid; it is suggested "that estrogen synthesis was one of the responsibilities of the granulosa from the earliest stages of vertebrate phylogeny and that this capacity developed in association with the synthesis of lipid materials present in the yolk; the high estrogen content of yolk in the ova of some species may represent their entire reserve of this hormone. With the variety and complexity of reproductive processes and controls found among fishes, it seems entirely reasonable that the same granulosa cells (in some fishes) may have specialised in hormone production to the point where corpora atretica become functional corpora lutea-even though they may synthesise estrogen rather than progesterone as their hormone." Chieffi's studies indicate how different the history of the pre and postovulatory bodies may be in closely related groups of fishes.

Accessory reproductive Structures and the Physiology of the Oviducts

In contrast to the males, secondary sex characters are inconspicuous in female fishes. The females are frequently larger, occasionally have specialised ventral fins (Ariidae), distinctive ventral surfaces or folds of unknown function on the abdominal wall (Bunocephalidae and Bagridae), tubercles on the head (Osteoglossidae), or an enlarged urinogenital papilla (Notopteridae, Cyprinidae). The genital papilla may be only slightly enlarged but in the cyprinid *Rhodeus* amarus-the European bitterling-it often extends well beyond the caudal fin as a specialised ovipositor.

Studies of the physiology of the oviduct have been almost completely confined to the Chondrichthyes. In this group, the oviduct (Miillerian duct) not only serves as an open tube for the collection of eggs from the abdominal cavity and their transport to the cloaca but also provides the secretions concerned with the formation of horny shelled eggs in the oviparous species and a site for the development of the young in the viviparous forms. In addition, this tube also serves for the reception of sperm and, at least in some species, for sperm storage and the dissolution of the hyaline cortex of the spermatophores. The oviduct of viviparous teleosts is also concerned with some of these sperm functions, but there are no systematic studies of its physiology.

The four regions of the Miillerian duct; the ostium or funnel, the oviducal or nidamental gland, the connecting isthmus, and the expanded posterior uterus. In some species the ostium is closely applied to an ovarian pocket from which the ova emerge to enter it directly; in other species the eggs are discharged at many points on the surface of the ovary and carried into the funnel by continuously beating cilia which line the abdominal cavity. The ciliation of the abdominal cavity has been described and illustrated by Metten. The absence of cilia in males and immature females and changes in the size of the ostium during the breeding season suggest a hormonal regulation of these structures; estrogens have been shown to stimulate the development of nidamental glands and other areas of the Miillerian tubes.

In all chondrichthyans, the oviducal gland secretes albumen and mucus; it is relatively larger in the oviparous species where it is also responsible for the formation of the shell. Two or three distinct glandular zones may be distinguished. In the oviparous species there is an anterior albumen-secreting area and a posterior shell-secreting zone; an intermediate mucus secreting area may be present, as in the ratfish, *Hydrolagus colliei*, or absent, as in the ovoviviparous species *Rhinobatus granulatus* or located caudad as in *Raja rhina*. The shellsecreting area of the nidamental gland also serves as a seminal vesicle in the dogfish where sperm are stored to fertilise eggs before or at the time of shell formation.

The wall of the uterus is smooth and covered with a flattened epithelium in the oviparous elasmobranchs where it serves only as a passageway for the eggs. In viviparous species, this portion of the Miillerian duct is variously modified through the formation of villuslike appendages which nourish the developing young. These specialisations are considered in the next section.

VIVIPARITY AND GESTATION

Evolutionary Considerations

If an animal species is to survive, each adult member of the population must on the average produce one reproductively active adult. Darwinian evolution rarely operates through one channel, and the fishes as a group have achieved reproductive success in many different ways. At one extreme, eggs and sperm are broadcast in sufficient numbers to balance the unusual pressures of the environment and satisfy the predators while, at another extreme, fertilisation is internal and the developing young are housed within the parent's body until ready for an independent existence. The provision of millions of unprotected gametes represents the primitive condition among fishes. Along the road to specialised viviparity many curious and successful devices have evolved to provide a measure of protection during incubation. Parental care must confer great biological advantages in reducing energy demands for production of eggs or by spreading this load over a longer period while the young are developing within the parent. Thus, it is perhaps not surprising that viviparity has evolved independently several times in both the vertebrates and the invertebrates.

Between those fishes which represent the primitive situation—scattering millions of unprotected gametes—and the highly specialised viviparous forms, there are numerous species which build nests and exercise parental care (Salmonidae and Gasterosteidae), other species in which incubation takes place either in the buccal (Cichlidae and Bagridae) or branchial chambers (Amblyopsidae), fishes which provide special devices for the attachment of their young to their bodies (cutaneous incubation in the male *Kurtus gulliveri* or the females of *Aspredo cotylephorus* and *Solenostomus laciniatus*), and the fascinating sea horses and pipefishes (Syngnathidae) in which the males have a ventral marsupial pouch where the young are incubated until ready for an independent existence. Bertin has provided a summary with many more examples and illustrations of these interesting and curious devices. Although there is a rich literature on the natural history and functional morphology, there are very few physiological studies beyond those which have shown a strong dependence of secondary sex characters such as the marsupium on the gonadal steroids.

Among the fishes, only elasmobranchs and teleosts have achieved a true viviparity. In each of these groups there is an array of species from the ovoviviparous-where the eggs have sufficient yolk for the nourishment of the young and the female provides only protection-to the truly viviparous species-where the yolk content of the egg is greatly reduced and the developing young establish a connection with the maternal tissues at an early stage to draw nourishment from them and to satisfy the respiratory and excretory demands.

The literature on viviparity in fishes has been reviewed several times. The following is a brief summary; since the phenomena are so very different in elasmobranchs and teleosts these two groups will be dealt with separately.

Viviparity among the Chondrichthyes

The Chondrichthyes produce relatively few eggs which vary greatly in their content of organic matter but are all of the yolky telolecithal type. In the truly oviparous groups (Scyllidae, Heterodontidae, Raiidae, and Chimaeridae), ova are housed in horny protective shells which provide protection throughout a long period of incubation. In the ovoviviparous and viviparous species, the young develop within the Miillerian duct (uterus) on which they usually depend for nourishment as well as protection. Ranzi did much of the pioneer work and his papers should be consulted for details; Needham gives a valuable English summary and bibliography. Budker summarises these important papers in French.

Internal fertilisation

Fertilisation is always internal among the Chondrichthyes. Leigh Sharpe in a series of classic papers described the modifications of the male pelvic fins which form the copulatory organs or claspers. These posterior extensions of the fins are stiffened not only by the cartilages of the metapterygia but also during copulation by erectile tissue. The relative contributions of cartilage and erectile tissues vary in different species. In all cases, an essential portion of the organ is the clasper groove formed by skin folds, the edges of which overlap to form a scroll-like tube along which the sperm are transported from the cloaca. Another constant feature of this apparatus is the presence of a clasper syphon or gland which contributes to the seminal discharge or provides a pumping mechanism for its release. In the sharks, the syphon takes the form of a blind muscular sac situated just under the skin anterior and lateral to the cloacal region; in the skates and rays a clasper gland takes the place of the hollow syphon. LeighSharpe concluded from his anatomical studies that the dilute fluid found in the syphon was mostly seawater and that during copulation the contractions of the muscular syphon wall pumped sperm through the clasper groove in a jet of seawater. It is by no means certain, however, that this is its main function. Botte *et al.* and La Marca among others have reported more recently on the physiology and histochemistry.

Mann has shown that the syphon of the spiny dogfish, *Squalus acanthias*, secretes an abundance of 5-hydroxytryptamine (serotonin) ; a pair of syphons in a mature male may contain as much as 20 mg of serotonin or 0.3% . The further demonstration of a marked stimulatory effect of serotonin on the isolated uterus of the spiny dogfish is strongly suggestive of a role for the syphon in the transport of sperm after transfer to the female. At this stage, generalisations are not justified since only traces of serotonin have been found in the syphon sacs of the smooth dogfish, *Mustelus canis*, and none was demonstrated in the clasper glands of *Torpedo* and *Raja.* The clasper glands of skates and rays produce a milky white semiviscous fluid which coagulates on contact with seawater; its function remains speculative and, indeed, there are still many questions concerning the physiological mechanisms of internal fertilisation in these fishes.

Ovoviviparity

The ovoviviparous species are far more numerous than the truly placental or viviparous ones; a true viviparity with yolk sac placenta is confined to certain species of two families of sharks-the Carcharhinidae and the Sphyrnidae, both belonging to the Galeiformes. Actually, the distinction between ovoviviparity and viviparity is a rather artificial one for the physiologist since the maternal contribution of the placental shark to the nourishment of the fetus is an intermediate one in a series ranging from almost zero in the primitive ovoviviparous species to an almost complete dependence in the highly complex ones. Moreover, in some sharks (*Sphyrna tiburo* and *Mustelus canis*) the placenta develops only after several months of an ovoviviparous existence. These facts are recognised by Budker who divides the Chondrichthyes into only two major groups-the oviparous and the viviparous-with the latter subdivided into the aplacentals and the placentals.

The rectangles to the left of the ordinate elsewhere in this chapter the extent to which the newborn fish receives nourishment from its mother while *in utero;* the rectangles to the right show the content of organic substances in the uterine fluids; the black bars show the relative reductions in size of maternal liver during development.

Although the presence of organic material in the uterus is universal, the organic content of the egg may be actually greater by 20-40% than the organic content of the animal at birth; in short, the mother has provided protection but probably does not provide organic nutrients to the fetus. At the other extreme, eggs with very little organic material (only 200 mg in *Pteroplatea micrura*) depend almost entirely on the maternal secretions for growth; at birth, *Pteroplatea* contains about 10 g of organic material, an increase of almost 5000%. In contrast, the change in organic substance from egg to newborn pup is only 840 and 1050% in two placental species*Carcharias glaucus* and *Mustelus laevis*, respectively. Selected tables summarising the biochemical data, largely the work of Ranzi, will be found in Needham and Amoroso. A host of research problems await the interested physiologist. It has now been almost 40 years since Ranzi's classic work on the morphology and biochemistry; most of the endocrinological work does not go beyond correlations between the cytology of the endocrine glands and the cycle of morphological events during gestation.

Histologically, the uterine lining varies from a mucus-secreting layer of cuboidal epithelial cells in species which depend on the egg yolk for nourishment through forms with moderately folded, serous-secreting linings, as in Torpedo, to those with uterine linings beset with villi or trophonemata of varying lengths and complexities and glandular surfaces which secrete an abundance of fat (8% of the organic substance in *Trygon violacea*). Amoroso refers to all these uterine secretions as "uterine milk" and provides a summary of the composition in various species. Budker describes them as mucous, serous, and lipid.

The yolk is mainly digested within the intestine of embryos which depend on it for nourishment. Te Winkel has described the process in the Squalidae. In *Squalus acanthias*, she found that the relatively enormous yolk sac established at an early stage in ontogeny decreased rapidly in size while an internal yolk sac (expansion of the yolk stalk) became relatively larger until the external one remained as only a remnant. Yolk granules are moved by cilia from the external sac through the yolk stalk into the internal sac; then they pass on into the intestine where digestion occurs. All of these structures, including the intestine, are ciliated. Enzymic activity is established in the gut when the embryo is about 60-70 mm long (size at birth about 250 × 300 mm). A small reserve of yolk still remains in the internal sac at birth.

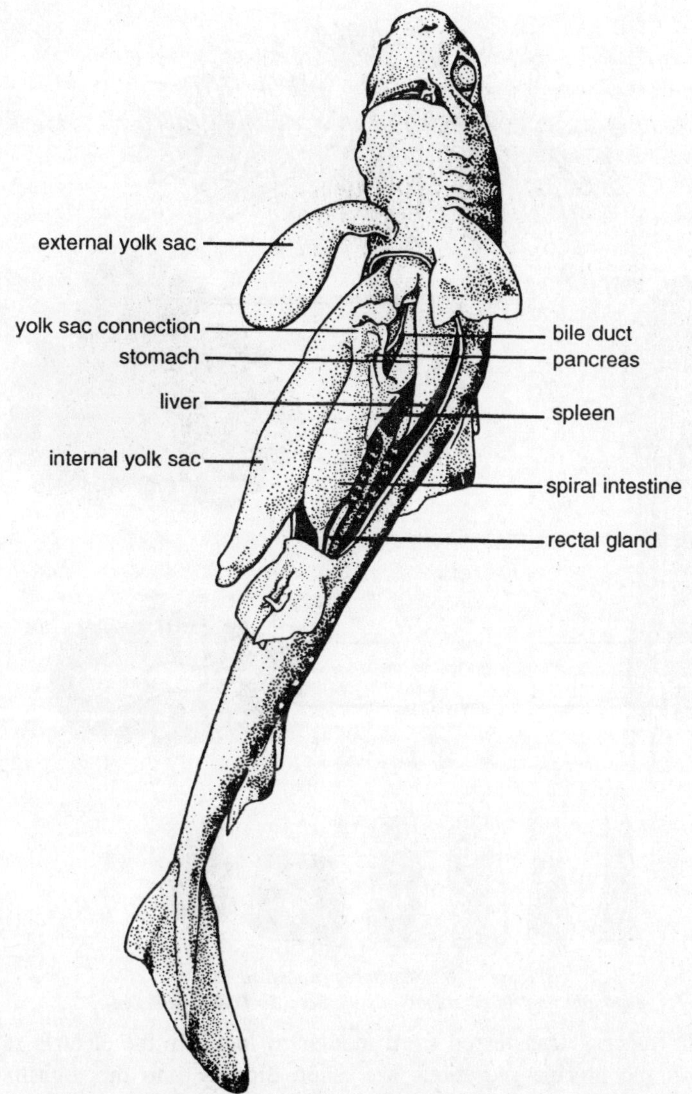

Figure 1.7 : Dissection of a 220-mm embryo of Squalus suckleyi.

The skates and rays display a variety of specialisations from those which are oviparous through a series of curious adaptations in ovoviviparity.

In the most highly specialised situations, nutrition by uterine milk or embryotrophe becomes more efficient than placentation in terms of

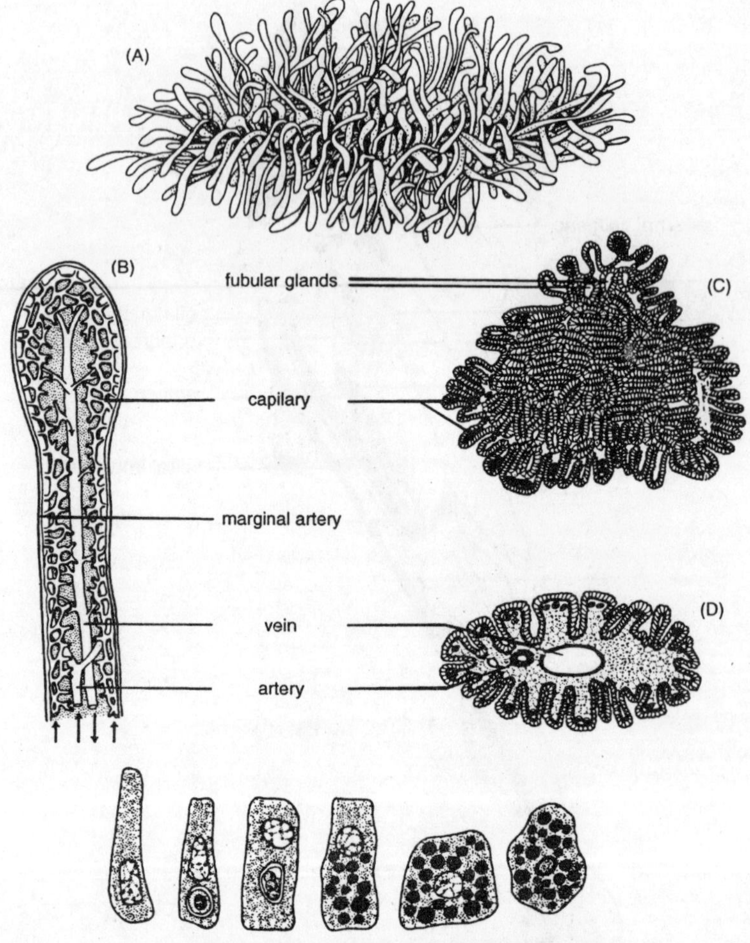

Figure 1.8 : Structures concerned with embryonic nutrition in Dasyatis violacea (=Trygon violacea).

organic material transferred from mother to fetus. In the electric ray *Torpedo* the uterine secretions are taken directly into the digestive tract through the mouth and spiracles; while yolk digestion is going on in the intestine, the uterine secretions are being digested in the stomach. In one of the stingrays *Pteroplatea*, the highly glandular trophonemata or villi enter through the spiracles of the embryo and extend down into the esophagus to release the secretions directly into the gut. In another stingray (*Trygon* or *Dasyatis*) and in the eagleray *Myliobatis* the villi are shorter but extremely numerous; they produce a secretion rich in fat which is aspirated into the gut mostly through the spiracles. The structure of the trophonemata *of Trygon*.

Shaun describes a very different mode of embryonic nutrition in the shark, *Lamna cornubica*. In this animal, the egg yolk is absorbed early in development and thereafter the developing pups depend almost entirely on the swallowing of immature eggs and degenerating ovarian tissues which pass down the oviduct packaged in a delicate secretion of the shell gland. As L. H. Matthews notes, some of these ovoviviparous arrangements are not very different from the oviparity found in the Holocephali, where only a small part of the yolk is enclosed in the yolk sac; the remainder breaks up into a thick milky fluid which is first absorbed by the filamentous external gills and later ingested through the mouth.

Viviparity

Descriptions of the placentae of the silky shark, *Carcharhinus falcif ormis*, and the bonnethead shark, *Sphyrna tiburo*, have recently been added to the literature of true viviparity cited in Amoroso's review.

The ova of *Mustelus canis*, on arriving in the uterus, become isolated into separate uterine compartments. These are formed by the growth of ridges on the uterine wall which eventually fuse so that each embryo is enclosed in a separate chamber ; although uterine compartments are not formed in most of the nonplacental sharks there are at least two exceptions (*Galeus canis* and *Mustelus vulgaris*) cited by Ranzi. *Mustelus canis* (also *S phyrna tiburo*) leads an ovoviviparous existence during this early period and may not establish a placenta for several months. During this time, while nutrients are obtained from the yolk, these embryos develop elaborate circulatory networks in the walls of their yolk sacs and the yolk sacs themselves become greatly folded. While these developments are taking place in the yolk sac, the uterine mucosa which is initially smooth, loose, edematous, and covered with a columnar or cuboidal epithelium also becomes folded. A placenta is established through the interdigitation of these two series of folds with a thinning of their epithelia to bring the maternal and fetal circulations into close proximity. The umbilical cord or stalk which attaches the embryo to the placenta is a modified yolk stalk. The interdigitations characteristic of the placentae of the two fishes just described do not seem to develop in *Carcharhinus*. The specialised yolk sac sits on a modified and extremely vascular discoidal patch of the uterus.

Detailed descriptions of the umbilical cords will be found in the literature cited. An interesting specialisation found in some forms

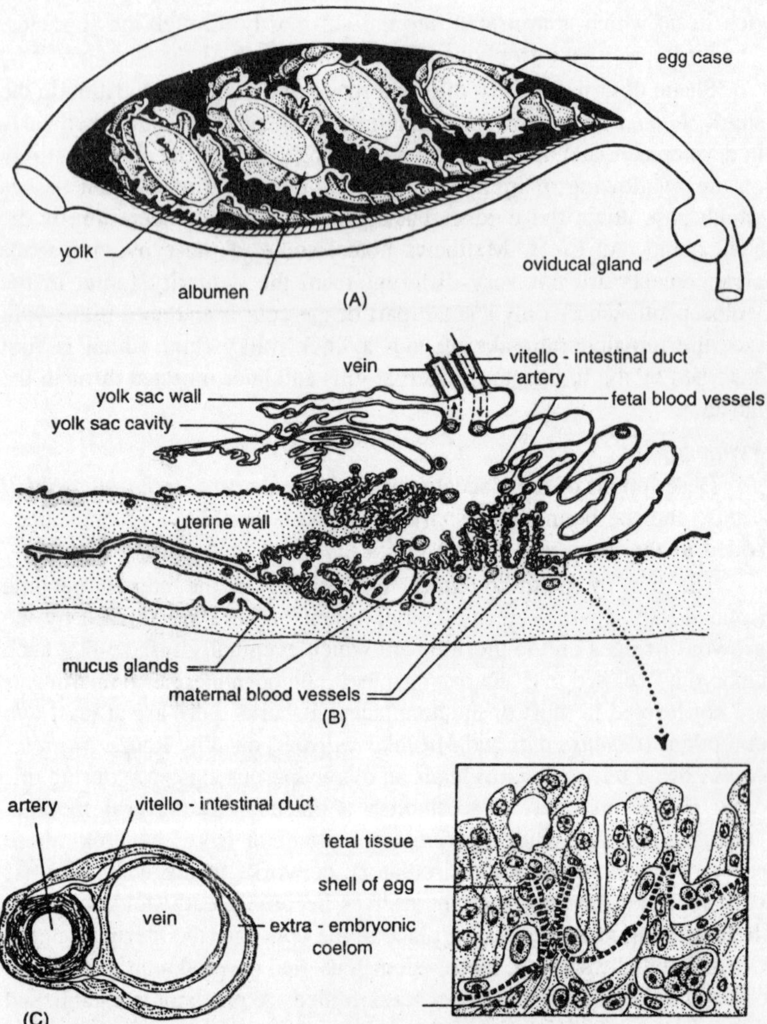

Figure 1.9 : Yolk sac placenta of Mustelus canis (=laevis).

(*Paragaleus*, *Scoliodon*, and *Sphyrna*) *is* the presence of numerous villi (appendiculae) which appear to be absorptive ; these suggest that the dependence on uterine milk was not completely lost in the evolution of viviparity among the sharks. In this and in several other features, it is clearly indicated that the phylogeny of viviparity in these fishes was via a highly specialised ovoviviparity and that the steps involved were not very long ones. Both conditions may be markedly developed within a single genus; *Mustelus vulgaris is ovovivip*arous while *Mustelus cans* is *viviparous*.

Viviparity among the Teleosts

Development of the young within the female occurs in only two orders of teleost fishes, the Cyprinodontiformes and the Perciformes. The nine families involved represent only a small proportion of the Teleostei. However, the diversity of mechanisms is as great as that of any other group of vertebrates which range from the oviparous, through several stages of ovoviviparity to highly successful viviparous forms. This, as well as a lack of correlation with particular habitats or geographical regions, argues for a separate evolution on more than one occasion. The viviparous teleosts include such geographically remote groups as the comephorids unique to Lake Baikal in Siberia and the goodeid fishes of the Mexican plateau ; both marine and freshwater species are represented.

Amoroso's review is the most informative of the recent ones and includes citations to the papers of C. L. Turner which are still the richest source of information on these fishes. Amoroso's summary is reproduced in Table I and lists the known families of viviparous teleosts. It will be noted that internal development in the teleosts always takes place in the ovary; the Miillerian duct, which houses the developing elasmobranch, is absent in the teleosts. Sometimes the young develop within the ovarian follicle to a very advanced stage before ovulation (Poeciliidae and Anablepidae) ; more frequently, embryogenesis occurs within the cavity of the ovary after fertilisation within the follicle. Although one might speculate that the ovarian type of gestation is more primitive, C. L. Turner believes that with one probable exception (*Zoarces*) fertilisation precedes ovulation in present-day viviparous teleosts. Bertin summarises the three general situations as follows where "0" represents ovulation, "F" is fertilisation, "H" is hatching, and "P" is parturition:

Type *Zoarces* O-F .... H .... P
Type *Jenynsia* F .... O.................. H .... P
Type *Gambusia* F .... O-H-P

Both follicular and ovarian gestation are characterised by ovoviviparous and viviparous types of development. As in the elasmobranchs, the total dry weight of the larva at birth may be less than that of the egg, indicating the complete lack of a maternal contribution to the nourishment of the embryo while in the more specialised viviparous types the egg is endowed with very little yolk and the developing embryo obtains almost all of its nourishment from its parent. Actually, most of the teleosts which develop within the ovary seem to draw nourishment from the mother and in this sense are not truly ovoviviparous. Only *Sebastes marinus* shows a decline in dry weight between fertilisation of the egg and parturition; most of the species show little change in weight indicating that the mother has met the metabolic costs of respiration in the developing tissues so that the total weight of the larva at birth is just balanced by the stored material in the original egg. Only two of the poeciliids (*Heterandria formosa* and *Aulophallus elongatus*) studied by Scrimshaw are truly viviparous, depending on the mother for both their respiratory needs and the anabolic demands for growth. The weight decrease of 34% in *Sebastes marinus is* essentially the same as the value 37% reported for the oviparous trout *Salmo fario*.

Internal fertilisation

As in the Chondrichthyes, internal fertilisation in the teleosts is not necessarily followed by a development of the young in the female. One of the most complex intromittent organs described in male teleosts (the priapium located on the throat of the Phallostethidae) serves only to fertilise the eggs as they are being laid.

In teleosts, the copulatory organ is usually an enlarged genital papilla or a specialised anal fin. Unlike the Chondrichthyes, the pelvic fins are rarely used directly in the transfer of sperm; the priapium of the phallostethid fishes is again an exception with skeletal elements derived from pelvic fins.

An enlarged genital papilla is sometimes referred to as a "pseudo-penis." Its size bears little relation to viviparity. In the Embiotocidae, a highly successful viviparous group, it appears only seasonally as a modest fleshy papilla; in the Cottidae, an oviparous group, it is often greatly enlarged and may contain erectile tissue. A pseudopenis is found in several other families both oviparous and viviparous. There are many strange modifications of the genital papillae of both sexes. In female *Orthonopias triacis* the protrusible oviduct is smeared with sperm during copulation and then withdrawn to effect fertilisation of the eggs which are laid one at a time. In *Apogan imberbis* the urogenital papilla of the female is enlarged and introduced into the male for the reception of sperm. As noted earlier, the elongated genital papilla of *Rhodeus amarus* transfers eggs to the gill spaces of a freshwater mussel. These enlarged genital papillae-both male and female-appear as secondary sex characters during the breeding season, and their development is probably always regulated by the gonadal steroids.

The copulatory organs of the cyprinodonts are specialised anal fins. The fins are only slightly altered in the Goodeidae with a foreshortening of the anterior rays to form a spermatopodium. In contrast, the anal fin of the Poeciliidae, referred to as the "gonopodium," is profoundly modified through an elongation and specialisation of several of its rays; the genital aperture is at its base, but during sexual activity the elongated bony fin rays and the web of tissue connecting these rays interact to form a transitory groove presumably associated with the transfer of sperm bundles to the tip of the gonopodium. The Jenynsiidae and the Anablepidae show a third stage in anal fin specialisation to form a large penis containing a permanent tube opening at its tip; at times the organ may be greatly extended; it is a naked structure in the Jenynsiidae but covered with scales in the Anablepidae.

Gestation within the Ovarian Cavity

Sebastodes paucispinis (family Scorpaenidae) undergoes an entirely ovoviviparous development within the ovarian cavity. Moser, who has recently studied this species, found numerous sperm singly or in clumps embedded among the epithelial cells of the maturing follicles or on their outer surfaces but never within the follicles; he believes that fertilisation occurs immediately after ovulation and this would place *Sebastodes* in the "Zoarces type" of Bertin's classification. *Sebastodes* embryos lack placentalike connections and depend entirely on their egg yolk for nourishment; they hatch just prior to spawning when the yolk has been largely absorbed and they are about 4-5 mm long; life in the ocean commences as a planktonic larva. Large females may produce broods of as many as two million larvae; the respiratory demands of this large mass of developing tissue must create one of the major problems in maternal physiology. A specialised dual vascular system to the ovary seems to be unique to this group of fishes and is evidently related to these special needs.

The Embiotocidae show an intermediate situation between the strict ovoviviparity of *Sebastodes* and the goodeid or jenynsiid fishes which usually form elaborate placentalike connections. *Cymatogaster aggregata* has been carefully investigated by C. L. Turner and Wiebe. Copulation occurs during the spring or early summer and sperm remain dormant in the ovary until fertilisation occurs about 6 months later. Eggs, fertilised within the follicle, are shed into the cavity of the ovary during early segmentation stages and develop there for 10-12 months to an advanced stage (sexual maturity in some males). The *Cymatogaster* egg contains relatively little yolk and depends mainly on the secretions of the highly glandular ovigerous folds for nourishment. By the time the larvae have reached a length of 15 mm, spatulate vascular extensions have developed on the vertical fins to assist in nutrition and respiration; these are resorbed just before birth. The alimentary canal begins to function later in gestation and long villi located in the gut are evidently concerned with digestion and absorption of foods entering through the mouth ; relationships are also established between the ovarian tissues and the embryonic gills.

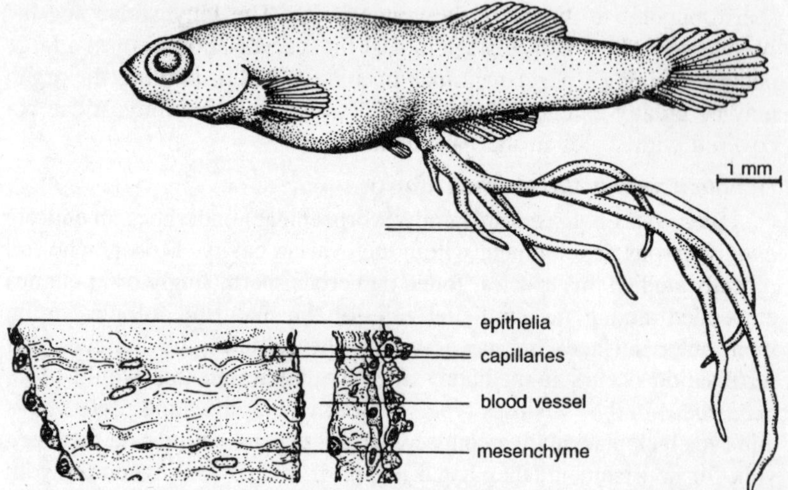

Figure 1.10 : The trophotaeniae of Zoogoneticus quitzeoensis during late gestation with a segment of a longitudinal section through one of the nutritive processes at the lower left.

More elaborate types of placental connections are characteristic of both the Goodeidae and the Jenynsiidae. Turner's investigations of the goodeids have revealed a spectrum of species with progressively diminishing yolk stores, a less significant role for the yolk sac and pericardial sac (mentioned below) and an increasing importance of highly vascular rectal trophotaenia of many diverse forms. These are bathed in the copious secretions of the ovarian cavity; the ovarian secretions are supplemented by degenerating embryos. The brotulid fishes have similar absorptive surfaces.

In the jenynsiids there is also an early ovulation following fertilisation, an initial dependence on the small supply of yolk, the development of a yolk sac and a pericardial sac. For the major portion of intraovarian development, however, an intimate association is established between the ovarian tissues and the pharyngeal and mouth cavities of the embryo; vascular folds of the ovarian epithelium are in close contact with the gills-an arrangement which Amoroso refers to as a "potential branchial placenta." These fishes also imbibe ovarian fluids as a source of nourishment. In several of the papers cited above, Turner discusses the progressive specialisation in ovarian gestation from the strictly ovoviviparous to these several curious placental connections and notes that this trend is evident in a single family-the Goodeidae.

Follicular Gestation

A prolonged follicular gestation is found only in the Poeciliidae and the Anablepidae. Again there is a series of forms from the ovoviviparous (*Poecilia*) to species with a specialised follicular pseudoplacenta; the complete array is evident within the poeciliids which remain in the follicle for the entire period of gestation.

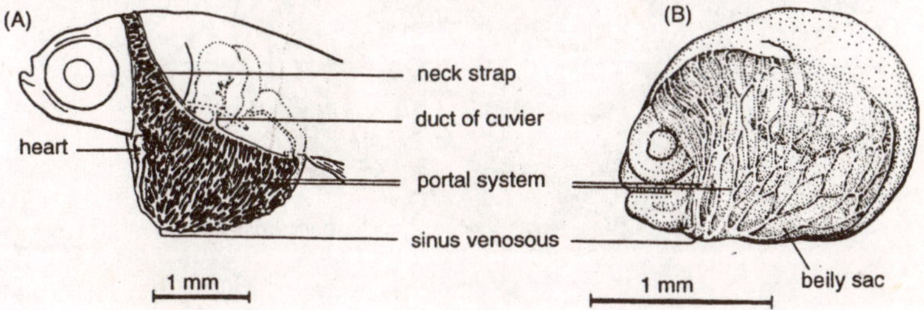

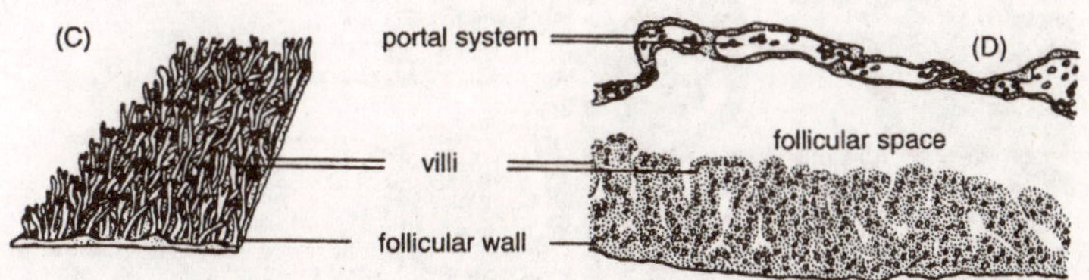

Figure 1.11 : The embryonic-maternal connections in several poeciliids.

The yolk supply is adequate for embryonic nutrition in several of the familiar poeciliids (guppy, black molly, and swordtail) and respiratory exchanges are effected through an expanded yolk sac which extends over the head as a "neck strap." Although it is generally assumed that the intimate association of the portal blood system on the yolk sac and the vascular wall of the ovarian follicles serves only for the exchange of gases and nitrogenous wastes, a transfer of nutrients has not been ruled out. In *Heterandria f ormosa* the yolk supply is meager and the yolk sac very small; the fetal-maternal exchanges are effected through a greatly expanded and vascular pericardial sac which is wrapped around the anterior part of the embryo; the wall of the follicle is smooth and makes intimate contact with the portal system of the embryo. A further stage of specialisation found in this group is an enlarged ventral expansion of the coelomic cavity to form a "belly sac" (only modestly developed in *Heterandria*) and the formation of finger-like villi on the follicular wall. Turner applies the term "follicular pseudo-placenta" to this complex of follicular wall, follicular space, and the portal system of the belly sac.

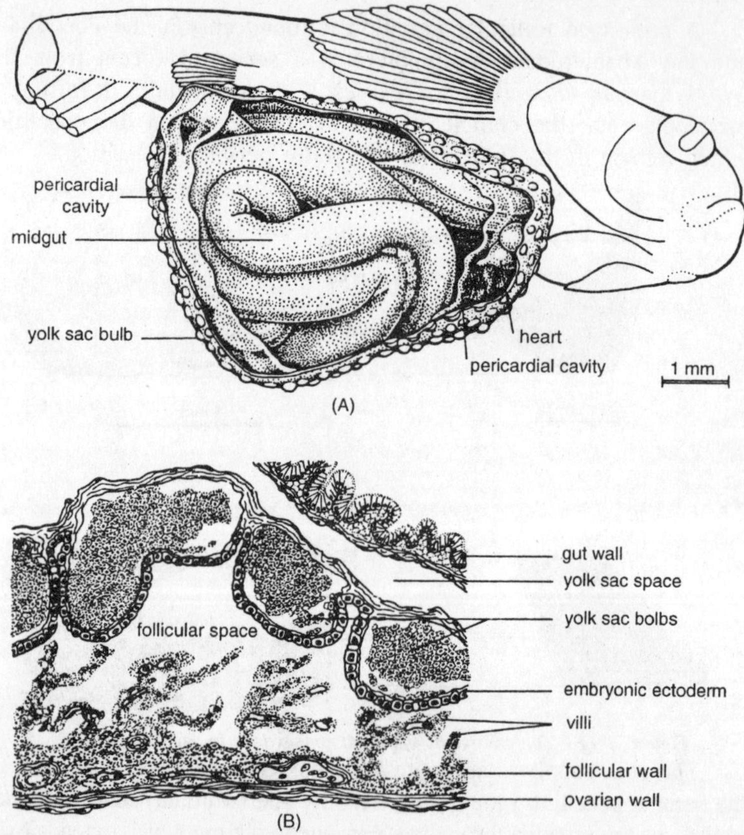

Figure 1.12 : Expanded midgut and maternal-fetal connections in Anableps anableps.

The anablepid fishes show still another stage in the specialisation of the follicular pseudoplacenta. Not only is the follicular wall clothed with complex villi but also numerous vascular bulbs appear in the portal circulation of the belly sac. In addition, there is a spectacular enlargement of the gut in *Anableps* (midgut in *A. anableps* and posterior gut in *A. doweri*), and this is evidently concerned with the digestion of follicular fluids taken into the alimentary canal; an involution of this specialised area of the gut occurs before birth. The enlargement of the gut and the presence of vascular bulbs is unique to the Anablepidae.

A multitude of fascinating problems awaits the physiologist who becomes interested in the viviparous fishes. Although the functional morphology has been systematically investigated there are relatively few studies of the physiology. Nutritional demands of elasmobranch embryos were investigated by Ranzi many years ago and there is a small amount of work on their nitrogen metabolism ; there are also a few studies of the nutrition of developing teleosts. However, to my knowledge, there are no careful investigations of the gaseous exchanges, the properties of maternal and fetal blood, or the ultrastructure of the maternal-fetal connections.

Supefetation

In some of the poeciliids, several broods of young may be developing simultaneously within the ovary (superfetation). C. L. Turner suggests that the condition has evolved through a shortening of the period required for the development and maturation of the ova; a second group of eggs reaches maturity and is fertilised before the previous group completes development. It must also be related to storage of sperm within the ovary. In many fishes the ovarian epithelium assumes a nurse-cell function and immobilised sperm, embedded in its epithelium, remain viable for many months. Superfetation seems to reach a climax in *Heterandria formosa* where live sperm have been found 10 months after a single contact with the male and as many as nine broods may be developing in the ovary simultaneously with births at intervals of about 10 days.

THE ENDORINOLOGY OF REPRODUCTION

The timing of reproduction, regulation of the associated morphological changes, the mobilisation of energy reserves for gonadal development, and intricate breeding behaviour are very largely dependent on the glands of internal secretion. Most of the endocrine organs are either directly or indirectly involved since this complexity requires profound metabolic adjustments, associated not only with gonadal development but sometimes also with changes in habitat during the breeding season. Only the pituitary gonadotropic hormones and the gonadal steroids are considered in this chapter. Chapters associated with the other endocrine glands contain pertinent references to their role in reproduction; the endocrinology of reproductive behaviour is considered separately in the chapter by Liley, this volume.

The literature on fish endocrinology has been reviewed several times during the past decade, while the reproductive hormones of fishes have been separately considered by Dodd and Hoar. The bibliographies of these reviews are a rich source of the earlier literature; the citations included here have been kept to a minimum.

The Pituitary Gonadotropins

Phylogeny of Gonadotropic Regulation—Studies in the Agnatha

It is now evident that the pituitary gland regulates certain aspects of reproduction in all vertebrate animals. It is equally true that this control is sometimes less precise and embraces fewer elements of reproduction among the fishes and, further, that there may be marked variations among the different groups of fishes.

Pickford and Atz have tabulated the early experiments involving hypophysectomy of fishes. By the midthirties it was evident that gonadal development and the maturation of the sexual products were dependent on the pituitary gland in all gnathostomes. Investigation of the gonadotropic activities of the pituitary in the Agnatha awaited the investigations of J. M. Dodd and his associates in the early sixties. Their studies and investigations of Larsen have clearly shown pituitary gonadotropic controls in the cyclostome Lampetra fluviatilis, one of the representatives of the most primitive living vertebrates. It would seem that the pituitary has dominated vertebrate reproductive processes since the Ordovician, a period of almost 500 million years. Dodd's studies, however, went further and revealed a less complete dominance among the Agnatha suggesting a certain phylogeny in the establishment of this physiological regulation.

Lampetra fluviatilis, hypophysectomised in the late autumn and winter, were followed for a period of 5 months and the changes in the gonads carefully described. At the beginning of the experiment, secondary sex characters were absent, the ovaries contained only small eggs, and the testicular ampullae a high percentage of spermatocytes. During the 5-month period the normal and sham-operated animals matured rapidly, secondary sex characters appeared, the ova increased in size, the testicular ampullae became filled with spermatozoa, and spawning took place in early April. On the contrary, the hypophysectomised animals remained immature in appearance while gonadal development was completely arrested in the females and markedly delayed in the males. Larsen finds continued relatively slow growth of the ovaries after hypophysectomy but confirms Dodd's findings in the male.

The effects on the gonads are particularly interesting. In the gnathostomes, hypophysectomy is followed not only by an arrest of ovarian development but also by a general atresia of the follicles (corpora lutea formation) while spermatogenesis is completely blocked in the testis. In the lamprey, spermatogenesis, spermiogenesis, and ovarian growth seem to be autonomous processes; they are retarded but not suppressed in the absence of a pituitary. The endocrine tissues of the gonads are evidently pituitary-regulated since the secondary sex characters which are thought to depend on them fail to develop in the operated animals.

Dodd *et al.* and Larsen's investigations of the adult *Lampetra fluviatilis* are the only long-term studies of the effects of hypophysectomy in the Agnatha. They have established the indispensable nature of the pituitary for reproduction. The Myxinidae have not yet been critically investigated in this connection ; these primitive animals should prove particularly interesting and may well increase our understanding of the phylogeny of gonadotropic controls among the vertebrates. On the basis of the lamprey work, it has been suggested that in phylogeny the metabolic controls involving yolk mobilisation and gonadal growth preceded those concerned with gametogenesis.

Chemical inhibition of the Action of Gonadotropins

Investigations of the pituitary functions of fishes have sometimes been hampered by the technical difficulties of hypophysectomy. This is true of some of the teleosts which are particularly popular with fish physiologists (*Gasterosteus* and *Cymatogaster*, for example) ; the mouth and opercular openings may be very small or the skull extremely deep dorsoventrally with much vascular tissue between the roof of the mouth and the brain or the fish may be difficult to handle postoperatively. Hoar *et al.* and Wiebe have used Methallibure as a chemical blocking agent of gonadotropic functions. This agent, prepared by Imperial Chemical Industries (33,828), is la-methylallylthiocarbamoyl-2methylthiocarbamyl hydrazine. It has been carefully tested in the homeotherms and is now finding a use in agricultural practice.

We have examined the reactions of four fishes treated with Methallibure either by injection or by addition to the ambient water. It appears to be particularly effective in *Cymatogaster aggregata*, and

Poecilia reticulata, but much less active in *Gasterosteus aculeatus;* the work with the goldfish, *Carassius auratus*, is preliminary. At present it is not known whether the variable response is because of species differences in reaction to the substance or species differences in the physiological actions of the gonadotropins; it is also possible that the effect varies with the environment-sea water vs. fresh water. It is of interest that some species of domestic animals are proving more responsive than others; the substance has been particularly effective in regulating the farrowing of pigs.

In spite of the variations in response, all species of fish studied show clear evidence of the blocking of gonadotropic action in a manner comparable to that which follows surgical hypophysectomy. Spermatogenesis is either completely suspended or greatly depressed; yolk deposition ceases, and, in the more responsive species, steroidogenesis comes to an end. The side effects of the compound appear to be minimal involving a slight stimulation of the pituitary thyrotrophs (because of a mild thiourea action) and some depression of the somatotrophs. In histochemical studies of *Poecilia*, Leatherland and Pandey have shown that the most likely locus for the block is in the synthesis of the gonadotropins; there was a sharp and significant decrease in both number and size of these cells and no evidence of suppression of secretory activity in the neurosecretory system of the hypothalamus. Ovine LH was found to partially counteract the inhibitory effects of Methallibure in the gonads of *Cymatogaster aggregata*.

The Poituitary-Gonadal Relations in Gnathostome Fishes

Pituitary-gonadal relations have now been investigated in several elasmobranchs and many teleosts. In general, the findings are consistent; the pituitary regulates both gametogenesis and steroidogenesis.

In the hypophysectomised male dogfish, *Scyliorhinus caniculus*, Dodd and his associates found that the effects of hypophysectomy were localised in the transitional zone between spermatogonia and spermatocytes. The zonated structure of the dogfish testis was particularly helpful in localising these effects. It is of considerable interest that the seasonal cycle-presumably regulated by the pituitarycreates an identical picture. Spermatocyte production ceases during the springtime or following hypophysectomy, but in both cases spermatocytes already formed continue to differentiate and become mature spermatozoa. Dodd has reached the tentative conclusion that the gonadotropin (s) are essential for the normal transformation which occurs when the spermatogonia (which

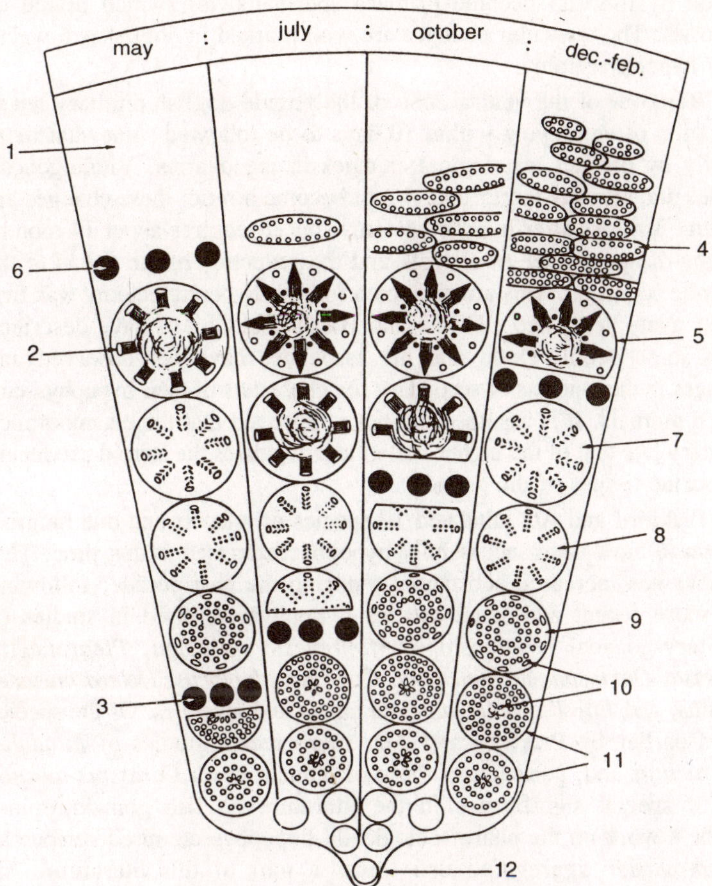

Figure 1.13 : Cycle of spermatogenesis in the dogfish, Squalus acanthias. The stages of gamete development are pictured in the circles representing ampullae. The four columns dividing the testis indicate four periods of the year. The filled circles indicate the position of the degenerate band, and they coincide with the absence of one stage and reduction of the neighboring stages of ampullae. Note the accumulation of evacuated ampullae to a maximum at the mating time (December-February) followed by a resting phase when breakdown of spermatogonia results in a new degenerate band (March-May). The rate of development from spermatocyte to spermatozoon is clearly seen using the degenerate band as a marker (7-9 months). 1, Epigonal tissue; 2, ampulla with sperm bundles and no Sertoli bodies; 3, origin of the degenerate band from breakdown of spermatogonia; 4, evacuated ampulla with Sertoli cell nuclei; 5, ampulla with ripe sperm and Sertoli bodies; 6, remains of degenerate band; 7, spermatid metamorphosis; 8, spermatids; 9, spermatocytes; 10, Sertoli cell nucleus; 11, spermatogonia; and 12, germ ridge.

divide by mitosis) become primary spermatocytes (which divide by meiosis). The testicular changes are well marked in dogfishes 6 weeks after hypophysectomy.

Removal of the ventral lobe of the female dogfish pituitary led to cessation of egg laying within 10 days to be followed somewhat more slowly by distinct histological changes in the ovaries. Vitellogenesis ceases and all eggs larger than 4 mm become atretic; these changes are evident 3 weeks after hypothysectomy, but it requires about 14 months for the disappearance of all yolk and the reduction of the gonad to the juvenile condition. This atresia which follows hypophysectomy was first noted many years ago ; Hisaw and Hisaw and Chief 1 have described it in some detail. Dodd and his associates have not observed any changes in the oviducts or secondary sex characters of their hypophysectomised animals. It is assumed but not yet proved that the elasmobranch pituitary like that of the higher vertebrates regulates the steroid-producing endocrine tissues of the gonads.

Pickford and Atz tabulated 17 species of teleosts and one lungfish known to have been successfully hypophysectomised at that time. This number has increased substantially during the past decade; following are more recent additional species hypophysectomised in studies of pituitary-gonadal relationships: *Mollienesia latipinna*, *Pleuronectes platessa*, *Ophicephalus punctatus*, *Couesius plumbeus*, *Hetero pneustes f ossilis*, and *Poecilia reticulate-both* adults and juveniles. Of the species listed earlier by Pickford and Atz, more recent studies of *Fundulus heteroclitus* and, particularly those of Yamazaki on *Carassius auratus* are of special significance in the literature on fish gonadotropins. Wiebe's work on the pharmacologically hypophysectomised surfperch, *Cymatogaster aggregata*, also forms a part of this literature. All investigators agree that the pituitary is required for gonadal maturation; in its absence vitellogenesis is suppressed with atresia of the larger developing oocytes, spermatogenesis is blocked at the spermatogonia-spermatocyte stage, and steroidogenesis does not occur in the gonadal endocrine tissues. Investigations of the endocrine tissues have been almost entirely confined to the testis.

Although these general findings appear to be consistent, there are still many unresolved details. As yet, it is not certain whether gonadotropin is required for (a) multiplication of the spermatogonia, (b) specifically triggering the reduction divisions, and (c) spermiation or ovulation, nor is it certain to what degree the Sertoli cells and other supporting tissues in the gonads are dependent on the pituitary. Barr's work suggests that some of the variations recorded by different workers may have a seasonal basis and depend on the variable stages of gonadal maturation at the time of hypophysectomy. The later stages of spermatogenesis and spermiogenesis, if well under way, seem to continue with production of mature sperm but spermiation is usually not observed in the absence of the pituitary and the mature sperm gradually disappear through phagocytosis or resorption. Spermiation has been reported in the hypophysectomised plaice and lake chub ; these differences may depend on the stage of sexual maturity at the time of surgery. It has already been noted that the pituitary normally triggers the secretions which are essential in thinning the seminal fluid prior to its discharge. Pandey's work shows that in the absence of the pituitary spermatophores form from the later spermatocytes, but they rupture and are resorbed without a discharge of the sperm. Pandey's work on the juvenile guppy has also emphasised the interplay of both pituitary gonadotropins and androgens in the maintenance of spermatogenesis.

The changes in ovarian histology which follow hypophysectomy have been much less frequently examined. However, the findings are consistent with those recorded for the elasmobranchs and indicate that after hypophysectomy a slow but inevitable resorption of ova may be expected; vitellogenesis comes to an end; new ova fail to form; and ovulation and spawning do not occur unless yolk deposition was complete prior to the surgery. Yamazaki found that relatively few goldfish ovulated even though the operation was performed only a few hours prior to the normal ovulation time.

A major component of the literature on fish endocrinology is concerned with the practical matter of regulating ovulation in fish culture operations. The work primarily involves the use of mammalian pituitary or chorionic gonadotropic preparations. Many citations will be found in reviews by Pickford and Atz, Ball, and Dodd. This topic is also discussed in the chapter by Liley, this volume. The activity of some of the preparations commonly used will be considered in the next section.

The endocrine-producing tissues of the gonads seem to be pituitary regulated in all of the vertebrates, and regressive changes may be expected in them after hypophysectomy. Atrophy of the testicular interstitial tissues has now been recorded in several teleost fishes and in lobule boundary cells. Regressive changes in the Sertoli cells have also been noted. The gonadotropic inhibitor Methallibure causes atrophy of both the interstitial cells and the Sertoli cells in the surfperch *Cymatogaster;* both of these groups of cells are actively concerned with steroidogenesis in this species. All these data are consistent with the concept of a pituitary control of steroidogenesis in the gonadal endocrine tissues. Regressive changes noted in the efferent duct system are probably secondary to the degeneration of the steroidproducing tissues.

Effects of Hypophysectomy on Gestation

Several workers have associated cytological cycles in pituitary secretory activity with the pregnancy cycle of viviparous fishes. Ranzi examined the endocrine glands in a part of his extensive studies of selachian viviparity and described pituitary hypertrophy and hyperemia with a decreased acidophilia in pregnant females. Chieffi and Della Corte and Chieffi cite confirmatory recent evidence for elasmobranchs. Likewise, pituitary changes have been associated with the reproductive cycles of viviparous teleosts. However, in both groups of fishes these pituitary cycles may only regulate the timing of reproduction and the development of the gametes as they do in the oviparous species. From present studies it must be concluded that the pituitary plays little or no part in the maintenance of pregnancy in the elasmobranchs but may be involved in some teleosts. However, there is still very little experimental work directed to relatively few species.

In the late thirties, Hisaw and Abramowitz reported that pregnancy was uninterrupted by hypophysectomy in the viviparous dogfish, *Mustelus canis;* animals hypophysectomised in early stages of pregnancy showed normal development of young for a period of 3½ months. These seem to be the only experiments of this nature recorded for the elasmobranchs, although Dodd *et al*. report that removal of the ventral lobe of the pituitary of *Scyliorhinus* has no effect on the oviducts which retain their normal secretory capacities.

Studies of the effects of hypophysectomy on gestation in the viviparous fishes are likewise preliminary. Ball found that 12 *Mollienesia latipinna*, hypophysectomised during pregnancy, gave birth to normal offspring; the embryos completed their development in the usual time and were as lively and survived as well as the controls. Chambolle, however, working with another viviparous cyrinodont *Gambusia*, recorded a highly significant mortality of embryos when females were hypophysectomised during the first days of pregnancy; in part, the mortality was thought to be related to a disturbance in water and electrolyte metabolism-perhaps associated with the loss of the pituitary-adrenal control. Thus, the evidence from two different species of viviparous cyprinodonts is contradictory but there are further indications from other studies that the pituitary is essential to successful gestation and parturition in some teleosts. Wiebe found that treatment with the gonadotropin-blocking agent Methallibure leads to an atrophy of the secretory ovigerous epithelium in *Cymatogaster*. Associated with this, Wiebe noted reduced growth and a 39% mortality of the intraovarian embryos. These results are preliminary and based on only three pregnant females; the effects may also be indirect ones mediated through the gonadal steroids. They are, however, indicative. In addition, there are several reports of premature release of young by pregnant cyprinodonts following injection of mammalian pituitary preparations. It is also pertinent that incubation and parturition of developing young in the male sea horse, *Hippocampus hippocampus*, is disturbed by hypophysectomy although gonadectomy is without effect. Finally, there is some suggestive evidence of a role for prolactin in the gestation of certain cyprinodonts and an intimation that the neurohypophysial hormones are active in parturition.

It is obvious that further research is required in this area of endocrinology. It may yet be necessary to modify Hisaw's conclusions—based on only one species—that the responsibilities of the fish pituitary end with ovulation and that an involvement in the maintenance of pregnancy developed at a later stage in vertebrate phylogeny.

Biochemical Nature of the Gonadotropin (s)

Two separate gonadotropins, the follicle-stimulating hormone (FSH) and the luteinizing hormone (LH), are physiologically distinct in all tetrapods. Although the nature of the fish gonadotropins is still not resolved, the data for a single factor seem to be accumulating steadily. In addition to the physiological literature summarised below, the biochemical evidence for a single protein increases as the preparations (salmon or carp) have been purified more carefully. Most of the studies-both physiological and biochemical-have been concerned with teleosts. The excellent discussion by Pickford and Atz is still valuable.

When fish pituitary extracts are tested in tetrapods, both the FSH- and the LH-like effects are frequently elicited. Witschi assayed pituitaries of sharks, garpikes and salmon-representatives of the three major groups of jawed fishes. Positive evidence for FSH was obtained by using the vaginal cornification test in the rat and for LH by using the weaver-finch feather reaction. The FSH content was very low; most of the gonadotropic activity appeared to be associated with LH. Ball has summarised these and other investigations-most of which provide suggestive evidence for the presence of the FSH-like effect and strong evidence for the LH action when fish pituitaries are injected into mammals. A series of papers by Otuska using partially purified salmon pituitaries injected into newts and mice are in agreement with evidence for two gonadotropic proteins with distinct physiological effects. However, not all of this evidence for an action of fish gonadotropin (s) in higher vertebrates is positive. Burzawa-Gerard and Fontaine prepared a carefully purified extract of carp pituitaries which appeared to be a single protein and was entirely inactive in mammals. The histophysiology of the pituitary and the question of two different gonadotrophs is discussed in the chapter by Ball and Baker.

Evidence concerning the nature of fish gonadotropin has also been sought in tests with the purified mammalian fractions injected into fishes. The preparations of mammalian FSH have consistently given negative results; LH, on the contrary, usually elicits both the gametogenetic and the steroidogenetic actions commonly associated with two different fractions in the higher vertebrates. The earlier literature is summarised in reviews previously cited. Among the more recent studies are those of Sundararaj and Goswami who report that LH alone, of several pituitary factors tested, would stimulate ovulation in *Heteropneustes;* Ahsan who found a partially purified salmon gonadotropin to have an almost identical physiological action to mammalian LH when injected into hypophysectomised *Couesius plumbeus;* Fontaine and Chauvel, Fontaine and Gerard, and Schmidt *et al.* who have reported strongly positive Galli-Mainini reactions (sperm release in frogs) with partially purified extracts of carp and salmon pituitaries; Ramaswami who has used enzymes which selectively digest or inactivate the gonadotropic factors in mammalian pituitaries (trypsin or pepsin for LH and ptyalin for FSH) and found convincing evidence for the LH-like factor but no support for FSH in *Heteropneustes fossilis*. Sundararaj and Nayyar's work with the latter species is confirmatory as is also that of Yamazaki and Yamamoto and Yamazaki for the acceleration of ovulation in hypophysectomised goldfish. However, it cannot yet be concluded that the fish gonadotropin is always physiologically similar to mammalian LH. By using the spermiation test in goldfish, Yamazaki and Donaldson have obtained strong reactions with a partially purified salmon gonadotropin and with human chorionic gonadotropin (HCG) but no reaction with either FSH or LH.

The mammalian chorionic gonadotropins are often used for experimental purposes and in fish culture work because of their ready availability. These factors, produced by the mammalian placenta during later stages of pregnancy, mimic many of the activities of the pituitary gonadotropins although they are definitely known to be different biochemically from the pituitary factors. Two chorionic gonadotropins are in common use:

one, prepared from the serum of pregnant mares (PMS), usually has biological properties similar to a combination of FSH and LH; the other, prepared from human pregnancy urine (HCG), acts more like LH. In mammals, the precise effects depend somewhat on the dosage used. When injected into fish, HCG has often been found to be quite effective and this is in line with its LH-like activity. Pregnant mare serum is sometimes active but has usually been less consistently effective than HCG. Sundararaj and Goswami report that 100 IU/fish of HCG will induce ovulation in *Heteropneustes* while 250 IU/fish of PMS are required to elicit the same reaction.

The evidence, both biochemical and physiological, now indicates a single proteinaceous gonadotropic factor in the pituitaries of teleost fishes. Although this may sometimes mimic FSH and frequently mimics LH when injected into tetrapods, it is clearly not identical with either of these factors. It more closely resembles LH but has some unique physiological properties and may be expected to differ from its counterpart in the tetrapods and even differ in its effects in different groups of fish. Investigations of some of the more primitive groups of fish should be rewarding. The lactogenic hormone (prolactin, luteotropin) is also gonadotropic in some mammals (the ovary of the rat), but no evidence has yet been presented for a comparable action in the lower vertebrates.

The Gonadal Steroids

In each of the major groups of fishes, development of the secondary sex characters prior to breeding has been shown to depend on gonadal steroids. This broad generalisation is now backed by considerable experimental evidence and there seem to be no exceptions to it. Following gonadectomy, secondary sex characters fail to develop or, if seasonal in occurrence, regress. Their differentiation is initiated or stimulated by a wide variety of androgens and estrogens-both natural and synthetic. This may be the only broad generalisation possible; in all the later events associated with reproduction (ovulation, spermiation, spawning, and breeding behaviour) the division of regulatory responsibilities between pituitary and gonads appears to be rather variable in different species.

The physiologically most active gonadal hormones of higher vertebrates are testosterone from the interstitial cells of the testis, estradiol-17b and its derivatives from the ovarian follicle, and progesterone from the corpus luteum. The biogenesis of all these compounds is from acetate via cholesterol, as the parent sterol, to testosterone through intermediate steps involving progesterone; the estrogens are derivatives of testosterone or closely related molecules and form the terminal part of the biosynthetic chain. Thus, the biologically active gonadal steroids are linked through common pathways and one may expect to find small amounts of any or all of them in tissues which are primarily concerned with the synthesis of a single important hormone. The biogenesis of the adrenocortical steroids is linked to the same chain.

These metabolic pathways are very ancient phylogenetically. Progesterone, estradiol-17β, and some other estrogens have been identified in the ovaries of invertebrates (echinoderms and mollusks) as well as the lower vertebrates. Estrogens are also well known in plant tissues. It is not then surprising that both androgens and estrogens of several biochemical sorts have been regularly found in the gonadal tissues and blood of many different fishes. However, the presence of a gonadal steroid is no evidence of its significance as a hormone and the much more difficult task is to identify steroids which are physiologically important in regulating the reproductive activities of fishes. At present, it is by no means certain that all the steroids recognised as physiologically active in higher vertebrates are biologically significant in the fishes. Different groups of fishes seem to react somewhat differently to some of the same steroids. Moreover, it seems certain that there are important biosynthetic pathways of steroidogenesis in fishes which have not been recognised in the higher vertebrates although the general scheme of biosynthesis is probably similar in both groups. In fishes, as in the higher vertebrates, some of the synthetic compounds are more active than the naturally occurring ones.

The localisation, identification, and analysis of synthetic pathways of fish steroids are being investigated in many places using a variety of techniques-physiological, biochemical, and histochemical. The specific histochemical techniques now available for the detection of enzymes involved in steroid biosynthesis (e.g., 3,8-hydroxysteroid dehydrogenase) have been particularly useful in localising the tissue sites of hormone synthesis. Bern and Chieffi have provided a useful and comprehensive bibliography of the steroid hormones of fishes while Nandi summarises the comparative endocrinology for the nonmammalian vertebrates.

The Androgens

The tissues of several different elasmobranchs have now been investigated, and there is good evidence that pathways recognised for biosynthesis of androgens in higher forms are operating in the elasmobranchs and that testosterone is an important end product in the testis. Chieffi and Lupo, using chemical methods, investigated the testicular tissues of mature dogfish, *Scylliorhinus stellaris*, and recorded progesterone (100 μg/kg), testosterone (50 μg/kg), androstenedione (70 μg/kg), and estradiol-17β (20 μg/kg). Idler and Truscott were the first to isolate testosterone from the blood of a male elasmobranch. They found values which were relatively high (7.4 μg/ 100 ml in *Raja radiata*) when compared with human males (about 0.56 μg/ 100 ml). Testosterone was also found in the blood of female *R. radiata*, but the mean values were much lower. Progesterone, androstenedione, androsterone, and other steroids have been isolated from the semen of the dogfish, *Squalus acanthias*, and testosterone biosynthesis was demonstrated by incubation techniques in the testis of this fish.

A number of different androgens have been isolated from the tissues and blood of teleost fishes. The studies of salmon tissues by Idler and his associates are particularly significant. Steroid values have also been recorded for testicular tissues of the teleosts *Morone labrax*, *Mugil cephalus*, and *Serranus scriba*; the latter is a hermaphroditic form.

Idler's meticulous chemical analyses have included gonadal tissues and blood from both the Atlantic salmon, *Salmo salar*, and the Pacific salmon-particularly the sockeye, *Oncorhynchus nerka*. Testosterone has been found in significant amounts together with several other steroids such as 17α-hydroxyprogesterone, recognised in higher forms as a step in testosterone biosynthesis, and androsterone—a product of testosterone synthesis. As might be expected from the known relationships of these substances, androgens have also been found in the tissues of female fish. Testosterone values are extremely variable (ranging up to 17 μg/100 ml of male blood) ; since tissue values must represent a balance between synthesis and utilisation it is always difficult to relate the actual amounts to the physiology. Three particular facets of Idler's studies require special comment: (a) the presence of 11-ketotestosterone, (b) the occurrence of conjugated testosterone, and (c) the marked variations associated with the life cycle of the salmon.

Idler *et al.* were the first to identify 11-ketotestosterone as a natural product and subsequently to show its biological activity as an androgen in stimulating the development of secondary sex characters in salmon and chickens. The isolation of 11-ketotestosterone from both males and females of Atlantic and Pacific salmon suggests that the synthetic pathways of the androgens in these fishes are somewhat different from the usually recognised ones. This steroid is present in amounts up to 17 μg/ 100 ml of blood and occurs along with several of the more familiar androgenic compounds including testosterone. Idler and Truscott found that testosterone and 17α-hydroxyprogesterone will serve as precursors for 11-ketotestosterone in the sockeye salmon and, subsequently Idler and MacNab demonstrated its *in vitro* synthesis from adrenosterone and testosterone in Atlantic salmon gonads and sperm; they suggest the probable pathways of biosynthesis. Arai and Tamaoki also studied the *in vitro* synthesis in Atlantic salmon gonads and sperm from adrenosterone and testosterone; they too suggest probable pathways of biosynthesis. Arai and Tamaoki have studied the *in vitro* biosynthesis of 11-ketotestosterone in rainbow trout, *Salmo gairdneri.*

Conjugated testosterone was first reported in fish blood by Grajcer and Idler. In higher vertebrates, the steroids are transported, at least in part, as conjugates with serum proteins and glucuronic acid. Grajcer and Idler obtained the release of testosterone in amounts of 13.7 μg/ 100 ml of male sockeye salmon blood after treatment with the enzyme β-glucuronidase. The "free" testosterone in these samples of blood was 1.7 μg/ 100 ml. The blood of female salmon also contained conjugated as well as "free" testosterone with relatively less of the conjugated form (7.6 μg of conjugated to 7.8 μg per 100 ml of free) even though the total amounts were about the same. Since these initial studies of testosterone glucuronoside, Idler and Truscott have reported on conjugated testosterone in the skates, *Raja radiata* and *R. ocellata;* these conjugates are probably as common in fish blood as they apparently are in the higher vertebrates.

Schmidt and Idler studied the quantitative changes in several plasma steroids (including 11-ketotestosterone and testosterone) during the migration of the sockeye salmon. Definite differences were recorded and the shift in ratio of 11-ketotestosterone and testosterone may be significant in the physiology and behaviour of the spawning migration. Changing levels of androgens have also been reported in females of the ovoviviparous elasmobranch, *Torpedo marmorata*, during the reproductive cycle; there was a moderate increase in androsterone at the end of the gestation period and a marked steady rise in dehydroxyepiandrosterone from pregestation to the midgestation period. These variations may indicate physiological changes in the demands for estrogens rather than imply a role for androgen in the gestating ray. In the stickleback, *Gasterosteus aculeatus*, where the development of sexual behaviour is cyclical and associated with strong agonistic behaviour in the males, Gottfried and van Mullem report that the dominant males have testicular androgen levels which are five to seven times higher than the nondominant individuals; testosterone could not be detected in the testes of the nondominant fish but was present in the dominants. These cyclical and seasonal changes in gonadal steroidogenesis are presumably triggered by variations in the gonadotropic activity of the pituitary; seasonal changes in the latter have been traced in both plaice *Pleuronectes* and the perch *Perca*.

The Estrogens and Progesterone

Estradiol-17β has been found in the ovaries of many fishes at all levels in phylogeny: the lampreys, *Petromyzon marinus*, dogfishes, *Squalus suckleyi* and *Scyliorhinus caniculus*, the ray, *Torpedo marmorata*, the ratfish, *Hydrolagus colliei*, the lungfish, *Protopterus annectes*, and many different teleosts. This wide distributiontogether with its presence in several invertebrate phyla-suggests that it will be found in at least small amounts wherever there is active steroid synthesis. Estradiol-17β has also been identified in the blood of elasmobranchs and teleosts and is probably the physiologically most important estrogen. Gottfried tabulates estradiol-17β values for ovarian tissues ranging from a trace to 120 μg/kg, with the most usual amounts varying from 10 to 20 μg/kg.

Estradiol-17β is recognised as the parent substance for several other estrogens known to occur widely in the animal world. Particularly frequent in occurrence are estrone (an oxidation product of estradiol-17β) and estriol which is derived from estrone by 16-hydroxylation and reduction at 17. Both estrone and estriol were found in the ovaries of many of the species listed above, but the proportions are quite variable and occasionally one of the compounds is absent. Again, this is not surprising with a series of related substances in which the active utilisation of at least one of them probably varies with the stage of maturity. Cedard *et al.* traced seasonal changes in the estrogens of the blood of both male and female Atlantic salmon. The total estrogen content in both sexes showed a five- to sixfold increase at the time of spawning, reaching 5—6 μg/ 100 ml of blood. Estrone was found at all seasons; estriol appeared in significant amounts only at spawning; and estradiol was only present in small amounts and seemed to disappear completely in the females at spawning and in the postspawning males. Again, it is difficult to interpret the findings in terms of physiological demands but results such as these emphasize the hazards of conclusions based on a few estimations of the gonadal steroids at only one season.

Progesterone has been found in the tissues of all vertebrates and many of the invertebrates. It is probably ubiquitous as a link in steroid biogenesis. As yet, however, there is no definitive evidence that progesterone is a physiologically active hormone with distinct endocrine responsibilities. Hisaw has discussed this problem at length and concludes that the role of progesterone as a hormone was established during the evolution of placentation in the mammals although it probably had its beginning among reptilian ancestors.

Chieffi has convincingly correlated the development of corpora lutea with the stages of gestation in the electric ray *Torpedo*. He has also demonstrated histochemically the biosynthesis of a variety of steroids-including progesterone-in the corpora lutea, while chemical methods have identified progesterone in *Torpedo* blood during pregestation, but again the question of its presence as hormone or precursor has not been settled. It appears, however, that active steroidogenesis occurs in the corpora lutea of some fishes; it seems equally evident that this is not the case in all fishes. Further work is required to clarify the endocrinological status of these structures.

Reproductive Cycles and Their Coordination

Reproduction is almost always a seasonal or cyclical phenomenon. An annual cycle of temperatures and photoperiods characterises the temperate and frigid zones; the rainy seasons may markedly alter freshwater habitats of the tropics. In these seasonally unstable environments, reproduction is geared to take advantage of seasons which offer the greatest opportunities for survival and development of the new generation. Even where conditions are relatively stable and the eggs or young are produced regularly throughout the year, there may still be a cycle of gonadal maturation imposed by the energy demands of maturing a batch of eggs or young.

These cycles of gonadal development frequently alter many aspects of metabolism as well as behaviour and reproductive physiology. Cycles of active feeding with storage of fat and long periods of starvation are characteristic of many species. The electrolyte metabolism may change seasonally both in species which inhabit the ocean waters of relatively constant salinity and in the euryhaline and anadromous forms. The basic cycle is probably the one imposed by the seasonal nature of reproduction, and these regular changes in metabolism are secondary to it.

The endocrine system forms the major link between the environment and the organs concerned with reproduction. Changing environmental conditions, operating through the sensory system and specific centers in the brain, trigger neurosecretions which in turn regulate the activities of the pituitary gland. The pituitary hormones have direct effects on gametogenesis, metabolism, and behaviour; these hormones also regulate the development of the gonadal endocrine tissues. Gonadal hormones take over some of these pituitary responsibilities and carry on the coordination of events in the production of gametes, sexual behaviour, fertilisation, and sometimes parental care. There are, in fact, several cycles within the gonads, but many of the details of regulation at the cellular level remain to be unraveled. Studies of the cyclical changes in the structure of the gonads and endocrine organs, as well as variations in actual secretion of hormones, form one of the most voluminous components of the literature concerned with reproduction. No attempt will be made to summarise it here; the reader is referred to the many existing reviews.

Although the pattern of regulation via the peripheral sense organs, the neural centers, pituitary, and gonads is general throughout the vertebrates, the details vary considerably even in closely related species. A definite breeding season is absent in the female of the spotted dogfish, *Scyliorhinus caniculus*, and in the male spermatogenesis is continuous with a regular progression in the ampullae from spermatogonia to sperm ; in comparison there is a limited breeding period of about 2 months in the spiny dogfish, *Squalus acanthias*, and during this period there is a maximum accumulation of semen in the efferent ducts which is followed by a pause in sperm production. This suspension of sperm production is the result of a cessation in the proliferation of spermatogonia several months earlier. Since a band of degenerating ampullae appears in the testis of the spotted dogfish following hypophysectomy, there seems little doubt that these cyclical changes in the sperm production of the spiny dogfish are regulated by the pituitary gonadotropins.

Similar examples could be drawn from many different groups of fishes. Occasionally, the details of regulation may even be somewhat different in the males and females. There are diurnal as well as seasonal cycles. In the cyprinodont, Oryzias latipes, there is a daily cycle of ovulation (between 1 AM and 5 AM), mating behaviour (4 AM to 7 AM), and the laying of fertilised eggs which follows soon after mating. Ovulation and oviposition are independent phenomena; the former depends on the temperature and the light cycle while the latter depends on contact stimuli associated with the sexual embrace. In another cyprinodont, Rivulus marmoratus, there is an internal self-fertilisation which is also timed by the daily light cycle. Rivulus marmoratus is hermaphroditic with functional ovotestes. The timing of events within these organs is so correlated that there is a peak frequency in ovulation; fertilisation occurs at dawn with a peak in oviposition at noon. Harrington also found evidence of seasonal changes in the cycles in accordance with light conditions.

The modifications in physiological controls are fully as numerous and diverse as those of an anatomical nature. In a sense, the reproductive system of an animal is independent of the other organ systems and can, perhaps, respond to evolutionary pressures more freely; although reproduction is indispensable to the survival of the species, it is not a matter of life or death for the individual. Perhaps, for this reason, the adaptive variations in anatomy and physiology are particularly diverse and curious in the biology of reproduction. At any rate they emphasize the opportunism, the compromise, and the adaptiveness of Darwinian evolution.

2

CULTURE OF FISHES

This chapter is intended to be a comprehensive source of information and a guide to methods of fish cell and tissue culture. Among the subject texts, Cameron presents the most information on fish cell and tissue culture, but the book is largely obsolete and inadequate for today's techniques and needs. There are specific requirements for culturing cells and tissues from fishes, but the basic principles apply to all animals. Parker, White, and Paul are single volume subject texts which are currently in their second or third editions. The newest and largest reference is the three-volume set edited by Willmer. Merchant *et al.* have prepared an excellent laboratory guide for both the student and technician or investigator; it has directness and simplicity. All the foregoing works reflect the fact that attention has been concentrated on mammalian and avian cells and that there has been much less work with the lower vertebrates and invertebrates.

Because the techniques have such widespread application, tissue culture literature is widely scattered and the interested reader must use several approaches in reviewing and keeping abreast of developments. A bibliography of cell and tissue culture covering was compiled by Murray and Kopech. The same workers undertook the tremendous task of compiling the subsequent literature; several issues of a serial publication were printed, but the endeavors were soon halted and fate of the project is unknown. Watson has compiled a bibliography which is restricted to invertebrate and poikilothermic vertebrate cell and tissue culture. In addition to the well-known biological indexes and abstracts, some of the current literature is abstracted by biological supply houses and published in serials. *Tissue Culture Abstracts* is an organ of Grand

Island Biological Co., Grand Island, New York, and *Tissue Culture Bibliography is* published by Microbiological Associates, Inc., Bethesda, Maryland.

The newcomer will find the catalogs from cell and tissue culture supply houses to be additional sources of useful information. Those from Grand Island Biological Co.; BioQuest, Cockeysville, Maryland, and Difco Laboratories, Detroit, Michigan, have "how-to-do-it" sections and informative descriptions and explanations.

History

Animal cell and tissue culture has advanced in a very short span of time from an esoteric art to a workaday tool of many uses. The dominant stimulus for this evolution has been the widespread need and the use of cultures in growing virus; knowledge of both animal virology and animal cell culture has grown apace.

Quite understandably, researchers working with diseases such as cancer, poliomyelitis, and a number of acute human, avian, and mammalian viral infections have concentrated efforts on the development of culture methods for homoiothermic animal cells and tissues. However, the techniques of cell culture had their origin not in mammalian tissue but first in chick tissue culture, next in frog, and soon after that in fish tissue.

The early work was quite simple and consisted of short-term study of bits of tissue in slide or watchglass preparations, and this all began in when Roux maintained chick neural tissue in warm saline. In, Arnold lured frog leukocytes into elder pith soaked in aqueous humor and observed cell movement and behaviour. It was also the frog which provided Harrison in with neural tissue which in clotted lymph showed outgrowth of nerve cell fibers-the first real "tissue culture."

Like its parent techniques, the evolution of fish cell and tissue culture follows a progression in time and complexity; at first, small bits of fully organised tissue were merely kept alive for a matter of hours in saline. Cover slip, embryo dish, and watchglass preparations gave way to flasks and culture tubes, while saline was fortified with inorganic ions and a wide array of sera and other natural fluids. Today's methods use a wide variety of glass and plastic laboratory ware. Media have become very complex, but the commercial availability has diminished the problems of preparation by the user.

Today, media may be purchased from a number of suppliers who offer it in ready-to-use form, in liquid concentrates, or in powdered or lyophilised preparations. Solution or reconstitution of the latter requires the use of high purity water. At the very pinnacle of today's accomplishments, some mammalian cell lines are grown in completely synthetic media.

The first report of fish tissue culture was made by Osowski who maintained fry and embryonal trout explants for 24 hr in both Ringer's solution and in frog lymph. In Lewis described a physiological medium based on seawater and employed bouillon as a source of nutrients. Subsequently, Dederer grew *Fundulus* embryo explants and kept heart beating for 10 days in Locke's physiological saline plus glucose and fish bouillon. In, Goodrich also used a similar medium for his work with *Fundulus*. Chlopin grew in cover slip culture a number of different tissues from lamprey, pike, and crucian carp using rabbit plasma diluted with fish spleen extract, and he specifically reported mitotic cells. He also cultured freshwater teleost and amphibian explants compatibly in the same vessel.

Except for an occasional paper during the and, fish were not in the mainstream of tissue culture. Pfeiffer cultured lamprey, *Petromyzon fluviatilis*, explants in Chlopin's medium. Lewis and MacNeal used a combination of fish and chick plasma for hanging drop cultures of pituitary from three marine teleosts and two marine elasmobranchs; the latter was the first work with this class of fish. They stressed the importance of isotonicity for the various animals' tissues and were able to maintain ciliary beating for 8-10 days and cultures for 3 weeks. In, Grand, Gordon, and Cameron cultured melanotic tumor tissue from freshwater teleost aquarium species with weekly changes of a medium composed of fish serum, chicken plasma, and embryo extract. They kept cover slip cultures "alive and active for many weeks." Prior as it was to the introduction of antibiotics, they were confronted with the problems of microbial contamination and found dilute Merthiolate to be a helpful disinfectant for use on the surface of their fish. Grand and Cameron used the same kind of medium for a similar study with pigmented tumor tissue of *Fundulus*, but they had the advantage of penicillin. The use of antibiotics was a major advance in animal tissue culture for it helped transform the techniques which at that time were almost a goal in themselves into a tool which could be easily and conveniently used by many.

Carrel introduced flask cultures in, but with few exceptions they were seldom used in early fish tissue culture. In time T-flasks and plastic flasks replaced the Carrel flask, but the culture tube became the standard vessel. Media too were improved, and semisynthetic preparations based on blood serum analyses were developed. Commercial marketing of such preparations began in, and this too was an important factor in speeding the advance and application of cell and tissue culture.

Although fowl pox had been grown in chick tissue culture some 20 years earlier, a major breakthrough occurred in when Enders, Robbins, and Weller grew poliomyelitis virus in nonneural human tissue· *in vitro*. Reciprocal stimulation had begun between virology and animal tissue culture, and the latter soon gave rise to the more sophisticated cell culture. During this transitional period, Schlumberger used roller tubes and mammalian-type medium in a tissue culture study of a neoplasm from adult goldfish, *Carassius auratus*. Some evidence suggested an infectious agent and although the etiology of this tumor has never been established, this was probably the first application of fish tissue culture in virology. Soon afterward, Sanders and Soret and Soret and Sanders reported growth of an arthropod-borne virus in fish tissue culture; in this case the "tissue" was actually intact *Gambusia* embryos grown in mammalian-type medium. The same methods permitted *in vitro* development of active young fish from early stages of eggs. Although it was not used for the bulk of the study, their work marked the first report of a synthetic medium designed for mammalian cells (medium 199) being used for teleost tissues.

Published two reports of a series of carefully conducted, detailed studies of fish tissue culture and its applications in fish virology. She cultured explants of embryonal juvenile and adult aquarium teleosts in diluted mammalian-type medium. Heart was kept beating and gut showed peristalsis for 2 weeks, and cultures in general could be kept for 3-6 weeks. Such cultures were employed in a study of viral lymphocystis disease and of carp pox. In the same year, Wolf used explants of adult trout tissue in mammalian-type medium in an attempt to determine whether virus was involved in bluesac disease of fish. The following year, Wolf and Dunbar reported the use of mammalian-type media for culture of adult trout and goldfish tissues. They reported no subculturing but kept cilia beating for nearly 2 months and goldfish heart beating for 10 months.

Grutzner made a major contribution by describing the first trypsinisation of fish tissue yielding cultivable cells which grew in monolayer. The methods which can be considered "second generation" followed those which had been developed for mammalian use, but temperature during digestion was 20°C or less and solutions were diluted to correspond to the reported freezing point of tench, *Tinca tinca*, serum. Grutzner's primary monolayers were successfully subcultured and thus *fish cell culture* had been established. Although its applications were to decrease, *fish tissue culture* was extended and used by others, Bargen and Wessing, Ghittino, Kunst, and Wolf *et al.* and the latter made the first isolation of a fish virus in cultures of trout fin explants.

Unaware of Griitzner's work, Wolf *et al.* employed a different procedure for use with cold-water teleosts. Using extended trypsinization at 4°-6°C and unmodified mammalian-type medium, they cultured cells of six freshwater teleosts, an amphibian, and a reptile. Subcultures were effected by mechanical dispersion as well as with trypsin or disodium versenate. Significantly, some tissues were stored at 4°C for 24 hr before use. Preparation of monolayer cell cultures from enzymically disaggregated fish tissues has been reported by a number of other workers who anticipated or made immediate application in fish virology. Clem *et al.* were the first to establish monolayer cell cultures from marine teleosts and understandably obtained their best results in commercial medium modified with about 0.07 M added NaCl.

Some generalisations seemed to be evolving. There was uniform agreement that teleost tissues could be trypsinized and grown with methods very much like those used with mammalian cells. Temperatures needed to be lower however (opinions differed as to degree), and all agreed that mammalian serum was at least as good and in most cases better than homologous serum. Embryonic and gonad tissues emerged as dependable sources of cultivable cells from fish and other lower vertebrates.

Continuously cultivable cells meeting certain criteria are termed "established" or "permanent lines" and might properly be considered a "third generation" in the advance of fish cell and tissue culture. Cell lines are usually derived from monolayer cell culture, but this is certainly not a requirement and the most noteworthy exception was the first-Carrel's line of chick fibroblasts, which were propagated for 34 years. Cell lines have a number of advantages, and as early as serial subcultivation of some tissues had been effected at the Eastern Fish Disease Laboratory; however, none of the early efforts resulted in a permanent cell line. With trypsinized cell preparations the likelihood of success was increased. RTG-2 cells of rainbow trout, *Salmo gairdneri*, gonad origin were one of several lines initiated at our laboratory in January,; they were the first permanent fish cell line to be developed. Within a year, Clem *et al.* initiated trypsinized blue-striped grunt, *Haemulon favolineatum*, fin cultures which provided GF-1 cells, the first line of marine fish origin. The next established line was initiated in it was an epithelial-like cell from the fathead minnow, *Pimephales promelas*, and designated FHM by its originators Gravell and Malsberger.

At Oregon State University, Department of Microbiology, Fryer made a very carefully quantitative study of requirements for dispersing tissues and culturing cells from embryonic Pacific salmon and rainbow trout hepatoma. The work is additionally noteworthy for it reports the establishment in of five different lines of cells from these fishes. Part of the work has been published, and equally important is the fact that the lines are still prospering.

Basic studies in fish cell and tissue culture continue to appear. Townsley *et al.* employed mammalian-type culture medium without osmotic adjustment and reported the successful culture at 5°C of explants from a number of cold-water Atlantic teleosts and one elasmobranch. Li and Stewart recently used short-term monolayer cultures of rainbow trout ovary for evaluating the growth promoting qualities of several natural substances. They found that a combination of serum and embryo extract gave maximum proliferation.

In the first such study ever to be made, Pilcher *et al.* compared growth and glycolysis in two established embryonic salmonid cell lines and a line of diploid human embryonic cells. Although incubating temperatures of necessity were different, they found one of the salmon cell lines grew almost as rapidly as the mammalian cell, but that rates of glucose utilisation and lactic acid formation were lower in the fish cells. They concluded that glycolysis in all cells was similar but that it differed quantitatively.

Understandably, most applications of fish cell and tissue culture have been in virology, but though they may be fewer in number, important applications have been made in other research areas. Cultured fish cells have been shown to produce interferon and in the same laboratory, Ortiz-Muniz and Sigel used organ cultures to demonstrate *in vitro* synthesis of antibody by fish lymphoid tissue. Roberts effectively used cell culture as a tool for determining the chromosome numbers of fishes. Because the *in vitro* approach provided a completely controlled endocrine environment, Hu and Chavin employed goldfish fin explants for a study of hormonal stimulation of melanogenesis. They found only ACTH capable of stimulating *in vitro* melanogenesis.

As with any system, things can go wrong and among those who work with mammalian cells there have been a number of disconcerting incidents of mistaken identity-possibly by mislabeling or "contamination" with another cell. To forestall such mixups among lower vertebrate cell lines, Levan and his co-workers showed chromosome idiograms to be useful in distinguishing certain cell lines. A simpler and more specific approach has been the cytotoxic antibody test.

Quite recently, there has been a revival of interest in *in vitro* culture of nonteleost fishes. Stephenson and Potter worked with a cyclostome while Wolf and Quimby used both a cyclostome and elasmobranchs.

Present Status

Many different kinds of tissues and cells from freshwater and marine teleosts can be cultured on a routine basis. As a general index of the state of the art, teleost cells are the second most numerous among the animal lines which have been developed-mammalian cells being the most numerous. Although all are heteroploid, there are at least 17 extant lines of fish cells.

Thus far, fish cell and tissue culture has been predominantly concerned with material from teleosts. Very little is known about culture of cyclostome cells, and while there is some indication that present methods and media may be satisfactory, there is also some evidence that both cyclostomes and elasmobranchs may have specific requirements different than teleosts.

The methodology for *in vitro* culture of fish cells closely follows that used with homoiotherm material; the major differences being, first, in temperature requirements and tolerances, and, second, in osmolarity of salines and media. For the most part, mammalian-type solutions are entirely satisfactory for many freshwater teleosts and possibly for cyclostomes. For marine fishes, best results are obtained after the osmolarity is increased.

Fish cells are comparable to mammalian cells in their response to freezing and storage at ultra-low temperatures, but like all poikilothermic animal cells, their rate of metabolism can be manipulated with temperature control. This attribute provides the researcher with a range of growth or activity rates for his studies, and at the lower end of temperature tolerance it permits long-term storage of cells without freezing.

To a certain extent, the present status of fish cell and tissue culture can also be described in negative terms.

PHYSIOLOGICAL SALINES

General

Physiological salines are fundamental to cell and tissue culture for they maintain pH, provide osmotic pressure, essential ions, and glucose as an energy source and are therefore in themselves adequate for handling, washing, manipulating, and short-term holding of living materials. More importantly, they provide the inorganic foundation on which media are elaborated. Ringer's solution, an empirical mixture of the essential ions Na, K, and Ca, was the prototype saline. Today's salines, balanced salt solutions (BSS) as they are usually called, are commonly patterned after the inorganic constituents of blood serum. Sera from the major vertebrate classes have basic similarities in their inorganic chemical composition, accordingly this resemblance is reflected in the physiological salines which are based upon them. In theory one should be able to use with fish the salt solutions which have been formulated for homoiotherms. In practice this has worked very well indeed, the one note of caution being that adjustment in osmotic pressure has usually been found necessary for marine forms. The existing mammalian-type balanced salt solutions then are in effect *vertebrate physiological salines*.

Understandably, the first physiological salt solutions were developed for use by physiologists, and different formulations were prepared for use with the various kinds of fishes. Young, who carried out studies with three classes of fishes, summarises fish physiological salines to that date. Lockwood prepared an extensive exposition on physiological salines for animals of different phyla including fishes. Curiously, it includes principal salines in use with mammalian cell and tissue culture, but there are no references on fish cell and tissue culture. The reverse situation is almost equally true; recent methods of fish cell and tissue culture have not employed the solutions used by the fish physiologists.

In the earlier literature on fish tissue culture one finds a considerable array of physiological salines. Reports and micrographs of tissue growth show that these solutions were adequate for their purpose. By present standards, however, they are primitive-being deficient in some components or lacking completely certain ions. Few had an energy source, and most were not adequately buffered. Undoubtedly one reason they were adequate is the fact that the tissues were not kept in saline for long periods. In addition, the culture systems usually employed rather large volumes of tissue for the volume of medium, therefore leaching could occur and the system would move to equilibrium. Most importantly, however, then as now, salines were usually only a part of the culture medium, and many deficiencies were met by serum, tissue extracts, or other natural products. The earlier physiological salt solutions are principally of historic interest. Today, virtually all fish cell and tissue culture employs the "mammalian-type" BSS usually without modification, but in some cases diluted or supplemented with additional NaCl (marine forms) or urea (elasmobranchs) or both.

Table 5.1 : Balanced Salt Solutions for Use with Fishes.

Constituents	**Earle's BSS-man (g)**	**Hanks' BSS-man (g)**	**Cortland BSS-brown trout (g)**
NaCl	6.80	8.00	7.25
$CaC1_2 \cdot 2H_2O$	0.27	0.19	0.23
KC1	0.40	0.40	0.38
$NaH_2PO_4 \cdot H_2O$	0.14	—	0.41
$Na2HPO_4 \cdot 2H_2O$	—	0.045	—
$NaHCO_3$	2.20	0.35	1.00
KH_2PO_4	—	0.06	—
$MgCl_2 \cdot 6H_2O$	—	0.10	—
$MgSO_4 \cdot 7H_2O$	0.20	0.10	0.23
Glucose	1.00	1.00	1.00
Water (ml)	1000	1000	1000
Mean freezing point (°C)	–0.58	–0.59	–0.58

Hanks' BSS is by far the most commonly used saline in fish cell and tissue culture, as indicated in about 80% of the reports published during the last 15 years. Earle's BSS has had limited use as a saline per se; it is however the most common base used in media for growing fish cells. Gey's BSS and the Cortland salt solution which was based on the determinations of Phillips *et al.* each have one report of use with fish. Hanks', Earle's, and Gey's solutions were formulated for mammalian cell use; the first two, and to a lesser degree the third, have won wide acceptance in homoiothermic cell culture. These salines are offered by cell and tissue culture supply houses and are available both in liquid and in powdered form. The Cortland salt solution was based on serum constituents of a freshwater teleost. A comparison with Hanks' and Earle's BSS shows their basic similarity.

Modifications in Balanced Salt Solutions for Use with Various Fishes

Freshwater Teleosts

According to Black, freshwater teleost blood has a freezing point of about -0.57°C (the range, -0.38 to -0.89°C). Lockwood cites somewhat narrower limits (-0.64°C to -0.45°C), but the mean is comparable (-0.54°C). One might conclude from the composition and freezing point of the foregoing salines that they are quite suitable for freshwater teleosts. Indeed, most workers have used them without alteration; others have found it desirable and even necessary to dilute the salines about 20%. Kunst initially found it desirable to dilute salines and media about 20%, but in subsequent comparisons found undiluted solutions to be better.

Osmotic concentration is usually reduced with water, this effects dilution without changing ionic ratios. An alternate method is to dilute with saline from which the NaCl has been omitted. Dilution alters the ionic ratio but maintains levels of other components. It is generally understood that slight hypotonicity is desirable for *in vitro* cell and tissue growth, but critical comparisons of osmotic concentrations have not been carried out. On the other hand, most freshwater teleost cell and tissue culture, including the establishment of permanent lines has been with salines and media having a freezing point of about -0.6°C.

Marine Teleosts

Clem *et al.* used both unmodified Hanks' and preparations which were supplemented with 0.07 M additional NaCl. They reported significantly better results with the latter and established a permanent marine teleost cell line. We have successfully used isotonically adjusted salines and medium (addition of 0.3 g % NaCl) for handling, establishing cultures, and subculturing cells of the spot, *Leiostomus xanthurus*, *a* marine teleost. If the serum freezing point has been properly determined for fish which are normal and healthy, the saline and medium should be matched accordingly. As an example, we established primary cultures of frog tongue fibroblasts in medium having a freezing point of about -0.59°C—considerably hypertonic for the frog. Several subcultivations were effected but cell number declined and cells became pyknotic. The medium was then diluted with NaCl-free Earle's BSS to a freezing point of about –0.47°C. Cell quality and vigor immediately improved, and a permanent cell line was established.

Supporting the thesis that a strongly hypotonic saline and medium is similarly unfavorable are the findings of Roberts. Although Atlantic herring cells grew in medium having a freezing point of –0.58°C, they were highly vacuolated, and Roberts considered this a possible consequence of hypotonicity.

Marine Elasmobranchs

A number of physiological salines have been developed for elasmobranchs, but only two have been used in tissue culture. The more complete of the two is Hanks' BSS modified with additional NaCl and urea.

Lockwood gives the composition of five different salines for different elasmobranchs; but none was used in cell or tissue culture. An additional reference is that of Pereira and Sawaya. All the formulations include the very high but completely normal and physiologically necessary levels of urea-the single component which literally characterises the serum of these fishes. Lewis proposed a culture medium based on a physiological saline. It is noteworthy for the fact that it has the necessary levels of both NaCl and of urea. We have tested salines and media with compositions intended for freshwater teleosts or marine teleosts and found them quite unsuited for marine elasmobranchs. Empirically, we determined the levels of NaCl and urea necessary for the most favorable response of shark cells and tissues. We used Hanks' BSS and 1.75-2.0 g % urea added as a small volume of stock concentrate. The stock solution was prepared by dissolving 40 g of urea in 50 ml of water. This was filter sterilised and had a volume of approximately 80 ml; accordingly each milliliter contained 0.5 g of urea. Sodium chloride alone did not provide proper physiological conditions even though the medium was isotonic with serum from our donor animals (-1.6°C). Hanks' BSS with about 0.5 to 0.6 g % additional NaCl (and the urea levels above) had a freezing point of about -1.5°C and seemed suitable for several species of sharks. There has been no culture work on freshwater elasmobranchs, but the animals have high urea levels, and this compound may be needed in the medium.

Cyclostomes

Unmodified Hanks' BSS has been found suitable for freshwater larval cyclostomes. From data presented by Urist and Van de Putte the inorganic constituents of teleosts and cyclostomes in freshwater are very similar. Conversely their data show that salines for marine cyclostomes would require higher levels of NaCl.

MEDIA

Not only do cell culture media provide nutrients, but also they are the reservoir for metabolic products. Considering nutritional requirements to mean knowledge of the qualitative and quantitative needs for specific elements, ions, vitamins, amino acids, etc., virtually nothing has been determined for fish cells. However, much of what is known about homoiotherm cells probably applies, at least broadly, to all poikilothermic vertebrate cells. Thus far, a "shotgun" approach has been effective in growing fish cells; they simply are provided with the inorganic salts, vitamins, and amino acids kr own to be needed by mammalian cells; for good measure, generous quantities of serum or other nutritive supplements are added. This method has certainly worked; it has permitted widespread application of fish cell and tissue culture as a tool, but it has contributed little to knowledge of nutritional requirements for fish cells *in vitro*.

Nutritional Factors

The first animal tissue culture media consisted of very simple salines plus serum, tissue extracts, or other natural supplements. Eventually, the proportion of natural components was reduced and the number of defined substances was increased. Thus the media became semisynthetic; they consisted in part of a defined basal solution of inorganic salts, glucose, vitamins, and amino acids. To sustain growth, such media required the addition of serum or other natural products. Today, practically all fish cell and tissue culture-and the great bulk of homoiotherm cell culture-uses media of this developmental level. The ultimate goal in cell culture has been to grow cells continuously in completely defined media. Today there are several formulations which supply all the requirements of some mammalian cells. These media quite probably will meet the requirements of at least some cells from some fishes. As formulations are improved, it is expected that other cell types can be grown in them. On the other hand, it is possible that synthetic media may have to be specifically tailored for certain cells from specific kinds of fishes.

As previously mentioned, the inorganic composition of mammalian and teleostean sera are quite similar, but too few data are available from fish to compare the organic constituents. The principal organic components may also be similar, for when the solutions are isotonic, teleost cells readily grow in media developed for homoiotherm cells. The same is true for cells from reptiles and amphibians. In discussing animal cell nutrition Paul has said: ". . . the most extraordinary thing about known nutritional requirements of different cells is their *general similarity*" (italics added). There are few published data, but it appears that cyclostome cell requirements could be like those of teleosts. The elasmobranchs also show a basic similarity, but there are significant differences, not only in the urea tolerance but also in the near absence of albumin.

Considering the fact that most fish cell and tissue cultures are now grown in semisynthetic media with natural supplements and that these have proved superior to older media, there seems to be little justification in discussing or enumerating the more primitive preparations. The interested reader will find partial reviews in Griitzner, Wolf and Dunbar, and Fryer.

Although their own definitive work employed a different medium, Soret and Sanders were the first to report use of a semisynthetic solution, medium 199, for fish tissue culture. Wolf and Dunbar grew tissues from several fishes in a mixture which contained medium 199, but in addition to high levels of serum they also used chick embryo extract. Griitzner also used medium 199; in part it was mixed with multiple supplements, but more importantly she was the first to use a synthetic medium with serum as the sole supplement-the mode of today's fish cell culture.

Cataloging the details of reports on fish cell culture shows that three synthetic media have been used far more than others; in order of decreasing frequency of use they are: medium 199, Eagle's minimal essential medium (MEM) and Eagle's basal medium (BME). Other synthetic media which have been reported as suitable for fish cells are: CMRL 1066 ; Leibovitz L-15 ; McCoy's 5a ; NCTC 109, ; and Puck's.

Eagle's MEM is the one medium that we suggest for routine fish cell and tissue culture. For our own use we prefer the formulation which uses Earle's BSS because it has greater buffering capacity. Admittedly, medium 199 has had a slightly greater frequency of published use, but from our own experiences and discussions with others, including people in the commercial supply houses, MEM has been effective and unquestionably is now used more frequently. Also, the cost of MEM is about 10% less than that of medium 199; in addition, MEM is used for culture of amphibian and reptilian cells and is widely employed in homoiotherm cell culture work. Other things being equal, it is likely that most if not all of the synthetic media originally intended for homoiotherms will prove satisfactory for culture of teleost cells and probably for cells from other fishes too. For exploratory work with new fishes or new techniques, we suggest a variety of media be used in order to determine which is the most satisfactory.

Serum Additives

With the above synthetic preparations, serum or other complex undefined organic materials are usually added—and probably needed

Table 5.2 : General Purpose Culture Media for Fish Cells and Tissuesa.

Culture type	*Synthetic component*	*Supplements* — *Required*	*Supplements* — *Optional*
Explant or monolayer in a closed system	Eagle's MEM with Earle's or Hanks' BSS, medium 199	5-20% serum (fetal bovine or calf serum suggested)	5% whole egg ultrafiltrate
Explant or monolayer in petri dish or other open system	Leibovitz' L-15	As above	As above

for sustained vigorous growth. Disregarding for the moment the quality of response, cost, and inherent risks, and considering all fish cell and tissue culture, homologous fish serum and calf serum have been used most often. These are followed by heterologous fish, human, human cord, and chicken serum. Horse, bovine, rabbit, fractionated calf serum, sheep, and swine serum have also been used. Some generalisations can be made about the results. Teleost cells often fare poorly in rabbit serum, and in 10% chicken serum they accumulate abnormal quantities of cytoplasmic globules which may be lipid. When chicken plasma is used as a clotting matrix and the culture medium contains another serum, the quantity of chicken serum released is too small to result in problems of cytoplasmic globule accumulation. Stephenson and Potter obtained excellent results with larval lamprey tissues grown in chicken serum. However, it is one of the more expensive sera. Human cord serum has uniformly been found to be excellent for culture of fish materials but limitations because of cost or problems in collecting and processing restrict its use. Human serum generally has been good to excellent.

Depending on the various workers, horse, bovine, and sheep sera have given good, fair, or poor results; there are too few data to support a generalisation. Fish sera, either homologous or heterologous, have also given mixed results. Among the reports of the more recent workers, some have found the growth response to fish serum to be excellent, but for others, heterologous or homologous fish serum has been inhibitory or toxic.

Compared to mammals, fish have a smaller blood volume. This can often be a problem in acquiring adequate amounts of fish serum for cell culture; of course, the problems are virtually insurmountable with the very small fishes. Perhaps the greatest deterrent to advocating use of fish serum for fish cell culture is the risk of introducing latent fish viruses.

Calf serum, be it the natural material from fetal or full-term animals, or chemically fractionated, stands out clearly as the serum of choice for growing teleost cells and tissues. It is also suitable for cyclostome cells and has been used for elasmobranch material but not completely satisfactorily. The problem with elasmobranch cells has been massive cytoplasmic accumulations of granular material, possibly albumin, which is almost a foreign protein to sharks, skates, and rays. Where budgets can afford it, fetal bovine serum is the serum of choice. It is notable for its lack of toxicity and for its growth stimulating properties. Without doubt, one of the most important reasons for its widespread use in fish cell culture is that much of the work has been in fish virology and fetal bovine serum is generally free of virus neutralising activity. In contrast, although the material may have been nonspecific, some lots of calf and human serum have been found to neutralise infectious pancreatic necrosis virus. Cross-reacting antibody may also neutralise virus, and to eliminate this possibility, y-globulins may be selectively precipitated; this results in the so-called agamma calf serum. Where viral neutralising activity is not a consideration, calf serum should suffice. Of course whether locally or commercially prepared it is always prudent to test any serum for growth promotion, toxicity, and viral neutralising activity prior to large-scale use. When serum from commercial sources is used and sufficiently large volumes are involved, the biological supply houses will usually provide a sample and reserve a specific lot or portion thereof pending the outcome of such testing. We use this system of pretesting serum and buy a year's supply at a time. It is stored at –20°C.

When used with MEM, medium 199, or comparable synthetic medium, the usual level of serum is 10-15%. Some fish cells will grow with as little as 2% serum, but the growth rate is slow. Some have found it necessary to use as much as 20 or even 30% serum levels, the latter however causes one to wonder whether such a high level was necessary, or whether the other components were adequate.

Other Addttives

A number of natural and partially processed materials have been used as nutrients for fish cell and tissue culture. In some instances these products are used in lieu of serum, in other work they have been used in addition to serum. The products are human ascitic fluid, bouillon, bovine amnionic fluid, fish, bovine or chick embryo extract, lactalbumin hydrolysate, serum ultrafiltrate, peptone, yeast extract, and whole egg ultrafiltrate. Bouillon is of historical interest and has not been used for about 30 years. Human ascitic fluid has been used with saline and chick embryo extract as a complete medium for very short term work. Greenberg *et al.* used 55% ascitic fluid with chick plasma and embryo extract and found it gave better results than a comparable amount of carp serum. Townsley *et al.* found that with medium 199, a 10% level of ascitic fluid gave excellent resultscomparable to the same amount of human serum. Li and Stewart used ascitic fluid in conjunction with 15% serum and 5% embryo extract in medium 199-the mixture gave a very rapid rate of growth. There is no doubt that ascitic fluid is a good source of nutrients; its cost and secondarily its pathological origin are factors which tend to limit its use.

Embryo extracts have almost been synonymous with tissue culture. They have been used as a stimulant to growth and as a source of enzymes essential to clotting of plasmas. For either use, fish, chick, or bovine embryo extracts can be used interchangeably. As a component of medium, the commercial products are expensive and there now seems to be little real justification or need for embryo extract in that role. Embryo extracts can be made locally quite cheaply, but it seems likely that similar benefit can be had from a comparable amount of good serum. In our own media we discontinued using embryo extract in 1959.

Ultrafiltrates, protein-free fractions of serum or of whole chicken eggs, are also used as sources of nutrients and growth factors. Though effective, there have been very few applications in fish cell and tissue culture, and the principal limiting factor has probably been cost. *In vitro* development of *Gambusia sp.*, from fertile egg to larva, has been carried out in medium composed of 20% ox serum ultrafiltrate and 80% Hanks' BSS. We have used a 10% level of whole egg ultrafiltrate in medium for developing a permanent line of amphibian cells and a 5% level during the development of several fish cell lines when it appeared that vigor was declining and continuation of the culture was threatened.

Bovine amnionic fluid (BAF), a natural product with physiological ions and a low level of protein, has had limited use in vertebrate cell and tissue culture as a saline substitute or serum extender. Under some local conditions it may provide a cheap component for medium, but commercial prices range from about half to twice that of calf serum. It has comparatively few advocates and its applications have been declining. When used at a 5-20% level it had neither significant advantages nor disadvantages. Grutzner obtained growth of tench, *Tinca tinca*, liver and occasionally of kidney cells in 70% BAF (containing lactalbumin hydrolysate and yeast extract) and 30% calf serum. However, this medium was later reported to be unsuited for tench swimbladder and gonadal cells. Kunst tried five media for growing carp, *Cyprinus carpio*, kidney cells and concluded that the best results were obtained with 74% BAF plus 26% calf serum when these were isotonically diluted (about 20%). Cells did not grow when the medium was not diluted. In later work with carp ovary cells the use of BAF was abandoned. It might be that kidney and liver cells better tolerate the urea levels present in BAF.

Lactalbumin hydrolysate (LAH), peptone, and yeast extract or lysate are low-cost sources of nutrients which have had limited use in fish cell culture. These ingredients have a history of use in the so-called maintenance type of media employed in homoiotherm virology. Jensen and Tomasec *et al.* used 0.5% LAH in growth media for fish virology. However, in several studies, which included an evaluation of media containing either these nutrients or serum, the authors consistently concluded that the best results were obtained with serum.

pH

The pH of medium necessary for good growth of fish cells does not appear to be particularly critical, and most cells seem to fare well in the range of 7.2–7.8. Primary cultures and low densities of cells will usually do better at 7.3–7.4 than at pH 7.8. Routine passage of some cell lines can be made at pH 7.8 or even 8.0, but the lag phase may be extended somewhat. At the other end, old cultures can have a pH as low as 6.8apparently without undue damage to the cells. Such cultures can be dispersed in fresh medium and growth will usually resume.

We have often heard of cultures becoming excessively alkaline soon after seeding, and this we believe has been erroneously attributed to an unusual cellular or tissue response. Invariably, these are media with a bicarbonate buffer and the rise in pH has simply been equilibration between the gas and liquid phases in the culture system. This alkaline shift occurs even when medium alone is added to comparable culture vessels. The pH shift is particularly marked if small volumes of medium are involved-as, for example, 1 ml or less per 16 × 125 mm culture tube. The remedy is to use CO_2 to lower the pH to 6.8 or 7.0. Upon equilibration, the final pH will be 7.4-7.6.

Storage of Prepared Medium

Since some of the growth factors in medium are labile, the medium should be refrigerated or frozen. We prefer not to store complete medium longer than 1 month at 4°C, but have found no untoward effects in some which were held for several months. We routinely prepare medium and store it at –20°C. Most lots are used within a month, but we have held some media for 4 years and found that they seemed to support growth as well as when prepared.

Antibiotics

The same antibiotics are used throughout vertebrate animal cell and tissue culture, but little quantitative work has been done with fish tissues or cells. Working concentrations of the more commonly used antibiotics have been within the same range and have been used with cells and tissues from a variety of fishes. At present, inhibitory or even toxic concentrations are largely unknown for cells and tissues from most fishes. The use of antibiotics involves selection of the drug and of the concentration that is to be employed. Antibiotics differ in the effects they produce on cells, and it is likely that cells of different fishes will vary in their response to a particular antimicrobial substance. Accordingly, the nature of the work must be considered; e.g., will it be longor short-term, a terminal use in virology, or a metabolic study? And of course the risk of contamination must be assessed. Routine passage of cells presents a much lower risk of contamination than establishing primary cultures of external tissues.

For routine purposes, many have used media containing 100 IU of penicillin, 100 μg of streptomycin, and 25 IU of nystatin per milliliter. We have used chlortetracycline at 50 μg/ml in lieu of the mixture of penicillin and streptomycin, but chlortetracycline forms a precipitate in the medium. In our experience kanamycin and amphotericin B are more inhibitory or toxic at lower concentrations than other antibiotics. We use them primarily when contamination risk is high or when it has already occurred and salvage of cells is necessary. If for example a mold colony is found in a bottle, it is carefully removed and the necessary amount of amphotericin B added to the medium. Kanamycin has similarly been used with light bacterial contamination in the presence of penicillin and streptomycin. Such cells are employed in terminal use only and subcultures are not propagated further.

Table 5.3 : Antibiotics Used in Fish Cell and Tissue Culture.

Antibiotic	Routine use	Remarks
Amphotericin B	2-10 μg/ml	May be toxic at <5 μg/ml
Chlortetracycline HCl	50 μg/m1	500μg toxic to BF-2 cells
Kanamycin sulfate	25-50μg/ml	100 μg halves growth rate of RTG-2 cells Neomycin 50-100 μg/ml
Nystatin	25-50 IU/ml	100 IU inhibitory for some cells
Penicillin G, potassium	50-100 IU/ml	2000 IU/ml has been used
Polymyxin B sulfate	No data	2000 IU/ml as brief bath
Streptomycin sulfate	50-100 IU/ml	2000 IU/ml has been used

Table elsewhere in the chapter is a suggested guide to the use of antibiotics with fish cells and tissues. It is based upon the values already reported and upon a limited amount of testing which we have not previously reported.

For general information on the use of antibiotics in tissue culture the reader is referred to a booklet by E. R. Squibb and Sons, New Brunswick, New Jersey, or to excerpts of the same which appear in the price and reference manual of the Grand Island Biological Company. The catalog of the Industrial Biological Laboratories, Inc., 451 S. Stonestreet Ave., Rockville, Maryland, 20850 contains a valuable table of both recommended and cytotoxic concentrations of antibiotics used in tissue culture.

METHODS

Preparation of Fish for Obtaining Tissues

In marked contrast to reptiles, birds, and mammals, the outermost layer of the fish integument is living, and if the animal is healthy and comes from a clean environment the microbial flora is very low in number. The skin may support protozoan and higher parasites. The kind of tissue needed, the habits and ecology of the donor fish, the nature and purpose of the work, and whether or not antibiotics are to be used will all be factors in deciding which decontaminating, disinfecting, or sterilising procedures should be used. No antimicrobial substance is without effect on the host tissue, and the rigors of decontamination should not result in death of the fish cells or tissues to be cultured.

Unless proper precautions are taken, the results of the most thorough disinfection or sterilisation of fish surfaces can be nullified quickly by contamination from draining feces or regurgitated stomach contents. This problem can be minimised by withholding food for several days before using the fish, but where this is neither practical nor feasible much of the feces can be manually stripped from the fish and the mouth thoroughly rinsed prior to disinfection.

External Tissues

The use of antibiotics in culture media has greatly reduced-but not eliminated—the problems associated with cell culture in growing external fish tissues with minimal microbial contamination. Fin, skin, barbels, cornea, and even caudad trunk portions of healthy fish may simply be washed in cold (preferably chlorinated) tap water and rinsed in sterile BSS. The BF-2 cell line was initiated from four small bluegills which were so treated. As specific tissues are removed they are rinsed in cold BSSeither in vessels or in a stream from a syringe and small gage needleand placed in a sterile covered container. We have used this procedure routinely for years with fin tissue and on occasion with corneal tissue. Gill tissue, which grows well, is often heavily contaminated with gramnegative bacteria, and is difficult to clean for culture. Polymyxin B is one of the few bactericidal antibiotics, and we have had some success with hour-long treatments of gill arches in BSS containing 2000 IU/ml. Grand, Gordon, and Cameron and Grand and Cameron used a 1: 10,000 solution of sodium Merthio!ate for 2-3 min followed by a wash of sterile physiological saline. Griitzner held guppies for 1-2 days in water containing 1000-1500 μg streptomycin per milliliter of water, then treated the freshly killed fish for about 5 min in 0.4% chloramine (sodium p-toluene-sulfonchloramide) in BSS. From 88 to 95% of her cultures were sterile. Babini and Ghittino washed yolk sac fry in 1:200,000 malachite green for 15 min followed by immersion in 1:10,000 Merthiolate for 10-15 min then a rinse in sterile water containing penicillin and streptomycin. Another method is to soak the tissue in Dakin's solution for 2-3 min with agitation, then wash it in three changes of sterile BSS.

The RTF-1 cell line was started from nonfeeding yolk sac rainbow trout fry from which we removed the yolk then simply washed the fry several times in cold sterile phosphate buffered saline (PBS). Fryer similarly found that rinsing embryonic or yolk sac salmonid fry was adequate decontamination. Such procedures will probably be adequate only if the fish have not yet begun to feed and thereby have yet to acquire an internal flora.

Where caudad trunk tissue is used for cell culture, external tissue which may be killed by sterilisation comprises only a small percentage of the biomass and therefore its loss will usually be negligible. In obtaining such tissue for establishing the FHM cell line Gravell and Malsberger immersed the donor minnows for 1 min in a filtered solution of 10% calcium hypochlorite and followed that with a rinse in 70% ethanol. The tissue was then given three 10-min washes in BSS containing penicillin and streptomycin.

Embryos

Sterile embryos may be obtained by surface sterilisation of either eggs or gravid females. Dederer immersed eggs for 1 sec in 95% ethanol and then transferred them to sterile water for aseptic removal of the embryo. Soret and Sanders immersed gravid *Gambusia* momentarily in Merthiolate (concentration not given) and followed with two washings in 70% alcohol. They dried the fish with sterile cotton and aseptically removed the embryos. We have used benzalkonium chloride followed by alcohol to disinfect sharks prior to surgical opening and collection of sterile embryos from developing eggs in the oviducts.

Internal tissues

Unless an animal is infected, and with the exception of the digestive tract, internal tissues of fishes are sterile and their aseptic removal is simple. Prior to opening the fish, the area of incision-or when feasible, the entire fish-is topically disinfected or sterilised. It is advantageous to remove scales from heavily scaled fishes. Wolf and Dunbar used procedures developed by Dr. S. F. Snieszko for bacteriological examination of fish. The fish is bathed for several minutes in a 1:1000 solution of benzalkonium chloride, and this is followed by washing in 70% ethanol. Isopropanol (70%) is equally satisfactory and we have since used 1:2000 household bleach (5.25% sodium hypochlorite) or about 500 ppm available chlorine interchangeably with benzalkonium chloride. Hypochlorite solutions have the advantage of wide availability and low cost.

For obtaining sterile internal larval lamprey tissue, Stephenson and Potter used a disinfecting technique developed for tadpoles. The ammocoetes were rinsed in tap water, washed in detergent (7X brand) and immersed briefly in 95% alcohol, then passed through two changes of BSS containing 40 IU of penicillin, 50 μg of streptomycin, and 20 IU of nystatin per milliliter.

Using sterile instruments, internal tissues are obtained by removing the lateral musculature covering the peritoneum and if the heart is needed, the pericardium. If necessary, soiled instruments may be wiped clean, boiled briefly in distilled water, then transferred to 70% ethanol. Sterile tubes provide protection for the instruments between use.

We have used a series of washes in sterile water to reduce external contamination and then held larval lampreys for 2 hr in sterile water containing 500 IU of polymyxin B, 500 μg of neomycin, and 40 IU of bacitracin per milliliter. The rate of contamination of external tissue explants was less than 5%

Preparation of Primary or Original Monolayer Cell Cultures

Freshwater Teleost

Disaggregation or partial "digestion" of tissues to obtain fish cells for primary monolayer culture follows the general procedures used with avian and mammalian tissues. An important exception however is temperature; to maintain viable fish cells throughout ' the procedures temperatures should not exceed 20°-25°C. There is, moreover, considerable evidence that lower temperatures are often desirable. Tryptic enzymes are employed most often, being used at a final concentration of 0.25% in a PBS having a pH preferably near 7.2-7.4 but not exceeding 7.6. As with the salines, some workers have diluted the PBS 20-30% but many do not-the decision has usually been based upon the serum freezing point of the fish involved. Digestions are either relatively short trypsinisations at temperatures between 15° and 20°C or extended digestion at 4°-6°C.

The tissues should be fresh, but with adequate precautions to prevent damaging dehydration, surplus tissues may be safely stored at 4°C for a day. On occasion, Roberts has had viable tissue after 48 hr storage at 4°C and once kept salmon ovary for 72 hr. Tissue is minced in BSS or PBS until the largest pieces are at most several millimeters in diameter. Fish embryos can also be mechanically disrupted simply by forcing them through a syringe several times. The tissue pieces are usually washed several times to remove debris and blood cells which are undesirable in the cultures.

Grutzner introduced modern methodolgy to fish cell culture; she used a relatively brief digestion period and cultured tench liver and kidney cells in media which were diluted from mammalian concentrations. Trypsin at 0.25% was used. in Dulbecco and Vogt's PBS which had been diluted 20% with water. After a 10-min mixing over a magnetic stirrer, the supernatant was discarded and fresh trypsin solution added. Harvests of separated cells were made at 30 min intervals, cooled with ice, and sedimented by gentle centrifugation. She cautioned that the temperature during digestion should not exceed 20°C. In later work mention was made that 4-5 hr digestion was so damaging to cells that they would not grow, and that overnight digestion at 4°C did not yield cultivable cells. Subsequently, however, tench gonads were trypsinized routinely at 4°C. Kunst used much the same procedure for preparing carp kidney cell monolayers; the temperature was 18°C and he used intermittent agitation instead of continual mixing for the tissue which was held in four volumes of digestant. More recently, Pfitzner and Froehlich extended the methods to gonadal cells from carp and goldfish.

Working principally with salmonids, Wolf *et al.* found that slight modification of the so-called cold trypsinisation used by Bodian for monkey kidney cells consistently gave cultivable cells from these and other fishes and also from an amphibian and a reptile. Dulbecco and Vogt's PBS was modified by buffering to about pH 7.2, and it constituted 87.5% of the digestion mixture. Trypsin solution (2.5% of 1:250) accounted for 10% and the remaining 2.5% was serum which was added to protect the released cells. Penicillin and streptomycin were present at 200 IU and 200 μg/ml, respectively. The ratio of tissue to digestion mixture was not critical; up to several volumes of minced tissue are used for each 10 volumes of digestant. The digestion itself is carried out on magnetic stirrers at 4°C. Cultivable cells are released during the first hour's treatment, but they are a minority and seldom worth culturing; they should be discarded.

Digestion is allowed to proceed until many cells and cell clusters are inn suspension. This takes several hours. The digestion can proceed overnight, then cells may be harvested and fresh trypsin added for further action. Because earlier work had shown mammalian osmolarity to be appropriate for trout and goldfish, neither the digestion mixture nor culture medium was altered.

Babini and Ghittino carried out cold digestions of carp, tench, and trout tissues in Dulbecco and Vogt's PBS which was diluted 20%. Others have successfully used the cold digestion in undiluted PBS. Kunst and Fijan trypsinized carp ovarian tissue at 4°C for both 3 and 20 hr and at 20°C for 0.5 and 2 hr. They found that the prolonged trypsinisation at 4°C gave the best results.

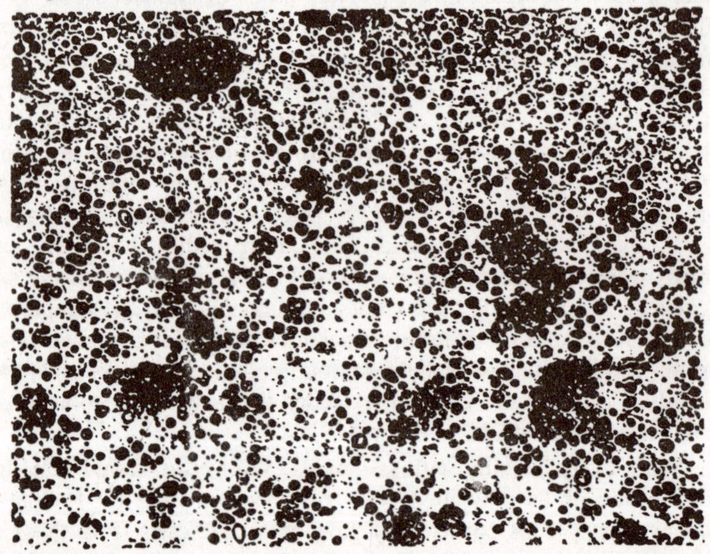

Figure 5.1 : An example of a trout kidney "digestion" at the time of harvesting. Note the many small fragments of nondispersed tissue including portions of tubules. Cilia in this preparation were highly active, and trypan blue dye exclusion indicated a viability of over 95%. We consider such preparations to be ideal for primary monolayer culture.

Fryer and Fryer *et al.* successfully used an intermediate temperature, 15°C, trypsin in pH 7.4 Hanks' BSS, and 30 min digestion periods for successful disaggregation of salmonid tissues.

Trypsinisation for primary cultivation of freshwater teleost cells has become routine and methodology is now omitted from some reports on fish cell culture.

Some teleost livers, kidneys, swim bladders, and ovaries possess ciliated cells. The student and newcomer who undertakes fish cell culture for the first time is urged to consider using these tissues for their beginning efforts. Positive assessment of viability can be determined at any step in the procedures simply by examining the released cells and tissue fragments for ciliary activity.

Although trypsin was used in the work thus far reported, other enzymes have been tried on fish tissues. Fryer tried collagenase, hyaluronidase, pancrease, papain, and *Vibrio comma* receptor destroying enzyme (N-acetylneuraminidase). While all were enzymically effective, trypsin had the least deleterious effect upon the cells. We have used pronase for digestions but find no particular advantage over trypsin; instead, it is more expensive. Pancreatin, a crude tryptic enzyme may be used interchangeably with trypsin, but it can be toxic to cells of some fishes.

Marine Teleost

The only detailed work on preparation of monolayer cell cultures from marine teleosts is that of Clem *et al.*. Their digestion mixture was prepared with magnesium-free Hanks' BSS. For some work the BSS was modified with about 0.07 M additional NaCl. Trypsinisation was carried out at room temperature with harvests at about 1 hr intervals. Roberts prepared primary monolayer cell cultures from a marine clupeid with procedures used for freshwater teleosts. He noted that the cells were highly vacuolated and considered the abnormal condition possibly a consequence of the hypotonic medium.

Marine Elasmobranch

There is no published information on preparation of monolayer cell cultures from the cartilaginous fishes. We have done a limited amount of work and have had limited success with spiny dogfish, *Squalus acanthias*, and the sandbar shark, *Carcharinus milberti*.

Digestion was effected with 0.25% pancreatin in pH 7.3 PBS which was modified by the addition of 1.75-2.0 g % urea and 0.5-0.6 g % NaCl. The freezing point (–1.59°C) closely matched that of spiny dogfish serum (–1.6°C). At different times, digestions were carried out at 26°, 15°, and 4°C, but the best results were obtained at the two lower temperatures. At 15°C, harvests were made at 1 and 2 hr intervals and oviduct cell cilia were still beating in the mixture after a total of 12 hr digestion. After the suspended material was decanted, undigested fragments were left in the residual fluid and kept at 4°C where cilia continued to beat for 76 days. The best results were obtained from digestions at 4°C with harvests at 4 hr intervals.

In order of productivity the following shark tissues yielded cultivable cells: embryo, oviduct, rectal gland, kidney, spleen, and heart. Ovary and pancreas were dispersed, but cells were not cultivable. Heat-inactivated homologous serum was toxic to the cells.

Seeding Density for Primary Monolayer Cultures

Following digestion of tissue, the cells to be cultured are harvested by centrifugation. It is generally agreed that 200 g for 10 min is both adequate and safe. Cells from many fishes readily tolerate centrifugation at 20°C or even higher, but frictional heating coupled with high ambient temperature may injure cells from cold-water fishes. The risk of heat damage may be avoided by chilling the cell suspension prior to centrifugation or by centrifugation in the cold. It is worth noting that cells from a wide variety of vertebrates are safely centrifuged at 4°-8°C.

Primary monolayer cultures of fish cells have most often been seeded with cell densities which have been established by counting, but to a lesser extent volumetric dilution of the harvested cell pack has also been used. There is a considerable range of cell numbers reported to be necessary to establish primary fish cell monolayers. A majority of the reports cite values well above those normally used for primary cultures of homoiotherm cells, but a minority found that 1 to 3×10^5 cells/ml were adequate. It seems likely that different tissues and even different species of fish could account for some of the differences. We believe however that other factors are more important and that the composition and properties of the medium, and especially the extent and conditions of digestion, temperature, physical forces, and possibly other factors of centrifugation critically affect the minimal threshold of cells necessary for primary monolayer cultures.

Griitzner found 2 to 2.5×10^5 tench liver or kidney cells to be inadequate and that 2, 3, or even 4 times as many cells were needed to establish monolayers. Her procedures included 30-40 min of centrifugation, but relative centrifugal force and temperature were not given. From marine teleost tissues, Clem *et al.* initiated cultures with 4 to 5×10^5 cells/ml when centrifugation was not used. Following 7 min at 5°C (g not given), they reported it necessary to use 8-10 $\times$ 10^5 cells/ml to establish primary cultures. Fryer and Fryer *et al.* considered cell concentration to be an "extremely critical factor" in establishing cultures. Cells were harvested with 20 min centrifugation (temperature and g not given), resuspended and washed in medium, and the minimal number necessary was about 6×10^5 cells/ml. For routine purposes however, the density was 1 to 1.5 × 10^6 cells/ml. Subculturing-which usually requires only one centrifugation-was successful with as few as 3 × 10^5 cells/ml though 6 × 105 cells/ml was preferred. The fewer cells required for subcultures could reflect enhanced survival with less centrifugation, but we recognise too that subcultures are usually more amenable to handling than are primary cells. Jensen used 10 min centrifugation (g not given) in *cold tubes* (italics added) and cultured trout ovary monolayers with 3 × 10^5 cells/ml. Li and Stewart harvested trout ovary cells with 35 g for 5 min in the cold and found 2 × 10^5 cells/ml to be adequate for establishing monolayers.

Volumetric dilution of centrifuged cell packs for seeding primary cultures is simpler and takes less time than counting. As determined by trypan blue dye exclusion but especially by ciliary activity we have had consistently high percentages of viable cells in digests of poikilothermic vertebrate tissues which have been stopped while small fragments of undispersed tissue were still present. The counting of cells in such harvests is tedious and the accuracy is unknown, accordingly we use volumetric dilution and suspend the cells in from 400 to 600 volumes of medium (1 to 3 × 10^5 cells/ml). We used two centrifugations but carried them out at 200 g for 20 min at 4°C. We have subsequently eliminated one centrifugation because it was found that washing the cells was not necessary; also, centrifugation time was reduced to 10 min. Babini and Ghittino also used volumetric dilution but at much lower ratios—1:25 to 1:100. Kunst used a volumetric dilution of 1:100 for carp ovary cells but in later work compared ratios of 1:300, 1:600, and 1:1200 and found the latter to be the the best. Although the relative centrifugal force and temperature were not given, they used two centrifugations each of 10 min duration.

Dispersion of Monolayer Cell Cultures

Monolayer cell cultures may be dispersed either mechanically or chemically for subculturing. At present, the use of a chelating agent with a tryptic enzyme is superior to other methods.

Fish cells differ in the tenacity of their attachment to each other and to the substrate. In part this reflects differences between cell lines; but media, culture conditions, and even culture age affect the ease or difficulty with which cells may be dispersed. In our experience old cultures are more cohesive than young cultures, and cells grown in medium with 10-15% serum are more cohesive than cells grown in 5% *serum.*

Dispersed fish cell lines can be easily heat-fixed for stabilisation and delayed counting. Preparations held at 57°-60°C for 30 min may be counted accurately through several days' storage.

Mechanical Dispersion

There is only occasional justification for using mechanical means to disperse cell sheets. The objections to this method are that even under the best conditions many cells are destroyed and the dispersion itself is usually incomplete-the sheet being broken into fragments rather than individual cells. This of course compounds the problems of accuracy in enumeration. The cell sheet is scraped from the surface-with a rubber policeman or glass culture scraper-into a sufficiently small volume of fresh medium to permit vigorous pipetting for additional dispersion. Estimates of cell loss have been as high as 50%.

Chemical Dispersion

Disodium versenate is a chelating agent for divalent cations. A 1:5000 solution is used to chelate calcium which is essential for cell cohesion and adhesion. Cell sheets are drained of medium and covered with ethylenediaminetetraacetate (EDTA) prepared in a calcium and magnesium-free salt solution. After 8-10 min, the action is stopped by adding an excess of calcium (the original medium is both effective and economical for this purpose) and the sheet fragments pipetted vigorously for additional dispersion. The cells must be sedimented by centrifugation, the supernate decanted, then the pellet of cells can be resuspended in fresh medium. Temperature should be maintained within the cells' tolerance. A 0.25% solution of trypsin or pancreatin in either BSS or PBS can be used in place of EDTA.

An alternate procedure eliminates the need for centrifugation. The cell sheet is covered with trypsin solution and watched carefully. When the very first cells are loosened-a matter of one to several minutes-the vessel is rotated 180° to invert the cell sheet. Most of the trypsin is decanted, but a small residual volume is retained and by gentle rocking the fluid is made to flow back and forth across the sheet until the cells are free. This usually takes 5-10 min and at that time a small amount of fresh medium is added and the cells dispersed by pipetting. The enzymic action is inhibited by serum, and the cells can be diluted as necessary.

In our experience, the best dispersion is achieved by a combination of 1:5000 EDTA (20 mg %) and 0.25% trypsin or pancreatin in calcium and magnesium-free salt solution used as described in the

Table 5.4 : EDTA-Trypsin Solution for Dispersion of Monolayer Cell Cultures.

Constituent	Quantity
Water	960 ml
NaCl	8.00 g
KH_2PO_4	0.20 g
KCl	0.20 g
Na_2HPO_4	1.15 g
Disodium EDTA	0.20 g

preceding paragraph. Cells so dispersed have had a high survival. Viability as determined by trypan blue dye exclusion has been about 95%.

Seeding Density for Cell Lines

The seeding density for subcultures of cell lines will vary with the cell, the medium, and the particular need. Where rapidity of growth

Table 5.5 : Effects of Seeding Density and Temperature on Growth of RTG-2 Cells.

Seeding density	*Days required to reach confluency (2.75 × 10^5 cells) at incubating temperatures of*					
	30°C	**25°C**	**20°C**	**15°C**	**10°C**	**5°C**
200,000	Death	1.8	1.5	2.4	3.5	8.0
100,000	Death	3.3	3.1	4.8	9.3	14.0
50,000	Death	6.1	4.8	7.3	14.0	—
25,000	Death	8.5	7.5	14.0	35.0	63.0
12,500	Death	14.0	10.5	20.0	—	70.0
Mean population doubling time (days)						
		2.5	2.0	3.5	7.6	13.2

is important and other factors are equal the lag phase will be shortest at high population densities. In routine use, fish cell lines are split at ratios of 1:5 to 1:20, but if a need arises well-established lines may be split 1:100 or even 1:1000. In terms of cells per milliliter of fresh medium, routine seeding densities would be from 10^4 to 10^5 cells/ml or more, but as few as 10^3 or even 10^2 cells/ml can be adequate under good conditions. Of course with the latter low numbers, the lag phase will be considerable.

Using initial seeding densities and different incubating temperatures as the variables, quantitative growth studies have been carried out with the RTG-2 cell line. Replicate culture tubes were seeded at densities of 1.25×10^4 and at twofold increases to 2×10^5 cells/ml and incubated at temperature increments of 5° from 5° to 30°C. Confluent cultures were found to have a mean population of 2.75×10^5 cells/ml. As shown in Table VI, cultures seeded most heavily became confluent in 1.5 days at 20°C, the cells' optimal temperature, while the lowest density required 70 days to become confluent at 5°C. The data clearly show that the cells grow more rapidly at 20°C than at 25°C.

Precautions to Be Taken with Fish and Other Poikilotherm Cells

It is a common practice to grow homoiotherm cells at 37°C in the presence of 5% CO_2, and many of the media use buffer systems which equilibrate in the physiological range under such conditions. Because the solubility of CO_2 is increased at lower temperatures, media which equilibrate at pH 7.4 at 37°C become acid when the temperature of CO_2 cabinets is lowered. It is a common mistake to use CO_2 cabinets in handling fish and other poikilotherm cells, and at the very low pH the cells are soon killed.

Excessive heat during enzymic disaggregation of poikilotherm tissues -especially those from cold-water species such as salmonids-is likely to damage many cells and may in fact kill all of them. Since digestions are usually stirred with a magnetic unit, the principal source of heat is from the stirring apparatus itself. Stirrers with a separate rheostat in the power line are strongly suggested. Although they are more compact, stirrers with a self-contained rheostat may transfer excessive heat to the cells. To prevent heat transfer, one may use insulation between the stirrer and the flask or keep the flask in a tray with crushed ice or cold water.

Cells in culture lack most of the defense mechanisms of the intact donor and are usually quite vulnerable to virus infection; this property makes them extremely valuable as a tool in virology. As a broad generalisation, primary cell cultures are usually susceptible to more viruses than are their derived permanent cell lines, in fact during their development, cell lines often lose susceptibility to some viruses. Primary cultures however can have disadvantages. Removing tissues from the host usually strips away defense mechanisms. If such tissues harbor a latent viruswhich sometimes happens-the virus can result in a fulminating infection which may destroy the culture. Latent IPN virus has been unmasked a number of times in primary trout cell cultures. Thus far, however, less obvious viruses which replicate without cytopathology have not been documented in primary fish cell cultures. The solution to the problem is to attempt to learn beforehand the health of the donor animals.

It is routine to carry out sterility checks on media as they are prepared for cell culture. For this purpose, aliquots of complete medium without antibiotics can be incubated as a test procedure. Bacteriological media such as thioglycollate or enrichment broths are usually used for testing salines and other solutions. On a number of occasions we have found bacterial contaminants which grew poorly in bacteriological medium but luxuriantly in cell culture medium; accordingly, we routinely use the latter as sterility test medium.

Mycoplasma, formerly called pleuropneumonia-like organisms or simply PPLO's, are microbial contaminants of significance in mammalian cell culture. Forms which are of human origin prevail, but mammalian serum sources are a secondary source. Thus far, no lower vertebrate cell line has been found contaminated with a mycoplasma. Whether or not this situation will prevail is not known, but maintenance of cell stock cultures by strict aseptic technique and without antibiotics is recommended.

Most fish, especially those from the wild, may have protozoan or metazoan parasites located internally or externally which can create problems for the fish cell or tissue culturist. One should be aware of them when planning for primary cultures. As an example, some female largemouth bass may have ovaries which are so heavily infested with cestodes that it is impossible to obtain unadulterated fish tissue. Kidneys, liver, and heart of many centrarchids are commonly riddled with metacercarial trematode cysts. Mincing and trypsinisation will release these minute worms and they will live for a week or more in cultures of the fish cells. While tissue from such infested sources may be useful for single usage short-term applications, it should not be used for critical work such as cell line development or other long-term applications.

CHOICE OF TISSUES FOR CULTURE

Critical comparisons have not been made, but cells and tissues from adult fishes seem to be cultured somewhat more easily than tissues from adult mammals. Such an impression is consistent with the biology of fishes for they generally continue to grow throughout life and have comparatively strong regenerative ability-as, for example, when fins are amputated. There is nevertheless a hierarchy among fish tissues; some are grown much more readily than others, and some-such as adult skeletal muscle—seem disinclined to grow *in vitro*.

With little doubt, embryonic fish cells are the most readily cultivated, and their state of active mitosis lends impetus to continued division in culture. Embryonic tissue has the added advantages of usually being bacteriologically sterile, of having a low probability of harboring latent virus, and of having relatively little cell storage products and a small complement of differentiated cells. The disadvantages of embryonic tissue are that the available mass is generally small. In addition, embryos are often inaccessible—as with pelagic spawning species; moreover, with many fishes embryos are available only during a limited portion of the year.

The second most easily cultured fish tissue is gonad preferably juvenile or immature organs because there is a greater proportion of undifferentiated tissue and more actual germinal tissue. Ovary is generally preferred over testes and may yield abundant cultivable cells at full maturity; the great mass of ovary at that time however is yolk and shell, and while it is acellular and noncultivable, it does not seem to interfere with the culturing of cells and tissue. The cultivability of gonadal tissue is a feature that applies broadly among the lower vertebrates. In our experience, the sole exception has been adult shark ovary which we have not been able to culture successfully. Townsley *et al.* failed to grow gonadal tissue from skate and a marine teleost.

Swim bladder, fin, mesentery, cornea, gills, heart, and skin generally culture well. Cornea and swim bladder are usually clean tissues which lend themselves well to neat explants. We have had trout swim bladder explant show persistaltic-like movement for over a year. Although it is usually cluttered with acellular debris, hematopoietic tissue such as spleen and kidneys usually fare well in culture. Strangely enough, circulating leukcoytes from teleosts may be induced to attach to culture vessel surfaces and form monolayers, but there is little if any mitosis. This may be a result of their being differentiated; if so, thymus and anterior kidney tissue—possibly taken by needle biopsy-may yield active stem cells. Cell culture thus could clarify some of the uncertainty existing in fish leukocyte phylogeny and nomenclature.

Liver tissue has responded erratically in culture. Griitzner apparently found tench liver to grow well, but Pfitzner noted it could only occasionally be cultivated. Others have also had mixed results with liver tissue. Babini and Ghittino and Fryer were unable to grow normal liver tissue. On the other hand, Fryer found that rainbow trout hepatomatous tissue readily yielded cultivable cells, and he established a permanent line of cells from such tissue. This probably reflects derepression of controls that exist for normal tissue, and tumor tissue almost traditionally has been easily grown in culture. A number of different fish tumors have been cultivated.

We have been unable to grow tissue from the corpuscles of Stannius, but Lewis and MacNeal grew pituitary and McLimans grew fish pancreas. As far as we know, no other fish endocrine tissues have been cultivated. In principle, tissue from the alimentary canal should respond favorably. We have explanted stomach and intestinal tissue a number of times but were unable to control the bacterial contamination.

STORAGE AND PRESERVATION

Freezing and long-term storage of living cells at ultra-low temperature has been practiced for about 15 years. The method that has been used so successfully for mammalian cells, works equally well with fish cells. Thus far, however, there are few published data on results of freezing and recovering fish cells. Cells from many poikilotherms not only tolerate, but also actually metabolise, at temperatures near freezing. The rate of metabolism is greatly reduced at low temperatures; thus, in the culture of cold-blooded animal cells one also has the advantage of prolonged storage without freezing.

Freezing

In general, cells are suspended in medium having 10% or more serum and 5-10% of either glycerol or dimethyl sulfoxide (DMSO) as a protective additive. The suspension is sealed in glass to prevent dehydration and to exclude CO_2 if refrigeration is to be with Dry Ice. After sealins, cells are allowed to equilibrate at 5°C for 30-60 min. For best results freezing should be done slowly by cooling 1°C/min to -25°C then transferring to -65°C or lower. For recovery, ampoules are thawed in less than 1 min and diluted for planting.

To date, the best single source of published data on freezing fish cells is the American Type Culture Collection, Registry of Animal Cell Lines Certified by the Cell Culture Collection Committee. The following tabulation lists fish cell lines that freeze well in the media given.

Cell line	Freeze medium
FHM (CCL 42)	Eagle's MEM (Hanks') 85%, calf serum 10%, DMSO 5%
RTG-2 (CCL 55)	Eagle's BME 80%, fetal bovine serum 15%, DMSO 5%
GF (CCL 58)	Eagle's BME with nonessential amino acids (Hanks' with 0.196 M
NaCl) 75%,	calf serum 10%, human serum 10%, DMSO 5%

During the past 4 years we have frozen fish and frog cells with the following procedures. Freeze medium consists of Eagle's MEM (Earle's BSS) 85%, fetal bovine serum 10%, and either glycerol or DMSO 5%. The medium is cooled to 4°C, and cells are added to a density of 2 to 6 $\times$ 10^6 cells/ml. One milliliter aliquots are sealed in 2 ml ampoules and cells are allowed to equilibrate at 4°C for about an hour. Ampoules are wrapped in several layers of flexible insulation, placed in an expanded polystyrene bead insulated container and moved to -80°C. Because of the insulation the heat loss is slow and good freezing is assured. For recovery the ampoules are thawed with rapid agitation in water not over 20°C, then the cell suspension is added to 6-8 volumes of growth medium and planted. Most of the viable cells attach within an hour. Following attachment, the freeze medium should be withdrawn and replaced with medium having no preservative. Storage at –80°C gave good recovery.

Although their immediate appearance following thawing and planting is not as good as with DMSO, cells frozen with glycerol survive longer in unopened ampoules at 20°C. We have held unopened ampoules of RTG-2 cells for 5 days and recovered live cells-a sufficient time to permit mailing. Using the above procedures we have successfully frozen, thawed, and cultured RTG-2, RTF-1, FHM, BF-2, and frog tongue fibroblasts. Cells from vigorously growing cultures are preferred.

Kunst and Fijan (1966) froze primary carp ovary cells in Eagle's BME with 10% calf serum and 8% glycerol. They used 60 volumes of medium per 1 volume of cells and stored the suspensions at –70°C.

Low Temperature Incubation

Cells from some but not all fishes have been grown at temperatures of 5°-10°C. It seems likely that the physiology of the fish and the quality of the medium are two of the factors involved, and there may be additional factors.

The RTG-2 cell line (from a "cold-water fish") grows well from 26° to 4°C and possibly lower, and rates of protein synthesis and glucose utilisation have been measured through that temperature range. Because of the reduced growth at low temperatures we have kept some cell stocks at low levels of activity for many months. Cultures can be incubated at 12°-15°C and subcultured at 3-4 month intervals. FHM cells are reported to grow at 14°C, but there is little activity at 4°C.

We routinely incubate some fish cells at 4°C. By seeding bottles at about 1.5 to 2.5 × 104 cells/ml, cultures of RTG-2 cells have been stored for 2 years at 4°C. The only attention given them is a visual check at 3 month intervals. It has recently been determined that RTG-2 cells at 4°C have a population doubling time of about 13.2 days. In contrast, the doubling time at 20°C, which is near the optimum, is only 2 days. The medium for this work has uniformly been Eagle's MEM (Earle's) 90% and fetal bovine serum 10%. While we have no quantitative data, it is our opinion that our stocks of FHM and BF-2 cells also metabolize and divide for months at 4°C.

Stephenson attempted culture of lamprey tissue at 5°C but reported that no outgrowth occurred. On the other hand, Townsley *et al.*, who were working with cold water marine teleost explants, chose 5°C as the most desirable temperature for their work.

FISH CELL LINES

General

The fish cell lines which have been established and those which are in development are all of teleost origin. Seven of the lines are from salmonids, six are from strictly freshwater species, and four are from marine fishes. With the exception of the lone aquarium species and the marine forms, all are from hatchery-propagated fishes of moderate to high economic value. The lines have been developed by people who are interested in the viruses and the viral diseases of poikilotherms, and most of the cells have been used for propagation of one or more fish viruses. Several of the lines support amphibian viruses as well, and FHM and RTG-2 cells are susceptible to some homoiotherm viruses.

Because of the work involved, chromosome numbers have been determined for only about half of the extant fish cell lines, and these are all now heteroploid. The fish counterpart of the mammalian diploid cell line may have existed, but it has been lost. An embryonic sockeye salmon line, Se E, remained diploid for 50 subcultures but could not be continued. Among cell lines from all animals, attainment of the potential for indefinite subculturing is usually accompanied by alteration to a heteroploid chromosome constitution. The modal chromosome number of such fish cell lines apparently is quite stable. The stability of karyotype is another matter and has been determined only for the FHM cell.

Sources of Fish Cell Lines

Starter cultures of fish cell lines may be obtained from the originator and if submitted, from the Cell Repository of the American Type Culture Collection, 12301 Parklawn Drive, Rockville, Maryland, 20852. Use of the latter spares the originator time, effort, and money but more importantly it provides Certified Cell Lines of known identity and of cultural quality possibly even exceeding that routinely available from the originator. Because of their widespread use, RTG-2 and FHM cells are available as starter cultures but also in production quantities from several of the cell culture supply houses.

One may of course wish to develop his own cell line. For this we would suggest starting with many culture units (we prefer tubes) and using several different media. When the response of the cells can be evaluated, the efforts should concentrate on the most promising-whether it be vigor, appearance, or uniformity. If feasible, one should consider working without antibiotics, or withdrawing them from at least part of the cultures after several subcultures. Initially at least, subcultures should be made 1:2 or 1:3. If possible, a portion of the primary culture and each tenth passage should be frozen for subsequent chromosome analysis. Once a line is established, the originators should submit it to the American Type Culture Collection for characterisation and accession.

SHIPMENT OF CELL CULTURES

Living fish cells may be shipped almost anywhere either in the frozen state with Dry Ice or as active cultures. Many hundreds of cultures have been sent from our laboratory by air and surface mail. With very few exceptions, they arrived safely. High temperature-at least for salmonid cells-and drying are probably the greatest hazards in shipping active cultures. Regardless of how a package is marked, one can expect that at times the labels will be ignored, therefore, cultures will probably be inverted and the medium drained from the cells. Cell sheets should be up to 50% confluent and grown either in tubes or small bottles. For shipment the vessels are filled with medium and tightly capped. Insulated containers such as expanded bead polystyrene are preferred for they provide both thermal and physical protection. When high temperatures may be encountered, either ice in plastic bags or commercial canned refrigerant is added for cooling. Dry Ice should not be used. For domestic mailing, packages should be marked "outside mail" and "special handling."

NEEDED DEVELOPMENTS

The needs of today's fish cell and tissue culture can be roughly divided into two categories: methodology and biology; separation of the two is sometimes difficult. Perhaps the greatest need is study of the physiology, nutrition, and metabolism of the cells themselves. Of course, one must specify the kind of fish cell. At this time we would suggest that it be of teleost origin because they are the dominant vertebrate class. Before such studies are undertaken, however, it would be most desirable to have the cells growing in a completely defined medium.

Although much of any medium in use today for growing fish cells is chemically defined, the serum fractions (5-20%), ultrafiltrates, embryo extracts, yeast extracts, and protein hydrolysates are almost completely undefined. The first need is to establish cells on completely defined medium. There are a number of defined media available commercially. One approach has been to grow cells in such media with serum then gradually reduce the concentration until none is required. This process of adaptation and selection is not difficult but it is time consuming. A more positive approach is to transfer cell lines to chemically defined medium or to establish primary cultures directly in completely synthetic medium.

The next need in the fundamental study is a standardised cell. Ideally, such a cell or cells should be as near normal as possible. Such a cell line does not now exist, therefore development of fish counterparts to the human diploid cell lines is one approach that could be taken. In the interim, another approach would be to select clones with stable karyotypes from existing cell lines. This is not a difficult feat, but it does highlight still another need-trained researchers who are principally interested in the cells themselves rather than in the applications of the cells to other uses.

Fish cells have not yet been grown satisfactorily in suspension culture. Large roller bottles which greatly enhance ability to grow many cells in a limited amount of space are helpful to the virologist, however, the biochemist and molecular biologist could also benefit by growing large batches of standard clones in chemically defined medium.

In the way of methodology, fish cell culture is notably lacking in techniques for growing leukocytes. Strange as it may seem, fish ought to provide an easily cultivable leukocyte of some sort and although we and others have attempted this, these efforts have not been successful. The cytogeneticists, of course, would make immediate use of the technique as a nondestructive means of determining the chromosome complement of particular fish. Perhaps use of the thymus or efforts to culture biopsy specimens from spleen and anterior kidney would be productive. Conceivably such studies would shed light on development of the various fish leukocytes, a currently disputed area.

There is also a real need for the application of organ culture to fish material. For example, the pseudobranch is a tissue which could be productively studied *in vitro*, and the corpuscles of Stannius-as yet with unknown function-may yield their secret *in vitro*.

Today's fish cell culture is largely *teleost* cell culture. Petromyzones and elasmobranchs have received little more than token attention while Myxini, Holocephali, and Dipnoi apparently have not yet been tried. Within the teleosts there are representatives of eight orders from which cells or tissues or both have been cultured *in vitro*, but there are also fishes such as the paddlefish, sturgeons, bowfin, and gar which have not yet been tried. Permanent cell lines have been established from representatives of three orders of fishes, but most of these lines have a fibroblastic morphology. The virologists certainly would welcome additional lines of fish cells with epithelioid morphology.

3

Embryonic Development

Fish eggs and larvae provide a relatively untapped source of biological material, increased by the recent improvements in techniques for rearing marine species. Apart from their intrinsic interest, experimentally based information on these early stages is required for further progress in the advancing fields of fish culture and fisheries research. General textbooks on ichthyology such as those of Lagler *et al.*, Nikolsky, Norman, and Marshall and on reproduction in fish by Breder and Rosen provide both general and some detailed information. Identification of eggs and larvae, apart from specialist papers, is possible from publications of Ehrenbaum, D'Ancona *et al.* through the current series of plankton sheets issued by the International Council for the Exploration of the Sea, and with the help of the extensive bibliographies by Dean and Mansueti.

Most species of fish pass through a larval stage before assuming the adult form at metamorphosis. Sometimes the newly hatched fish is called a "prolarva" (or alevin in salmonids) until the yolk is resorbed, and then a "postlarva" (or fry). The term "larva" is used here for all stages to metamorphosis in marine fish, although alevin and fry may be used when referring to salmonids or other freshwater groups.

THE PARENTAL CONTRIBUTION

Apart from the more obvious genetical effects on differentiation, rate of development, body form, size, and behaviour, the parents, and especially the female, have an important influence on the viability of the offspring both on a species and individual level, in ιerms of (a) the conditions for incubation, (b) fecundity, and (c) egg size.

Conditions for Incubation

Differences of spawning season and time and of spawning sites and substrate mean that incubation can take place in a great variety of conditions which influence the early development and physiology of the offspring.

Eggs Single, with No parental care

(a) Buoyant, planktonic—most marine fish, e.g., gadids, clupeids, flatfish, and deep-sea fish.

(b) Nonbuoyant, loose or attached to substrate-a common freshwater characteristic, e.g., cyprinids, pike *Esox*, or in littoral species, e.g., blenny *Blennius*, bullheads *Cottus*, sand eels *Ammodytes*; also found in some marine species, e.g., herring *Clupea harengus*, capelin *Mallotus villosus*, catfish *Anarhichas*, and American flounder *Pseudopleuronectes americanus*. Tendrils for attachment are found in many oviparous elasmobranchs, in the hagfish *Myxine*, smelt *Osmerus*, saury *Scomberesox*, and flying fish *Exocoetus*.

(c) Nonbuoyant, buried in sand or gravel-many salmonids, grunion *Leuresthes tenuis*, and lamprey *Petromyzon*; in peat or mud *Aphyosemion* and *Cynolebias* where the eggs undergo diapause during the dry season.

Eggs Single, Special Environments

The bitterling *Rhodeus amarus* lays eggs in the gills of the freshwater mussel and the lumpsucker *Careproctus* under the carapace of the Kamchatka crab.

Eggs Single, with Parental Care

(a) No nest, but eggs protected-found in many littoral forms, e.g., the bullheads Cottidae, blennies Blenniidae and gobies Gobiidae.

(b) Nests, often with parental protection and ventilation-also found in littoral species, e.g., blenny *Ictalurus*, sticklebacks *Gasterosteus*, and in other freshwater species such as sunfish Centrarcidae, bowfin *Amia*, lungfish *Protopterus* and *Lepidosiren*, and in the Cichlidae. Bubble nests giving good aeration are found in tropical or swamp species, e.g., Siamese fighting fish *Betta splendens*.

(c) Parents carrying eggs-sea horses *Hippocampus* and pipefish *Syngnathus* have brood pouches and the sheat fish *Platystacus a* specially modified area of "*spongy*" skin. Marine catfish Ariidae, cardinal fish Apogonidae and *Tilapia* are mouth brooders, and *Tachysurus* incubates the eggs intestinally.

(d) Ovovipiparity and viviparity—elasmobranchs include picked dogfish *Acanthias*, smooth hound *Mustelus vulgaris*, electric ray *Torpedo*, stingray *Trygon*, and the nurse hound *Mustelus laevis*. Teleosts include redfish *Sebastes*, *Heterandria*, *Anableps*, poeciliids such as *Xiphophorus* and half beaks *Hemirhampus*.

Eggs Massed

Angler fish *Lophius* and yellow perch *Perca flavescens* have massed but unprotected eggs; in the lumpsucker *Cyclopterus* and butterfish *Blennius pholis* the eggs are protected by the male.

Fecundity and Egg Size

In higher latitudes the spawning season is often short and the eggs are liberated over a brief period of perhaps hours (clupeids) or over periods of some days, probably at certain times of the day or night (flatfish and gadids). Where the seasons are less marked spawning may occur over a much longer period or be intermittent throughout the year, especially where the time between generations is only a matter of weeks or months. Fecundity may be considered as the number of eggs produced in one year by a female although this may be very difficult to determine where spawning is protracted.

In general, fecundity is high where the eggs are liberated into open marine waters; it is lower in freshwater species and where there is parental care. There is also a strong tendency for fecundity and egg size to be inversely related.

Apart from enormous interspecific differences, there are also considerable variations of fecundity within a species. Many authors have found that fecundity increases with length, weight, or age, the relationship usually being of the form $F = aL^b$, where F is fecundity, L is length, and a and b are constants. Year-to-year differences resulting almost certainly from environmental effects are also well established. For instance, sea temperature was correlated by Rounsefell with the fecundity of pink salmon, *Oncorhynchus gorbuscha*, higher temperatures apparently resulting in lower fecundities. (Here the effect of temperature on growth is a complication.) Bagenal reported density-dependent factors operating in Scottish flatfish, high densities being correlated with low fecundity. Anokhina found that fecundity in Baltic herring could be related to feeding conditions, high fat content of the female being related to high fecundity. Experimental studies by D. P. Scott indicated in rainbow trout, *Salmo gairdneri*, that an insufficient diet caused a reduction in egg number: in the guppy *Lebistes* fewer offspring were also produced when the females were kept on short rations. Extensions of this type of work are badly needed.

Differences within a species resulting from latitude, area, race, or season are no doubt interconnected.

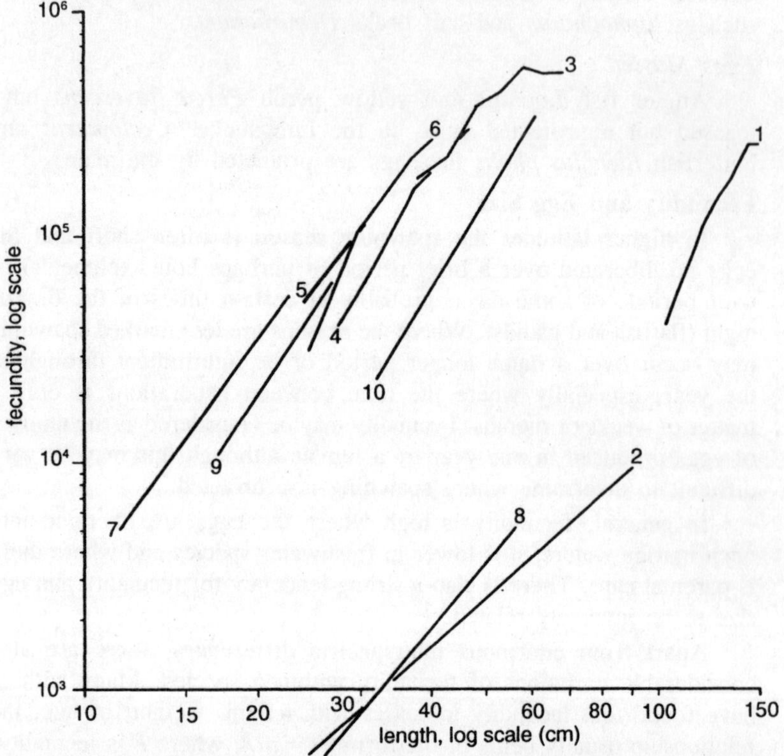

Figure 3.1 : The relationship between fecundity and length within a species. 1. Acipenser stellatus; 2. Salmo salar; 3. Cyprinus carpio : 4. Pleuronectes platessa (Clyde); 5. Clupea harengus (northern North Sea) ; 6. Melanogrammus aegle finus; 7. Osmerus eperlanus; 8. Salvelinus fontinalis; 9. Sardinops caerulea ; and 10. Seabstes marinus.

Considerable differences of this type were reported for plaice, *Pleuronectes platessa*, for herring, for species of *Oncorhynchus*, and for *Salmo salar*. An interesting characteristic of certain species is a difference in fecundity of the left and right ovary. The left gonad contains more eggs in *Oncorhynchus*, *Salmo*, and smelt *Osmerus*. The significance of this is not clear.

Egg size varies at the interspecific level, with larger eggs being especially associated with freshwater species like the salmonids or where fecundity is very low, as in many elasmobranchs. At the intraspecific level, differences of egg size as a result of area or river were noted by Rounsefell in *Oncorhynchus*, *Salmo*, *Cristivomer*, and *Salvelinus* species. In *Salmo salar* there are differences in egg diameter related to length, fecundity, and river. In *Tilapia* egg weight may increase 2–4 times or even more depending on the size of the female ; in the flounder, *Platichthys flesus*, large females, or females from low salinities, have larger eggs. Larger eggs are also found in larger females of the spur dogfish *Acanthias*. There are differences in average dry weight of the order of four times among the various races of herring and small differences between very young first spawners and repeat spawners. In two darter *Etheostoma* species with long spawning seasons, the egg diameter tends to be greater in the cooler winter months.

The connection between egg size and fecundity in related species may be correlated with the conditions for incubation. For example, *Tilapia tholloni*, a substrate brooder, has 500-3000 eggs depending on length. On the other hand, *T. mossambica* and *T. macrocephala*, which are mouth brooders, have less than 500 eggs that are considerably greater in weight. Garnaud reported two species of *Apogon*, one, A. *imberbis*, with an egg diameter of 0.5 mm and fecundity of 22,000, and the other, A. *conspersus*, with an egg diameter of 4.5 mm and fecundity of 150. The relationship between egg size and egg weight within a species is shown for the different races of herring in Fig. 2. The winter-spring spawners have a low fecundity and large eggs, an adaptation to poor food for the young, but a low predator population. In summer-autumn conditions fecundity is high and egg size low, presumably an adaptation to good food supplies and many predators. Intraspecific differences in fecundity and egg size deserve further study in terms of a link between ecological conditions and the physiology of maturation of the ovary.

A general pattern seems to emerge of marine fish with many, small buoyant eggs, a short incubation period, and vulnerable larvae. Freshwater fish have larger demersal eggs, a long incubation period and larger, less vulnerable larvae, while the littoral forms exhibit protective devices to prevent losses in this particularly difficult environment. The initial conditions of development determined by the genotype of the parent and reproductive behaviour must have a considerable influence on the viability of the young and its physiology, in that environmental conditions such as temperature affect the speed of development, salinity presents problems of osmoregulation, and oxygen must be obtained for respiration. The egg is susceptible and yet cannot make defensive responses to mechanical shock, drift by current, toxins, light, and predators. The larva should live in conditions where food can be obtained and protective behavioural devices can be practiced.

EVENTS IN DEVELOPMENT

Fertilisation

The physiology of fertilisation in fish ,with special reference to the extensive japanese work, has been fully reviewed by Yamamoto from which much of the present account is taken.

There is some evidence for the action of gamones in *Lampetra* and in teleosts; these activate the sperm and serve as chemical attractants toward the egg, while other gamones are known to paralyse or agglutinate sperm. In the bitterling species *Acheilognathus* and *Rhodeus* sperm aggregation and activity have been noted in the micropyle region of the chorion.

The chorion or egg case is relatively tough with a funnel-shaped micropyle at the animal pole. Within the chorion a plasma or vitelline membrane [also called a pellicle or surface gel layer] surrounds the yolk and cytoplasm (ovoplasm) of the egg. Fertilisation, which requires the presence of small concentrations of Ca or Mg ions, is normally monospermic in teleosts, the micropyle being too narrow to allow more than one sperm to pass at a time. The ovoplasm and chorion separate as the egg is activated by the sperm and a plug forms in the micropyle, further sperm being rejected. Where polyspermy occurs, as in some elasmobranchs, only one sperm fuses with the egg nucleus, the rest probably being resorbed and used as nutrient. Removal of the chorion permits polyspermy in teleost eggs; it seems that polyspermy is usually prevented by rapid changes at the micropyle, rather than over the egg cortex. In salmonids water activation (not to be confused with activation by sperm) takes place. When the egg is released into hypotonic solutions like river water the vitelline membrane becomes opaque and there are changes in its permeability. If sperm are not immediately available these changes may also affect fertilisability.

Following fertilisation, the prominent alveoli in the egg cortex of salmonids, acipenserids, and lampreys disappear. In the medaka, *Oryzias latipes*, these alveoli break down progressively from the animal pole. The separation of the cortex from the chorion leads to the appearance of the perivitelline space. The chorion is permeable to water and small molecules, but larger molecules of a colloidal nature are retained in the perivitelline fluid. In *Oryzias* and *Lampetra* these colloids maintain an osmotically based tension within the chorion. It seems likely that the colloid is derived from polysaccharide material in the cortical alveoli so that the formation of the perivitelline space is partly owing to a decrease in volume of the ovoplasm, as the alveoli release colloid, and partly due to an osmotic distension of the chorion. According to Ginsburg, polyspermy is blocked in sturgeon and trout eggs by the discharge of the cortical alveoli in the micropylar region. Chemicals, such as urethane, which cause polyspermy apparently retard the secretions of the alveoli, while removal of the perivitelline fluid in trout eggs allows the penetration of many sperm.

The chorion also hardens thus protecting the embryo in the early, more vulnerable stages. Probably the inner layer of glycoprotein is mainly responsible for this and it has been suggested that the alveolar colloid, Ca ions, phospholipids, or hardening enzymes also play a part. In salmonids, Zotin reported hardening of the chorion because of an enzyme in the perivitelline fluid; Ca ions affect the enzyme rather than the chorion itself. Ohtsuka considered that a phospholipid was liberated from the cortex (not from the alveoli) in *Oryzias* eggs. In *Fundulus* the chorion hardened with oxidising agents but not with reducing agents. The soft chorion appeared to be impregnated with protein containing SH groups. Hardening resulted from oxidation of SH to SS groups by means of aldehydes produced from polysaccharides with α-glycol groups. Zotin distinguished between the initial enzyme action and subsequent hardening processes which last much longer and where the enzyme is no longer functioning. Thus the initial enzyme reaction is blocked when Ca ions are bound by citrate or oxalate or by the use of NaCl or other chlorides. Later hardening is not susceptible to many of these factors.

The eggs of *Oryzias*, *Gasterosteus*, and *Lampetra* can be activated by pricking. Other activating agents are surface-active chemicals and lipid solvents (perhaps emulsifying the cytoplasm at the animal pole) and thermal shock, electric fields, ultraviolet light, and high frequency vibrations. This type of artificial parthenogenesis usually leads to irregular cleavage, but stringent precautions are needed to prevent contamination with sperm in such experiments.

Of interest is the ability of eggs and sperm to retain their fertilisability after leaving the parent. According to Yamamoto fish eggs lose this capacity after a very short time, but this can be increased if they are retained in isotonic Ringer's solution. While this is true of some freshwater eggs, presumably as a result of water activation, in seawater the capacity for fertilisation is retained for much longer-certainly for hours in the herring. Nikolsky states that sperm motility is short lived where spawning takes place in fast-flowing water, for example, 10-15 sec in *Oncorhynchus*. In slower flows, sturgeon sperm is motile for 230-290 sec, and in the sea herring sperm may be motile for hours or days. Observations of short-lived activity are difficult to make; thus, some of these figures must be considered approximate.

Storage of gametes is a useful technique in fish farming to allow controlled fertilisations in the laboratory and to obviate the need for transporting eggs in the susceptible pregastrulation stages. Salmonid gametes are best stored dry below 5°C ; those of the herring may be held in buffered egg-yolk diluents, but are also best kept dry at about 4°C. While this permits storage for, at most, a few days long-term techniques are also possible. Herring sperm, but not eggs, were kept for some months in a diluent consisting of 12.5% glycerol in 3% salt solution (diluted seawater) at –79°C and crosses made successfully between spring and autumn spawning races. Sneed and Clemens also succeeded in holding out-of-season carp sperm immotile for 30 days at 3°–5°C in frog Ringer. Carp sperm could also be frozen and stored at –73°C in isotonic Ringer containing 6–12% glycerol with some survival when thawed after 60 hr of storage. More recently, Truscott *et al.* have shown that salmon sperm can be stored for 1–2 months at temperatures of –3° to –4.5°C using diluents such as 5% ethylene glycol or 5% dimethyl sulfoxide, retaining 70-80% fertility. Horton *et al.* obtained alevins from salmon eggs fertilised with sperm frozen in liquid nitrogen with dimethyl sulfoxide as a protecting agent, but the fertility rate was low. Hoyle and Idler have also obtained fertile salmon sperm after storage in liquid nitrogen using ethylene glycol with added lactose or serine. Slow freezing produced the best results. Mounib *et al.* succeeded in storing cod sperm for up to 60 days using 17-24% glycerol and temperatures of –79° and –196°C. Initial experiments suggest that faster cooling rates gave the better results with cod sperm.

Incubation (Fertilisation to Hatching)

The progress of cleavage, formation of layers, and morphogenesis have been described in a number of standard textbooks such as Rudnick, Waddington and Smith, with Oppenheimer and Devillers stressing structural changes from the viewpoint of experimental embryology. More detailed information is limited mainly to freshwater species like the trout *Salmo trutta*, killifish *Fundulus*, medaka *Oryzias*, goldfish *Carassius*, and to the dogfish. New gives information on the problems of culturing *Fundulus*, *Oryzias*, and *Salmo* eggs for the purpose of experimental embryology.

Most fish eggs are round, although in the anchovy *Engraulis* and bitterling *Rhodeus* they are ovoid, and in certain gobies pear-shaped. Most species have telolecithal eggs with yolk more concentrated at the vegetative pole; some marine species have oil globules of varying size and number. Before fertilisation the cytoplasm may be mixed with or separate from the yolk. The extent to which polarity exists at this stage has not been described in many species. After fertilisation (as the cortical alveoli release colloid, the perivitelline space develops and the chorion hardens) cytoplasm migrates to the future blastodermal region, most of it arriving by the first cleavage. The remaining cytoplasm forms a "halo" or periblast under the blastoderm.

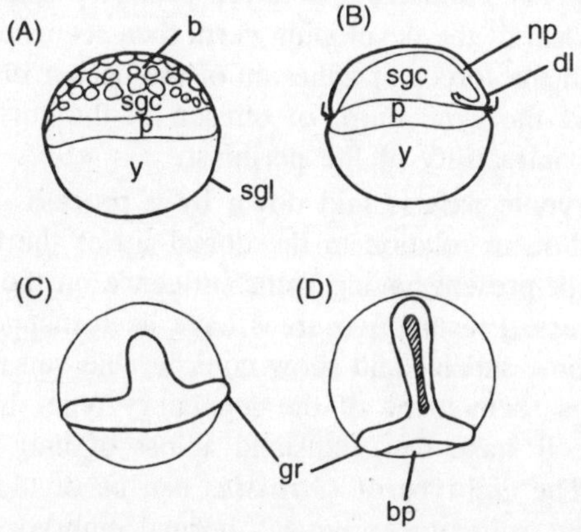

Figure 3.2 : (A) Transverse section of early blastula showing adhesion of blastomeres to the surface gel layer and attachment of blastoderm to the periblast at the periphery only. (B) Sagittal section of later blastula showing gastrulation; epiboly shown by arrows. (C) and (D) Surface view of eggs in later stages of gastrulation. Key: b, blastoderm; bp, blastopore; dl, dorsal lip of blastopore; gr, germ ring; n, notochord; np, neural plate; p, periblast; sgc, subgerminal cavity; sgl, surface gel layer; and y, yolk.

In lampreys cleavage is holoblastic but with the formation of microand macro-meres. In hagfish, elasmobranchs, and teleosts it is meroblastic. Other groups such as bowfin *Amia*, gar *Lepidosteus*, and sturgeon *Acipenser* have intermediate features. With some variation, the meroblastic group possess a blastodermal cap of cells at the animal pole after the initial stages of cleavage. In *Fundulus* the surface gel layer overlying the blastoderm, being sticky on its inner surface, serves to hold the outer blastomeres together. Usually cleavage is not complete, and in the deeper layers the periblast becomes syncytial and is involved in mobilising the yolk reserves. There are substantial cohesive forces between the developing blastomeres and the surrounding periblast which are important in the subsequent morphogenetic movements.

The blastoderm now commences to thin and overgrow the yolk (epiboly) and at the same time invaginate at its periphery. The periblast seems closely connected with this spreading tendency of the blastoderm, which at the junction may be thickened to form a germ ring. The syncytial periblast seems to have the property of autonomous spreading, and it is likely that cell proliferation in the blastoderm is relatively unimportant. Devillers suggests that the periblast acts as an intermediary between two "non-wettable" components-the blastoderm and yolk. As epiboly proceeds the blastopore contracts as does the surface gel layer over the yolk.. In *Fundulus* this layer probably solates and passes inward. The form of the developing germ then seems to be controlled by a balance of the forces of adhesion of the deeper blastomeres with each other and the syncytium, of tension in the surface (yolk) gel layer and of contractility of the periblast.

The embryonic axis is laid down by a process of convergence and concentration in relation to the dorsal lip of the blastopore, the quantity of yolk present having some influence on the time at which this event occurs. Presumptive areas have been mapped in the early gastrulae of some species and show considerable variation. In earlier stages the eggs seem to be of the regulatory type. In *Fundulus* the 2-cell and 4-cell stage can withstand a loss of half the number of blastomeres. The embryos of *Carassius* can be divided at the 8-cell stage, each part sometimes giving a normal embryo. Up to the 16-cell stage two embryos can be fused resulting either in twinning or an oversized single embryo. Removal of the yolk from the blastoderm before a critical stage is reached (8 cell stage in *Carassius*, 32 cell stage in *Fundulus*, blastula in *Salmo*) brings development to a halt. Incomplete removal of the yolk before the critical stage may, however, not prevent further development. It is likely that organiser substances rather than nutrient material diffusing from the yolk are more important. At later stages even the embryonic shield of *Fundulus* may be isolated and cultured to a fairly advanced stage, with the development of primitive axial organs, ears and eyes, and even with cardiac contractions and independent movement.

Hatching

The time to hatching is both a specifically and environmentally controlled character with temperature and oxygen supply exerting a considerable effect. Hatching results from a softening of the chorion because of enzymic or other chemical substances which are secreted from ectodermal glands usually on the anterior surface or from endodermal glands in the pharynx. In the sturgeon *Acipenser* the latter are innervated by the palatine nerve. The activity of the larvae, which may be enhanced by increase of temperature or light intensity or by reduction of oxygen tension, assists in breaking through the chorion.

The biochemical aspects of hatching are dealt with by Hayes, Smith , and Deuchar. The enzymes have apparently been identified in a number of species, but certainly in salmonids Hayes' work shows there is doubt about their mode of action. The chorion, which resists digestion by trypsin and pepsin, appears to be of "pseudo-keratin." Kaighn measured the amino acid and carbohydrate components in *Fundulus*. Cystine comprised only 1%, compared with 12% in keratin, It is therefore unlikely that disulfide links play a role in stabilising chorionic protein as they probably do keratin. The hatching enzyme works best under alkaline conditions, pH 7.2-9.6 and temperatures of 14°-20°C having been reported as optima. Very little hydrolysis takes place and Hayes speculates that the enzyme may be a reducing agent which liquefies the chorion. In *Oryzias* the enzyme is probably a tryptase.

In *Fundulus*, Kaighn obtained purified chorionase and concluded that digestion of the chorion was mainly a proteolytic process, although he could not duplicate its action with other proteases. Whatever the mode of action of the enzyme, it is likely that a considerable part of the nutrient material in the chorion can be utilized by the embryo via the perivitelline fluid and the losses at hatching may not be too serious.

The Larva

At hatching the larva is usually transparent with some pigment spots of unknown function. Notochord and myotomes are clear with usually little development of cartilage or ossification in the skeleton. A full complement of fins is rarely present, but a primordial fin fold is well developed in the sagittal plane. The mouth and jaws may not yet have appeared, and the gut is a straight tube. Although the heart functions for a considerable period before hatching, the blood is colourless in the majority of species and the circulation and respiratory systems poorly developed. The yolk sac is relatively enormous with, presumably, hydrodynamic disadvantages. Pigmentation of the eyes is very variable, but where the eye is not functioning at hatching it very soon develops. The kidney is usually pronephric with very few glomeruli. Very little is known about the endocrine glands, gonads, and other organs of the body cavity at such an early stage.

As the yolk is resorbed, the mouth begins to function, the gut and the eyes develop further, and the larva becomes fitted for transfer to sources of external food. One of the earlier systems to develop is that responsible for locomotion and support, the primordial fin being fairly soon replaced by median fins and the skeleton laid down. This is one of the better known aspects of later development because it is a system less easily damaged in such delicate organisms and because of the importance of meristic characters in racial studies of fish. Branchial replaces cutaneous respiration as the gill arches and filaments appear. The swim bladder may or may not be present during the larval phase. It is possible that this and the eyes, which are potentially dangerous in making the transparent larva visible, are silvered in such a way as to render them inconspicuous.

Metamorphosis

A clear change or metamorphosis from the larval to adult form is to be found in many species. In others there may be a number of less marked metamorphoses, e.g., in salmonids and eels. The most obvious signs are the laying down of scales and other pigmentation and often the first appearance of hemoglobin in the circulation. The swim bladder and lateral line may also develop first at this stage. In flatfish there is rotation of the optic region of the skull and the change in the normal orientation of the body so that they eventually come to lie on one side. There are often concomitant changes in distribution and behaviour such as schooling. Barrington gives detailed consideration to the physiological changes associated with metamorphosis in salmonids, eels, and the lamprey, especially from the aspect of thyroid activity and osmoregulatory functions.

The time to reach metamorphosis may be a matter of days in tropical species, a few weeks or months in the majority of fish from temperate latitudes, or periods of years in the sturgeon *Acipenser* and eel *Anguilla*. It is controlled not only genetically but also by temperature and food supply, which may affect the rate of growth, and possibly by social (hierarchical) factors as well.

Timing

To give some idea of the timing of the events just described, examples of the approximate duration of different stages under natural conditions are given in Table elsewhere in this chapter. The modification of these times experimentally or by fluctuations in environmental conditions is discussed in the succeeding pages.

METABOLISM AND GROWTH

Rate of Development

Obvious specific differences in time to hatching may be masked by variations in ambient temperature, which is one of the most potent influences on rate of development. Detailed observations of temperature

Table 3.1 : Duration of Events in the Development of Some Species

	Weeks from fertilisation to			
Species	**Hatch**	**First feeding**	**Meta-morphosis**	**Temp. range (°C)**
Scomberscombrus (mackerel)	0.8-1.5	1.3-2.0	11-13	9-15
Roccus saxatilis (striped bass)	0.25	0.75	4-5	17
Osmerus eperlanus (smelt)	2.0-4.0	2.5-5.0	8-10	4-14
Acanlhurus triostegus (convict surgeon fish)	0.15	0.7	?	26
Clupea harengus (Clyde herring)	2.5	3.5	16-18	7-10
Pleuronectes platessa (plaice)	2.5	4.0	10-12	7-11
Oryzias latipes (medaka)	1.5-2.0	2.0-2.4	Not clear cut	20-25
Salmo salar (salmon)	20-22	26-28	Gradual	1-7
Squalus acanthias (spur dogfish)	ca. 104	ca. 104	< 104	4-12
Scyliorhinus caniculus (spotted dogfish)	24-32	28-36	28-36	4-12

effects during development are scarce. Fliichter and Rosenthal, however, showed between 3.5° and 9°C a doubling of heart rate in the embryos of the blue whiting, *Micromesistius poutassou*, and more rapid embryonic movements. The effect of temperature on time to hatching, a commonly used criterion. Low temperatures retard hatching and at a theoretical low temperature (the biological zero) the incubation period will be infinite. The product of incubation time (*D*) and temperature (T)—day-degrees—was originally thought to be constant, i.e.,

$$TD = k \quad (1)$$

This was modified to use the temperature, not from zero, but from the biological zero (T_0), i.e.,

$$(T - T_0)\,D = k \quad (2)$$

There has been increasing criticism of the concept, for example, by Kinne and Kinne and Garside, on various grounds. Plots of *1/D* against *T* are *curvilinear* over wide ranges of temperature, simple linearity only applying over a narrow range. This invalidates the formulas given above. Furthermore, there may be inflections even of the curvilinear relationship at extreme temperatures. The biological zero, which is usually given between 0° and -2°C and most often around -1.5°C, may be below the freezing point of water or the body fluids themselves. In addition, abnormalities may occur at less extreme temperatures which are not necessarily lethal in the strict sense.

Improvements in describing the mathematical relation between *D* and T arise in later work. For example, Blaxter used the equation

$$(T - T_0)\,(D - D_0) = k \quad (3)$$

for development of Clupea harengus, where D_0 is the theoretical time to hatching at infinite temperature, not in itself a very satisfactory additional constant. Lasker, working with the eggs of Sardinops caerulea, used the equation

$$D = aT^b \quad (4)$$

where *a* and b are constants, and Braum, using the eggs of whitefish *Coregonus* and pike *Esox lucius,*

$$D = D_x + 1.26^{T_x - T} \quad (5)$$

where D_x is the minimum possible incubation time at the maximum permissible temperature T_x.

The van't Hoff values over different temperature ranges reflect the nonlinearity of plots of log *D* against T, the values being higher at lower temperatures. Thus in *Enchelyopus cimbrius* the $Q_{}$, varies between 6.5 and 1.5 over the temperature range 5°-23°C and between 6.5 and 2.0 over the range 3°-18°C in herring. The value of this type of theoretical consideration may lie in establishing criteria for optimum conditions of development. Thus the optima may be where van't Hoff values lie between certain limits. Certainly the day-degree concept is useful as an approximation for predicting events in normal hatchery practice.

Other environmental factors influence the rate of development. Low salinities may accelerate or retard the time to hatching, while oxygen lack has a retarding effect on development, especially at higher temperatures. Laale and McCallion found that the development of the zebra fish, *Brachydanio rerio*, could be arrested before gastrulation by the use of the supernatant of homogenates produced from other zebra fish embryos. This arrest, which could be reversed, appeared to be an effect at the cellular level, the nuclei of the arrested embryos all being in interphase.

Yolk Utilisation

The efficiency with which yolk is transformed to body tissue and the effect of the environment on utilisation is important in that larger larvae may be expected to be stronger, better swimmers, less susceptible to damage, and less liable to predation. Efficiency at any time may be expressed as a percentage:

$$\frac{\text{dry weight increment of body}}{\text{dry weight decrement of yolk}} \times 100$$

More often efficiency is measured from fertilisation to final yolk resorption (or to maximum weight attained on the yolk reserves), This is gross efficiency, i.e.,

$$\frac{\text{dry weight of final body}}{\text{dry weight of original yolk}} \times 100$$

or from fertilisation to intermediate stages as

$$\frac{\text{dry weight of body}}{\text{dry weight of original yolk} - \text{dry weight of remaining yolk}} \times 100$$

or more precisely

$$\frac{\text{dry weight of body}}{\text{dry weight of body} - \text{dry weight of yolk used for maintenance}} \times 100$$

The difficulties of measuring efficiency by dry weight lie in the need for taking samples of an egg population at different stages with the accompanying problems of initial differences in egg weight. Utilisation of material from the chorion or losses of excretory products are also difficult to allow for, as are the possibilities of uptake of organic matter from the environment. Another serious problem when comparing, for example, environmental effects such as temperature on efficiency is the question of making dry weight measurements at "equivalent" stages. Both hatching and maximum weight (attained on the yolk) can be questioned for staging; hatching at different temperatures can result in larvae of quite different appearance, while full yolk utilisation is often not complete when maximum weight is reached, some yolk remaining in the yolk sac. Furthermore, the disappearance of the yolk sac is no certain indication that all the yolk has been used as it may be present in storage spaces within the larval body, for example, in the subdermal spaces of cod and plaice larvae. D. H. A. Marr adopted the ratio

$$\frac{\text{dry weight of body}}{\text{dry weight of body + remaining yolk}} \times 100 \quad \text{(as \%age)}$$

as a criterion for ***equivalent staging***, comparing in *S. salar* the efficiency

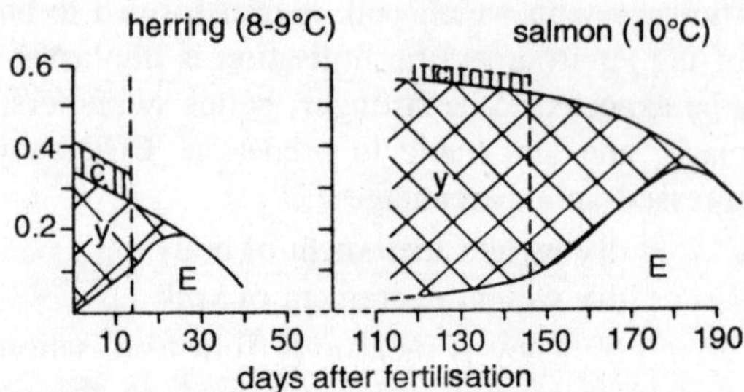

Figure 3.3 : The relative proportions of yolk (Y), embryo (E), and chorion (C) during the development of a small egg and a large egg. The vertical dashed line represents hatching. Note the difference in the scale of the ordinates.

of development at different temperatures between the 15% and 80% stages. Ryland and Nichols used the ratio

$$\frac{\text{rate of growth in length}}{\text{rate of yolk disappearance}} \times 100$$

for equivalent staging when comparing the efficiency of development at different temperatures during the yolk sac stage of *P. platessa*. The use of maximum weight as a stage for making comparisons still remains, however, a useful criterion and one with immediate meaning when deciding on the optimum conditions for hatchery practice.

Efficiency over the whole process of yolk utilisation is mainly between 40% and 70% although clearly cumulative efficiency must decrease as growth proceeds and the maintenance requirements increase. Experiments with temperature indicate certain optima for maximum efficiency. Other influences on efficiency are the original egg weight at the intraspecific level in *C. harengus*, and light conditions, contour of the substrate, and turnover of water in the photonegative alevin of S. *salar* living within the interstices of the spawning redd. Here highest efficiency is achieved under dark conditions, on a grooved substrate, with a rapid turnover of water.

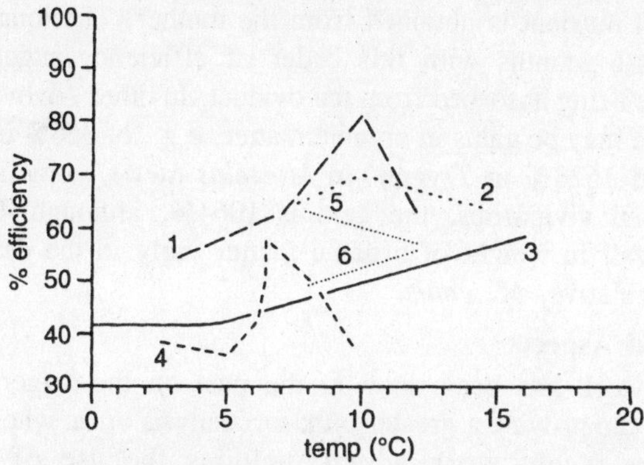

Figure 3.4 : Efficiency of development, at different temperatures. 1. Salmo trutta (fario)-yolk sac period; 2. Salmo salar-alevin stage; 3. S. salar-early alevin ; 4. Pleuronectes platessa-yolk sac larva ; 5. Clupea harengus-small eggs, yolk sac period; 6. Clupea harengus-large eggs, yolk sac periodb.

A word of caution is required where larval feeding may occur well before final yolk resorption. High efficiency may result from low activity, a high proportion of yolk being used for growth; if this is reflected in low feeding activity it could be a very undesirable trait.

Viviparity

Amoroso, who gives a comprehensive review of this subject, points out the rather indistinct barrier between ovoviviparity where the young develop within the female only on their yolk reserves, and viviparity, where the nutrient requirements are obtained from the mother. In the first place, there are a wide variety of methods of obtaining these nutrients: by absorption through simple external surfaces, by swallowing, or by "placental" connections. Perhaps all these may be considered as viviparous. In the second place, initial development may be ovoviviparous (on the yolk supply) with viviparity superimposed later. This is likely to be the case in the very early stages of all viviparous fish, but in the smooth dogfish, *Mustelus laevis*, and goodeid teleosts, for example, there is a change from one to the other rather later. Details of the functional morphology of viviparity are dealt with in the chapter by Hoar, this volume.

The changes of weight found during the development of various species can be very striking. Some of the ovoviviparous ones, where maternal nutrients are scarce, show decreases of organic matter between the fertilised egg and final embryo, for example, of 23-34% in *Torpedo spp*. This gives an efficiency of 66-74% which puts this species very much in the same category as oviparous forms although it is rarely certain what nutrient is obtained from the mother. The long gestation period of 4-6 months with this order of efficiency suggests some nutrients are being absorbed from the oviduct. In other "ovoviviparous" species there may be gains in organic matter, e.g., of 356% in *Mustelus vulgaris* and 1628% in *Trygon*. In *Mustelus laevis*, sometimes more strictly called viviparous, the gain is 1064%, although Te Winkel reported a fall in weight of organic matter early in the development of a close relative, *M. canis*.

Biochemical Aspects

Much work has been done in the past on the larger salmonid eggs which can provide a greater bulk for analysis or on whole ovaries. Some more recent work which includes the use of isotopes, chromatography, and histochemistry is mentioned by Deuchar and Williams.

Water Relations

The swelling of eggs, with the formation of the perivitelline space as a result of water uptake, is a general phenomenon signifying fertilisation. The chorion at fertilisation is permeable to water and also to urea, glucose, salts, and certain dyes. It seems likely that the colloidal material liberated from the cortical alveoli cannot escape through the chorion and creates an osmotic pressure which draws in water. This effect can be inhibited by high osmotic pressure in the outside medium. The chorion then hardens, a process taking a matter of a few hours and the osmotic forces become matched by the resistance of the chorion. In hypertonic solutions the eggs of salmonids and of *Mullus barbatus* lose water only from the perivitelline space and not from the embryo. Some loss does, however, occur in acipenserids. The use of D_2O on water-activated and early fertilised eggs of *Oncorhynchus tshawytscha* seems to confirm the view that only the perivitelline space is penetrable by water. Subsequent use of ^3H_2O, $^{22}NaCl$, and $Na^{131}I$ on the water-activated eggs of *Salmo gairdneri* has shown a definite but limited permeability of the vitelline membrane to anions, cations, and water. Unfortunately, this was not done on fertilised eggs. Recently, however, Potts and Rudy have confirmed with ^3H_2O that the vitelline membrane of fertilised eggs of S. *salar* has a high permeability before laying and during water hardening. Subsequently permeability is low until the eyed stage. The use of ^{24}Na showed that sodium exchange is confined initially to the perivitelline fluid but accumulation within the embryo occurs during the eyed stage. Terner reported that during the eyed stage of *Salmo gairdneri* external substrates such as ^{14}C-labeled pyruvate and acetate were apparently taken up and metabolised as judged by the presence of $^{14}CO_2$ in the respiratory CO_2. Wedemeyer also found that ^{65}Zn was taken up by developing eggs of coho salmon, *O. kisutch*. Almost all was bound to the chorion, but 26% was found in the perivitelline fluid, 2% in the yolk, and 1% in the embryo. Mounib and Eisan found that both ^{14}C-labeled pyruvate and glyoxylate could be utilised in the form of an exogenous substrate by salmon (*S. salar*) eggs. Lactate was produced and any carbon atom in these compounds could be incorporated by the eggs into organic acids, lipids, nucleic acids or proteins. The formation of ^{14}C-amino acids indicated the presence of an active transaminase system.

The use of freezing point measurements on the yolk of herring and plaice eggs gives an alternative picture. In the herring, which has a demersal egg, the yolk is not regulated osmotically until after gastrulation when it is covered by cells. However, in the pelagic floating egg of the plaice, regulation occurs from fertilisation, indicating the ability of the vitelline membrane to osmoregulate. This seems necessary from the buoyancy aspect.

If Gray's measurements of wet and dry weights of the embryo and yolk of trout are generally true of salmonids, then the yolk as a high density nutrient (41% dry weight) requires considerable quantities of water for transformation to the relatively watery embryo (16% dry weight). Obtaining this water may be a problem if the permeability of the vitelline membrane remains limited throughout much of development. Smith suggests that growth may be retarded until hatching occurs and water becomes more readily available. It is likely that highly desiccated yolk is only feasible in the demersal egg; in *Sardinops caerulea* the water content of the larval yolk is about 91%.

Chemical Composition

The chemical composition of the eggs of two species. Further data are given by Phillips and Dumas, who showed in particular that there was no difference in the constituents of different sized eggs of *Salmo trutta*.

There is considerable difficulty in deciding on the sequence of utilisation of various materials for energy production. Analysis of the change in chemical components may be unreliable where substances like carbohydrates are being synthesized. Heat production and respiratory quotients are difficult to determine accurately in small eggs, and CO_2 liberation may be masked by a buffered external medium. Amberson and Armstrong found the RQ's of *Fundulus* eggs over the first 6 days of development were 0.90, 0.78, 0.77, 0.76, 0.72, and 0.72 suggesting very early carbohydrate metabolism. In

Table 3.2 : Analyses of Eggs

	Percent of wet weight						
Species	***Dry weight (%)***	***Protein***	***Total fat***	***Oil***	***Phos-pholipid***	***Carbo hydrate***	***Ash***
S. gairdneri (*irideus*) (rainbow trout)	33.8	20.2	—	3.6	3.8	0.2	1.3
Sardinops caerulead (California sardine)	29.3	21.0	3.8	—	3.2	<0.3	2.1

Oryzias, Hishida and Nakano found 0.75 after fertilisation, 0.92 at gastrulation, and 0.70 later. Another problem to overcome is the retention of nitrogenous excretory mattern within the egg which gives values of protein metabolism which are too low. The original concept of Needham that materials were used up for energy in the sequence

carbohydrate—protein—fat

has to some extent been supported by Devillers. However, Haye was of the opinion that the sequence in salmon eggs was

hatch
↓
fat—protein—fat—protein

and Smith that it was

hatch
↓
phospholipid—protein + carbohydrate—protein—phospholipid—triglyceride fat

Carbohydrate Metabolism

Since carbohydrate is present in small quantities and synthesis is continuously occurring, its role is particularly difficult to ascertain by bulk analysis. Hayes reported a gradual increase in carbohydrate level during salmon development with a temporary fall in glucose level at hatching. Glycogen is probably synthesized near hatching and is present exclusively in the embryo, being stored in the liver near the end of the yolk sac stage. Glucose, in aqueous phase, is present in relatively greater quantities in the embryo than in the yolk due to its higher water content. Smith described falls in total carbohydrate after gastrulation (linked with the establishment of the circulatory system and therefore a higher potential for metabolism) and a further fall in glucose level at hatching, perhaps because of an interruption of glycogen synthesis. He did not report the general increase found by Hayes. Terner has more recently found that the ova of *Salmo gairdneri* have a free store of glucose which increases during development and only falls at hatching. Carbohydrate is certainly being synthesized in development for use in the final stages of energy production and for mechanical and osmoregulatory work.

Protein

Protein is the dominant raw material in the yolk and the main source for tissue formation. The use of labeled amino acids in *Oryzias* shows a slow passage from the yolk to a low molecular weight pool, a more rapid uptake occurring when the vitelline circulation is established. The proportion of yolk used for energy requirements is not easy to determine since the egg retains some nitrogenous metabolities (shown by an increase of nonprotein nitrogen with age) and only small amounts of ammonia are excreted. Of particular interest is the finding by Read of ornithine carbamoyltransferase and arginase in the embryos of *Squalus suckleyi* and *Raja binoculata*, suggesting an early functioning ornithine urea cycle with the retention of urea for osmoregulatory purposes. The use of protein for energy seems to decline after hatching, while the amount used over the whole of early development depends on whether calculations are made when some yolk remains or whether, at a later stage, the larva starts to consume its own tissues. While it has been suggested that 40% of the original protein may be metabolised, more recent detailed studies using isotopes on *Oryzias* indicate that all protein being resorbed from the yolk in the 9 days subsequent to gastrulation was being transformed into new tissue and none lost in combustion.

Regarding different types of protein it appears that in *Oryzias* 40% of phosphorus is incorporated as phospho-protein; it falls after gastrulation and the loss is very rapid at hatching, especially in S.

gairdneri (*irideus*). Deuchar, reviewing analyses of amino acids present in the same species, reports that aspartic and glutamic acid are always present, valine and leucine appearing after gastrulation.

Fat

There is some ambiguity about the role of fat in energy production during different phases of development. Fat is undoubtedly used as a fuel, possibly 70—80% being consumed over the whole period of development. Hayes reported the main loss of fat to be in the fourth week after hatching in the salmon. It seems that the triglyceride fats are the last to be utilised before food is required from external sources. The energy requirements before hatching may also be met by fat, according to Hayes, or by fats in combination with protein.

According to Phillips and Dumas fat is synthesized from protein in the eyed stage of the eggs of brown trout *Salmo trutta*. Terner *et al.* found that ^{14}C-labeled acetate in the incubation medium of the eggs of *S. gairdneri*, *S. trutta*, and *Salvelinus fontinalis* was incorporated into free fatty acids and other lipids of the egg, which were probably the substrates for endogenous respiration. In addition, labeled acetate was found in the egg phospholipids, presumably as an intermediate stage in the synthesis of complex fats. These authors suggest that the lipids of the embryo are not transferred directly from the yolk but are resynthesized by the embryo after the breakdown of triglyceride fats in the lipid pool of the yolk.

Respiration

In the sense of obtaining oxygen, respiratory problems vary with the environment of the egg and larva, but they are probably rather rarely limiting with pelagic eggs. Eggs deposited on or in a substrate may act as an oxygen "sink" the oxygen level at the egg surface always being less than that of the surroundings even in high velocities of current. Hayes *et al.* calculated a flow of 1.2×10^{-2} μl/cm^2/cm thickness/min through the chorion of salmon eggs. Sensitivity to low oxygen tensions or anoxia varies with both species and stage of development. In salmonids, cleavage may continue in anoxic conditions; perhaps this is possible because requirements are low and oxygen is stored within the egg, especially in the perivitelline fluid. Gastrulation is more easily blocked by lack of oxygen or by respiratory poisons such as cyanide or azide. Development can also be retarded by blocking oxidative phosphorylation with dinitrophenol. During oxygen lack there are often increases in lactic acid production but, in the shortterm, retardation of development to anoxia is reversible. De Ciechomski succeeded in keeping the demersal eggs of *Austroatherina* for 17-18 days in "Vaseline oil." Development was at first normal, but no heart beat or movement was observed. Transfer to water resulted in the heart starting to beat and some movements occurred, but abnormal pigmentation developed, and no embryos hatched. Eggs of three pelagic species were killed by immersion in the oil.

Various adaptations are found which assist in obtaining oxygen. The spawning redd of the salmonid fish has a rapid current of water passing through it; parents guarding their eggs may ventilate them (e.g., the stickleback, *Gasterosteus aculeatus*) while the lungfish *Lepidosiren* has special pelvic "gills" for oxygenating the eggs in its nest; viviparous species obtain oxygen as well as nutrients from the female. After hatching functional gill filaments are often lacking; in herring they do not appear for some weeks at a length of about 20 mm. Oxygen is obtained over the body surface and almost certainly via internal body surfaces such as the pharynx and gut. The ultramicroscopic corrugations described by Jones *et al.* and Lasker and Threadgold on the epidermis might possibly be directing the flow of water over the body surface. In less favorable environments external gills may be found, for example, in the bichir *Polypterus*, in the lungfish *Protopterus* and *Lepidosiren*, in the loach *Misgurnus fossilis*, and in *Gymnarchus niloticus*. In newly hatched trout, *Salmo trutta*, respiratory currents are produced by the pectoral fins which also keep the body clear of silt. After a short time mouth and gill movements replace those of the fins and then clogging of the gills by silt is prevented by a coughing reflex or aggregating the silt with mucus.

Intake of oxygen has been measured in a number of species. In the egg stage this may be very low and variable before fertilisation. In *Oryzias latipes* there is no sharp rise until a few hours after fertilisation, although it seems likely that the respiratory rate becomes steadier at this stage. Before any movements occur within the egg, relative values of oxygen uptake between different stages and species are quite valid, but subsequent to movement being possible the oxygen consumed can be dominated by bursts of activity. Oxygen uptake may be used to express metabolic functions, total or general metabolism consisting of two components, active metabolism (owing to activity) and basal metabolism (owing to maintenance functions). Standard metabolism applies to practically motionless fish under experimental conditions; in the fry of *Salmo salar*, for example, the basal metabolism was 64-69% the standard rate. For active fish metabolism is proportional to KV^n, where V is velocity and K and n are constants. The change in oxygen uptake with varying velocity is elsewherein this chapter. The function KV^n is called the "scope for activity" and has been estimated for a number of species.

Some values of oxygen uptake expressed as Q_{O_2} (μl/mg dry weight/ hr) based where possible on subjective criteria of activity and inactivity. Without the type of precise control as demonstrated in Fig. 7A these are the best available data at the present time. The standard Q_{O_2}'s show some similarity considering the variety of the material used. Active Q_{O_2}'s will naturally fluctuate widely, depending on the type and extent of activity permitted. The use of certain anesthetics to control activity may also help in establishing at least some of the factors controlling oxygen uptake.

Cumulative oxygen uptake has been measured over short periods as well as fairly long periods of development. In *Fundulus heteroclitus*, the killifish, it was 80 μl/egg from fertilisation to hatching

Table 3.3 : Scope for Activity.

Species	*Stage*	*Range of increase of Q_{O_2} owing to activity*
Salmo salar (salmon)	Eggs	3
Salmo salar (salmon)	Fry	7-14
Sardinops caerulea (California sardine)	Larvae	32
Clupea harengus (herring)	Larvae	9-10

In salmon eggs the uptake was about 0.2 μl/egg/hr at fertilisation rising to 3.4 td/egg/hr at hatching, the cumulative total between these stages amounting to 1400 μl/egg. Privolniev, quoted by Devillers, found only 850 μl/egg from fertilisation to an age of 120 days. Such values may be compared with the short yolk sac period of the California sardine, *Sardinops caerulea*, where only 12 μl/larva were used from hatching to 180 hr subsequently. Obviously cumulative uptake and uptake per unit time will vary with size, age, and temperature. Devillers considers that increases in oxygen uptake with age may be shown by the exponential equation

$$Q = ae^{kt} \tag{6}$$

where t is time and a and k are constants. Q_{O_2}'s will also vary with environmental conditions. In herring eggs and larvae the Q_{O_2} for oxygen uptake (Q_{O_2}) was 2.0 between 5° and 14°C. Q_{O_2} also increased with a salinity shock but not with constant rearing salinities. There were only slight effects because of light intensity and starvation. In salmonids oxygen uptake falls at low ambient oxygen tensions, but the young stages of fish will resist oxygen lack quite well for a time.

This raises the problem of oxygen debt in such experiments and indeed in any respiration measurements where activity is intense. The accumulation of such a debt can be tested by continuing respiration measurements after activity or anoxia. Hayes *et al.* found no oxygen debt after periods at reduced oxygen tensions in salmon eggs, suggesting there had been a reduction in metabolism under these conditions. Salmon fry, however, showed an oxygen debt amounting to 48% of general metabolism after a period of activity. Fry of bleak, *Alburnus alburnus*, had a negligible debt when stemming a water velocity of 2 cm/sec but one of about 15% in 6.42 cm/sec.

An overall review of respiration rates, including that of larval fish, allowed Winberg to deduce that the relationship

$$Q_S = 0.3\ W^{0.8} \qquad (7)$$

represented the best fit for the data on standard metabolism in microliter per hour at 20°C for organisms of wet weight W mg. He provides conversion factors for other temperatures but stresses that such relationships are designed to evaluate experimental results rather than replace them. Blaxter, recalculating the data of Holliday *et al.* for herring, concluded that in larvae of mean wet weight 1.53 mg respiration amounted to 17 μl O_2/day at 8°C. Using Winberg's relationship a value of 23 μl O_2/day is arrived at, fair agreement considering his relationship is operating at its limit.

Deductions can be made about the type of metabolism involved in respiration. The consumption of 1000 μl O_2 corresponds to the utilisation of different materials as follows:

$$1000\ \mu l\ O_2 \equiv \begin{cases} 1.05 \text{ mg protein} \\ 1.23 \text{ mg carbohydrate} \\ 0.50 \text{ mg fat} \end{cases}$$

1 mg of protein yields 4.25 cal, 1 mg of carbohydrate yields 4.15 cal, and 1 mg of fat yields 9.45 cal;

therefore,

$$1000\ \mu l\ O_2 \equiv \begin{cases} 4.5 \text{ cal from protein} \\ 5.1 \text{ cal from carbohydrate} \end{cases}$$

$$\frac{4.7 \text{ cal from fat}}{4.77}$$

The value 4.77 cal/ 1000 μl O_2 is sometimes called the oxycaloric coefficient. This makes it possible to calculate the requirements for growth and metabolism from measurements of growth rate and respiration. On the basis of 1000 μl O_2 being equivalent to 4.77 cal, and 1 mg wet weight of growth corresponding to 1 cal, Ivlev concluded that young Baltic herring of wet weight 66.5 mg required 16.2 cal/day for growth and maintenance.

Table 3.4 : Calorific Intake in Some Species.

Species	*Wet weight* (mg)	*Daily itake* (Cal)	*Temp.* (°C)
Engaraulis japonics (achovy)	1500-6300	78-426	14-20
Alburnus alburnus (bleak)	250	3.3	19-22
Clupea harengus (herring)	66.5	16.2	(?) 15-17
C. harengus (herring)	1.53	0.18	8

Smith reviews the work on growth, heat production, and respiration. High respiration and high rates of growth appear to be correlated, at least at some phases of development. In particular, specific growth rate and heat produced show very similar trends at various stages of development of the rainbow trout, *Salmo gairdneri* (*irideus*). Up to hatching, plots of heat production on specific growth lie on a straight line with positive slope passing through the origin; after hatching the line passes through an intercept on the ordinate. There are two components in heat production-one resulting from growth and the other from maintenance. Temperature may affect the relative requirements of growth and maintenance thus giving differences in efficiency of development with temperature optima.

Growth

Growth is influenced in the early stages by the ratio between embryo and yolk weight, especially by the problems of efficient yolk utilisation and the need to mobilise the nutrient reserves. Higher maintenance requirements per unit weight in small organisms, surfacevolume ratios, change of diet with age, feeding and searching potential are also important. Growth has often been shown in terms of length but there is a need for uniformity, i.e., whether the caudal fin or fin fold is included, and length gives no information on the condition (fatness) of the organism or of the changes in body proportions with age; this is especially true in the yolk sac stage and at metamorphosis. Similarly, wet weights introduce the error of varying water content; thus, dry weight is the most satisfactory measure.

Examples of the relative quantities of yolk and embryo in a large and a small egg at different stages of development. The increase of weight of the embryo and decrease of the yolk tend to be logarithmic in nature. It is clear that in the earlier stages relatively less yolk is being used for maintenance than for growth, giving a higher efficiency of utilisation. Without external feeding the larval weight begins to fall before all the yolk is resorbed. Farris divided the growth in length of four species of pelagic larvae after hatching into three phases: an early rapid phase after hatching, a slow phase near the completion of resorption, and a subsequent negative phase if no food was available. This shrinkage in terms of length was also described by Lasker in the larvae of *Sardinops caerulea.*

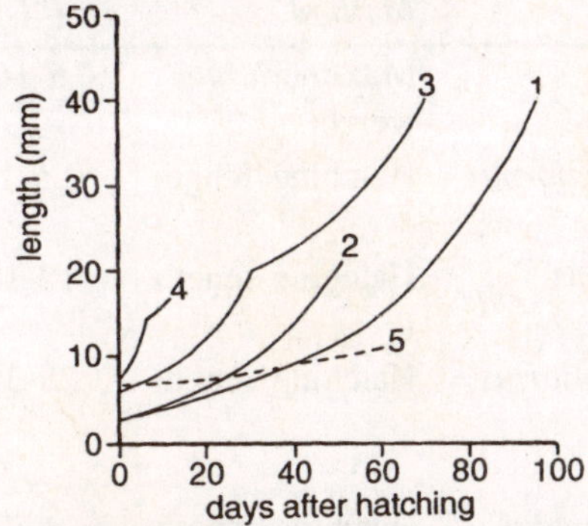

Figure 3.5 : Length of different species related to age under natural conditions. 1. blelanogrammus aeglefinus; 2. Scomber scombrus; 3. Clupea harengus; 4. Esox lucius ; and 5. Pleuronectes platessa.

Growth of salmonid and other freshwater fry in hatcheries has received considerable attention from an economic point of view, to provide maximum growth at minimum cost. Growth in marine conditions, where sampling of the same population is difficult owing to mortality, migration, or changes in net avoidance with age.

Some estimates of survival of fish larvae depend on a knowledge of growth rate and this is often inadequate or based on flimsy experimental data. Comparisons from tank experiments are difficult owing to mortalities, size-hierarchy effects, or other possible aquarium artifacts. Showing the enormous range of size which may be found from an initially fairly uniform stock. Thus Shelbourne, after a series of rearing experiments on plaice, found the length ranged from 7.5 to 37.5 mm. In dense populations there were also relatively larger numbers of small larvae. Magnuson tested the effect of food supply on *Oryzias* kept at high densities. As long as food was adequate, growth was not retarded, social hierarchies did not develop, and aggressive behaviour was not manifested. Territorial behaviour only occurred when food was limited spatially. Whether size hierarchies occur in natural conditions is not known. It seems they must stem from inherently different growth rates or from some type of dominance hierarchy—the first could occur in the wild, but the second is more likely to occur in crowded tank conditions.

Table 3.5 : Effect of Temperature on Size, Using Yolk Reserves Only.

Species	*Method*	*Temp. range (°C)*	*Corresponding size range (mm)*
Salmo trutta (brown trout)	Maximum wet weight	5-16	135-95 (mg)
Coregonus clupeaformis (whitefish)	Hatching length	0.5-10	13-8.8
Osmerus eperlanus (smelt)	Hatching length	12-18	5.2-4.6
Cyprinodon macularius (desert minnow)	Hatching length	28-35	5.3-3.7
			4.2-3.6
Coregonus wartmanni (whitefish)	Hatching length	1-7	11-9
Gadus macrocephalus (Pacific cod)	Hatching length	2-10	4.1-3.5

Food supply and temperature are the most potent environmental factors in controlling growth in the later stages, and there are few data available on this except for salmonids. In the egg stage low temperatures often produce longer larvae at hatching, although Lasker found the maximum length of *Sardinops caerulea* larvae at intermediate temperatures. There are then probably temperature optima for yolk utilisation which partially give rise to these effects, but often the range of temperature used may be insufficient to show them. On the assumption that high temperatures are nonoptimal other conditions producing smaller larvae, such as low oxygen tensions and current velocities for salmonids, as well as high salinities in the desert minnow, *Cyprinodon macularius*, shown by Sweet and Kinne and in cod, *Gadus macrocephalus*, *by* Forrester and Alderdice, may also be considered nonoptimal. Combinations of temperature and salinity suggest that the optimum conditions for growth in length *over the ranges used* were 26°C for 70%$_0$ salinity, 28°C for 35%$_0$ and 33°C for freshwater in *Cyprinodon*, and 6°C and 19%$_0$ for *Gadus macrocephalus*. The mortality, length, and maximum weight of salmon alevins (on the yolk supply) was greatest when they were reared in the dark, on a grooved surface, where presumably activity was minimal.

Endocrines, Growth and Metamorphosis

Russian, Canadian, and other work on the role of the endocrine system in fish larvae is fully reviewed by Pickford and Atz and the role of the thyroid in metamorphosis by Barrington. The main techniques used were histological, histochemical, or immersion in dilute solutions of thyroid hormone or antithyroid drugs. Evidence from a number of sources suggests that the pituitary is inactive until metamorphosis, for example, in herring *Clupea harengus*, eels *Anguilla anguilla*, bream *Abramis brama*, milkfish *Chanos chanos*, and sturgeon *Acipenser stellatus*.

The thyroid varies in its histological "activity" in the early larval stages but at metamorphosis in sturgeon, flatfish, eels, herring, pilchard *Sardina pilchardus*, and bone fish *Albula vulpes*, there are signs of secretion of colloid into the lumen of the thyroid follicles. Whether this is storage, or release into the circulation is taking place, can be satisfactorily tested only by measuring hormone levels in the blood. Earlier, thyroxine may retard growth and accelerate morphogenesis (e.g., in salmon, sturgeon, *Misgurnus fossilis* and *Lebistes*) at a concentration of 1 ppm or less. Rate of oxygen consumption may also be increased. In the developing eggs of *Scyliorhinus canicula* the functioning of the thyroid seems highly dependent on temperature. At 8°C use of ^{131}I showed the thyroid concentrated and bound iodine, but there was no evidence of thyroid hormone formation. At room temperature thyroid hormone reached a measurable level. Thiourea at concentrations of about 100 ppm may delay hatching, retard yolk resorption, and inhibit growth in salmonids, inhibit morphogenesis in sturgeon larvae, and decrease oxygen consumption in salmon and sturgeon larvae by as much as 15%. Thiourea seems to improve the ability to utilise oxygen at low tensions, although it is not clear whether this is because of its antithyroid action. Thiourea might well improve survival in suboptimal conditions of oxygen concentration that might occur, for example, during transportation. The effect of thyroxine on promoting metamorphosis was tested on the eel. There was an accelerating, but not a sudden, effect with an indication that an increasing threshold controls the sequence of events in metamorphosis. In other words, the later stages of metamorphosis require higher concentrations of thyroxine. Other experiments on lampreys showed no effect of thyroxine, thyroid extract, iodide, or iodine on metamorphosis.

It may be concluded that there is a suggestion of the thyroid playing a role in growth and differentiation in the larval stages but further work is required, especially at the important metamorphosis stage. If the pituitary is not functioning until metamorphosis the thyroid must be relatively independent of pituitary control in the early stages of life.

FEEDING, DIGESTION, AND STARVATION

Often the mouth is not completely formed at hatching, but rapid development in many marine fish larvae leads to the possibility of taking external food before the yolk is finally resorbed. Without success in feeding there is eventually self-metabolism and loss of weight. Yolk may be transported and stored within the body, especially in the base of the primordial fin and other subdermal spaces, and fat may be stored in the mesenteries of the larval gut. Unlike demersal freshwater species there is no vitelline circulation in most pelagic larvae although *Scomberesox*, *Trachypterus*, and a few other species are known to be exceptions. Usually there is a yolk sac sinus in connection with the heart and with lateral branches to the subdermal spaces; in the species with a vascularised yolk sac these spaces are much less inflated.

After final resorption of the yolk the larvae retain their potential to feed for some days depending on species, egg size and temperature. The concept of a "point of no return" was introduced by Blaxter and Hempel for herring larvae. This point, after which the larvae are still living but too weak to feed, may vary from 15 days after fertilisation for summer spawners in warmer water to 45 days for winter spawners. In the larval cisco, *Leuchichthys artedi*, it was 27-

36 days after hatching at 3°-4°C. The activity of larval *Solea solea* drops from about 70% of the time active at first feeding to 20% some 5-6 days later when, if feeding is unsuccessful, inanition occurs. Information on the time to point of no return in relation to spawning and food supply may help to predict the probability of survival in different broods.

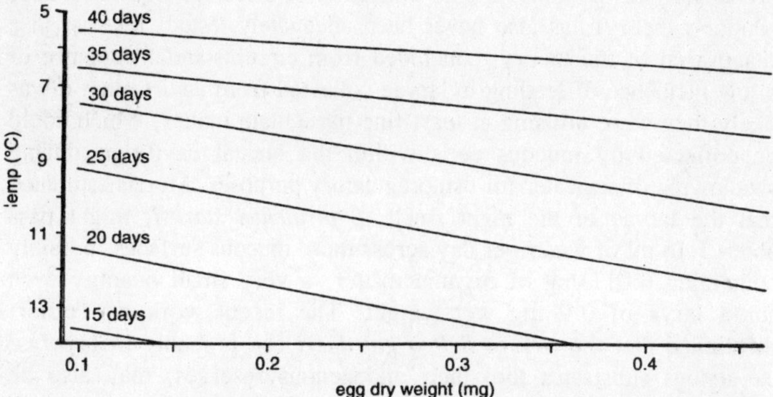

Figure 3.6 : The concept of "point of no return." The time from fertilisation to the point where the larvae of Clupea harengus *are too weak to feed is given in the form of a nomogram based on original egg dry weight of different races and the temperature.*

Most fish larvae are predatory with a large mouth and well-developed eyes. In herring the gape of the jaw increases by 50% during the yolk sac stage: the gape of 0.3-0.4 mm at first feeding depends on the length of the larvae and, therefore to some extent, on the original egg size. An elastic ligament at the articulation makes it possible for even larger organisms to be taken. Some species have very small gapes which is partly a function of their small size (e.g., lemon sole *Microstomus kitt*, sprat *Sprattus sprattus*, pilchard *Sardina pilchardus*, sardine *Sardinops caerulea*, and many gadoids). These must require very small food organisms for their early survival, a factor which is of considerably less importance for many freshwater fish.

The forward darting movements of fish larvae in order to engulf the prey are described in a number of species, for instance, in *Coregonus* and *Esox* by Braum and in sole, *Solea solea, by* Rosenthal. Iwai found three types of feeding in the larvae of *Plecoglossus altivelis:* a predatory snapping, a respiratory current with a sieving action by the gill rakers, and a ciliary current associated with the olfactory pits.

The extent to which fish larvae are herbivorous and might make use of such currents is not clear. Green food in the gut may be incidental to the swallowing of water or due to fecal material from prey species, and its presence is no evidence of utilisation. In the anchovy, *Engraulis anchoita*, however, phytoplankton in the guts of young stages can be correlated with the formation of gill rakers which presumably act as a sieve. The utilisation of dissolved organic matter (Putter's theory) has also never been adequately tested. Morris, in a reappraisal of the theory, concluded from circumstantial evidence of a low incidence of feeding in larvae collected from nature, that it was likely they were utilising at least fine particulate matter, which could be collected by mucous cells within the buccal cavity or during swallowing of seawater for osmoregulatory purposes. Morris estimated that the larvae of the night smelt, *S pirinchus starksi*, might pass about 1.15 ml of water per day across these mucous surfaces, possibly containing 0.011 mg of organic matter, a very small quantity even for a larva of 0.9 mg wet weight. The recent work of Terner, mentioned that the ova of *Salmo gairdneri* can probably incorporate exogenous substrates into their endogenous reserves, may also be significant.

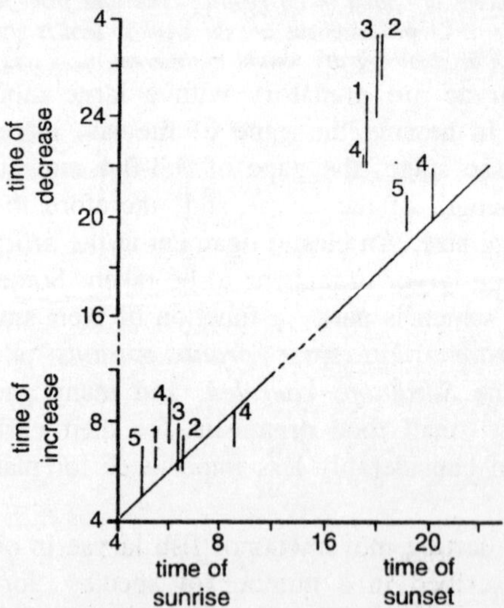

Figure 3.7 : The time period over which the gut contents of various species of larvae increase and decrease related to the time of sunrise and sunset. Note the lag at sunset caused, presumably, by the time taken to digest food taken earlier. 1. Pleuronectes platessa ; 2. Pleuronectes platessa; 3. Ammodytes ; 4. Clupea harengus ; and 5. Salmo salar parr.

The guts of many pelagic marine larvae are often empty, a puzzling feature when one considers the fairly high food requirements in the early stages. Apart from a lack of suitable food this may result from the capture by net of the weaker feeders only, rapid digestion, defecation during capture, or diurnal feeding rhythms. Examples of such rhythms, which are given in Figure elsewhere in this chapter, do indicate a decrease in feeding at night and an increase at dawn.

Table 3.8 : Main Range of Light Intensity over which Feeding Becomes Reduced.

Species	***Range of light intensity (me)***
Oncorhynchus keta (chum salmon)	10^{1}-10^{-3}
O. gorbuscha (pink salmon)	10^{1}-10^{-3}
O. nerka (sockeye salmon)	1O'-10^{-3}
O. kisutch (coho salmon)	10°-10^{-4}
Coregonus wartmanni (whitefish)	?-10°
Esox lucius (pike)	?-10^{-1}
Clupea harengus (herring)	10^{2}-10^{-1}
Pleuronectes platessa (plaice)	10^{2}-10^{-2}

The role of vision in feeding has been tested experimentally and thresholds of light intensity measured. The fall off in rate of feeding corresponds with the dusk and dawn periods. Some species can take food in the dark, e.g., cisco, especially when it is present in high concentrations. Some, like sole, take food in the dark for most of the larval life and others, like plaice, only at later stages around metamorphosis. Where vision is important the daily feeding period must vary considerably with season, latitude, clarity of water, and even average cloud cover. Ivlev calculated, from the food requirements of young Baltic herring, their rate of feeding and the density of food, that they needed to feed 15 hr/day in August, in other words nearly all the hours of daylight. Blaxter estimated that much younger stages of herring had 10 hr/day to feed in the southern winter spawning groups and up to 24 hr/day in the more northern summer spawners. Variations in feeding time affect searching power, and therefore survival and growth, and these may be further influenced by the temperatures prevailing in different seasons.

Success in early feeding, feeding drives, and learning factors have been studied. At very early stages Braum found only 3-8% of feeding movements in *Coregonus* were successful, but 30% in *Esox*. In the yolk sac stage, herring larvae take food successfully in 3-10% of their feeding movements but later in 80-90%. In plaice, feeding is 60-80% successful throughout larval development. In herring larvae from 20 to 40 mm in length the feeding drive depends on factors such as satiation an especially on the activity of the prey. Live organisms, although more difficult to catch, seem to increase the feeding drive and are probably very important in the young stages, where the main visual process may centre round movement perception. Larval cisco, however, showed a much higher rate of feeding on dead than live *Cyclops*.

The gut is usually a straight or simple tube at first feeding and the food is often digested near the anus. This is an interesting reverse of the normal situation in adult vertebrates. It seems likely that the gut is unspecialised at this stage and digestive enzymes are secreted along its length, although no critical work has been done on this. Harder gives a detailed account of the changes in the gut during the early growth of clupeids and engraulids; Ryland gives a rather simpler account for the plaice. Movement along the gut is mainly by peristalsis, although Iwai found cilia in the gut of ayu, *Plecoglossus altivelis*, especially in positions posterior to the liver. Backwardly directed ciliary currents and forward peristalsis seemed to cause a circulation of the gut contents, perhaps to aid digestion. In further reports use of the electron microscope showed intermingling of ciliated cells and columnar cells with microvilli in both *Plecoglossus* and *Hypomesus olidus*. Reduction of cilia in later stages suggested a transition from cilia to microvilli. Whether cilia are to be found in many species is still open to investigation.

Rates of digestion have been measured by rate of disappearance of the gut contents (the gut wall and body being transparent) or by interposing differently coloured food in the normal diet and observing the first signs of coloured feces. Kurata, using herring 12 days old, found that the gut clearance rate ranged from 12 to 19 hr or more at 9°C depending on the extent of food intake. The rate of digestion varies from 9 hr at 7°C to 4 hr at 15°C judged by transparency of the gut contents in this species.

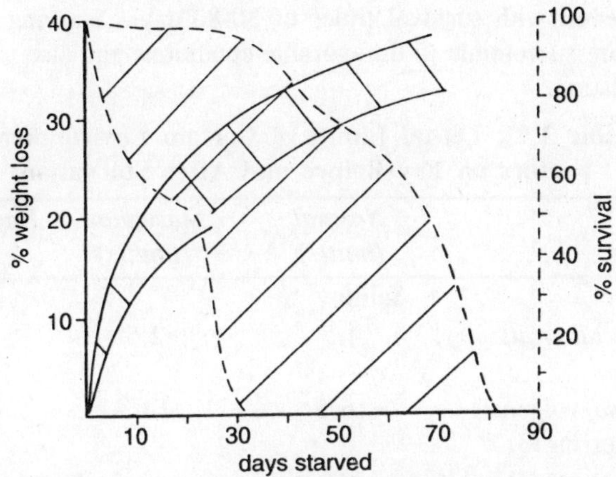

Figure 3.8 : Weight loss (— and left-hand ordinate) and percent survival (- - - and right-hand ordinate) as "envelopes" for 12 species of fish larvae during starvation.

Reports of larvae in poor condition or dead in net hauls may be due to selective capture or poor washing and are not certain evidence for mortality. Shelbourne estimated the condition of plaice larvae in the North Sea and found more signs of inanition early in the year. Measurements of condition factor [weight/length^3 × 1000] in herring larvae compared with starving larvae in tanks suggested that larvae in the sea were often very near the point of starvation although, over a number of years, it was not possible to correlate high condition factors with a good food supply. Ivlev looked at various aspects of starvation in young catfish, *Silurus glanis*, and bream, *Abramis bran*. With complete starvation the survival time was 34-46 days corresponding to a weight loss of 33-39%. With rations below the maintenance requirement survival was prolonged, but even on a full maintenance diet, the weight remaining constant, the fry died after 126-151 days, indicating that a "need" for growth may be characteristic of these young stages and that merely to maintain uniform weight is physiologically or developmentally inadequate. The current velocity stemmed by young carp *Cyprinus carpio* and roach *Rutilus rutilus* for 5 min was reduced by 10 times, from about 100 to 10 cm/sec, over 52 days of starvation. Age had a considerable influence on survival time without food. In catfish, roach, and bleak *Alburnus alburnus*, it ranged from 3 to 6 days at yolk resorption and from 109 to more than 180 days when 100 days old. The loss in weight and percentage survival of 12 species of larvae 25-30 days old during starvation is elsewhere in this chapter. Losses of weight of 30-35% were possible before death, with survival times of 30-80 days. Starving fish were much more susceptible to unfavorable conditions and also to infection and predation.

Table 3.9 : Lethal Limits of Certain Environmental Factors on Fry Before and After Starvation.

Species	*Normal limit(s)*	*Starvation limit(s)*	*Days starved*
	Salinity %		
Alburnus alburnus (bleak)	4.5	2.5	20
Caspialosa volgensis (Volga herring)	16.5	14.5	20
	pH		
Tincatinca (tench)	4.0-10.8	5.7-8.3	50
Perca fluviatilis (perch)	5.3-8.2	6.3-7.8	30
	O_2 PPM		
Cyprinus carpio (carp)	*0.7*	2.1	50
Alburnus alburnus (bleak)	2.8	3.1	20

SENSE ORGANS

Vision

The newly hatched larvae of many species have unpigmented, and presumably nonfunctioning, eyes (e.g., sole *Solea solea*, mackerel *Scomber scombrus*, whiting *Merlangius merlangus*, pilchard *Sardina pilchardus*, and sardine *Sardinops caerulea*); others have pigmented eyes (e.g., plaice *Pleuronectes platessa*, cod *Gadus morhua*, herring *Clupèa harengus*, and salmonids). Schwassmann examined sections of the *eye* and brain of larval *Sardinops caerulea* after hatching. On the first day there was no pigment and little differentiation, but by the second day pigment with a stratified retina and visual cells had developed. At 5 days the optic nerve was myelinated and the optic chiasma showed some interdigitation. Considering there are only about 40 individual fibers in the optic nerve when feeding commences at 3 days, it seems probable that the eye at first feeding is only capable of coarse movement perception. The larval eye is also limited in other respects. In newly hatched *Lebistes* and *Oncorhynchus*, and in *Clupea harengus* and *Pleuronectes platessa* and in a number of other species up to metamorphosis, there is a pure-cone retina, although the adult has both cones and rods. Lyall examined the developing eye of the trout, *Salmo trutta*, and concluded that the rods might develop from single cones or by an outward migration of bipolar cells, the latter view being supported by Blaxter and Jones.

Retinal pigment migration and other retinomotor responses associated with dark and light adaptation only develop when the rods appear and are therefore absent in herring and plaice until the larvae metamorphose. This may be true of many other marine species. Thus the larvae are, visually, poorly equipped, with a single type of visual cell, no ability to dark or light adapt, and with a coarse retinal mosaic. *Oncorhynchus* species acquire rods and retinomotor responses very rapidly after hatching compared with the marine fish which have been examined. The range of light intensity over which light and dark adaptation takes place is then 10^{1}-10^{0-1}, which corresponds fairly well with the thresholds for feeding.

The acuity of the larvae is poor compared with the adult. Although the retinal cells are small and fairly closely packed, the eye is also small, as is the focal length of the lens. Baerends *et al.* found that young cichlids, *Aequidens portalegrensis*, could be trained to distinguish stripes 1.5 mm apart at 3 cm body length; this improved to 0.3 mm at 11 cm body length. Blaxter and Jones calculated the acuity of the eyes of larval herring (minimum separable angle) to be 200 minutes at 1 cm body length improving to 50 minutes at metamorphosis.

Thresholds and spectral sensitivity were measured by behaviour techniques in herring, plaice, and sole larvae. Using a negative phototaxis as criterion the visual thresholds were about $10^{-5}10^{-6}$ mc, the spectral sensitivity curves being plateaulike with peaks which might correspond to different cone populations.

Distances of perception and visual fields have been measured by a number of workers.

Neuromast Organs

It has probably not been fully appreciated that many species of fish larvae have free neuromast organs. Iwai reviews earlier reports and his own work on *Tribolodon hakonensis*, *Tridentiger trigonocephalus*, and *Oryzias latipes*. These organs are usually situated in lateral rows along the body with the small hump of sensory cells being the main element visible especially in sea-caught or fixed specimens. Rearing, with the use of phase contrast microscopy, makes it possible to retain and see the cupulae which are relatively enormous-0.05 mm long in *Oryzias* but 0.1-0.4 mm long in some other species. In the adult it is likely that they are more often sunk into pits. The ability of larvae to avoid capture before the eyes fully develop is almost certainly a result of these neuromast organs.

ACTIVITY AND DISTRIBUTION

Phototaxis and Activity

Much of the published work on phototaxis in fish larvae is anecdotal. Generalisations that marine fish larvae are photopositive and demersal freshwater larvae photonegative do not stand up to a detailed investigation using different temperatures, light intensities, species, or ages. Changes of the sign of the phototaxis may be related to temperature, especially where a sudden change is applied, e.g., in *Coregonus*. Herring larvae show a positive phototaxis at high light intensities which becomes negative below a certain threshold.

Activity levels are frequently associated with diurnal light changes and phototaxis. Hoar, Ali , and Heard review much work on activity, rheotaxis, schooling, and migration of young salmonids. Emergence of the alevins of sockeye salmon, *Oncorhynchus nerka*, seems to be controlled by incident light on the gravel and can be delayed by artificial light at night. The younger fry of pink salmon, *Oncorhynchus gorbuscha*, for example, are negatively phototactic and hide among the stones of their home stream by day. They become active at night and may rise to the surface and swim downstream or get displaced by the current. After schooling both pink and chum salmon *O. keta*, prefer lighted conditions and presumably substitute schooling for hiding as their protection against predators. They show then a positive phototaxis, stem the current by day displaying a cover reaction only with abrupt changes of light, and again become displaced downstream at night when the visually controlled rheotaxis phases out. Coho salmon fry, *O. kisutch*, are rather different in behaviour; they are strongly territorial and generally more active, but their responses to light and their diurnal changes in activity are much less marked. The behaviour of the fry of the different species of *Oncorhynchus* can, in fact, be related rather generally to their migratory habits. Pink and chum fry migrate to the sea quite soon after hatching, whereas sockeye fry tend to move into lakes, and coho remain in the home stream for a year or so. Loch trout, *Salmo trutta*, also show incipient territorial behaviour at an early age. Woodhead, in a laboratory study of the larvae of *Salmo* species, related changes in photokinetic activity levels to ambient light intensity and age. Between 0.005 and 100 me the photokinetic activity of *S. trutta* gradually increased above a basal dark level, and this species and *S. gairdneri* (*irideus*) and *S. salar* all showed increased activity with age. A positive photokinesis, together with a negative phototaxis, would work together to keep the larvae within the stones of their home stream. Woodhead and Woodhead found a similar increase in the activity of herring larvae with light intensity above 0.3 me, and there was a basal level of activity in very low illumination as there was in the salmonids.

Vertical Distribution

Diurnal variations in distribution are linked to activity, phototaxis, and brightness discrimination. Some earlier work on changes of vertical distribution did not take into account the need for opening and closing the plankton nets at defined depths and for preventing differential net avoidance by day and night. A general lack of larvae by day, especially larger specimens, partly results from this latter factor, as emphasized by Bridger. With slow tow nets it can be difficult to estimate the relative effects of net avoidance and vertical distribution when comparing day and night catches. Strasburg used slow tow nets to catch tuna larvae. Some species such as skipjack, *Katsuwonus pelamis*, were *rarely* caught by day at all, while catches of all species were always much higher by night. He concluded that there was some net avoidance, but the main diurnal difference was owing to vertical migration. Stevenson used high speed nets for larvae of Pacific herring, *Clupea pallasii*. The catches at night were far higher and the average size of the larvae was greater, from which he concluded that the larvae were concentrated at the surface by night and spread over a fairly wide stratum by day. There seemed to be net avoidance, even of high speed nets, by the larger larvae in the daytime. Ryland found a reverse distribution of larval plaice and sand eels *Ammodytes*, which were spread over a wide depth range from the surface to about 35 m by night and much more concentrated between 5 and 10 m during the day. The use of high speed nets in this study and in that of Colton on haddock larvae, *Melanogrammus aeglefinus*, prevented diurnal variation in size and presumably overcame the net avoidance factor.

Buoyancy and Pressure

There are a number of devices which provide buoyancy in pelagic fish eggs and larvae. In marine species, found in high salinities, maintenance of the body fluid concentration below that of the environment acts as a buoyancy mechanism. In the early stages lack of a skeleton also keeps the specific gravity fairly low. In freshwater the eggs are often demersal so that the much greater problem of maintaining buoyancy does not exist. In fact, in salmonid eggs the specific gravity may be very high in the early stages because of low water content of the yolk. It is sometimes said that a large perivitelline space is a mechanism for buoyancy. However, this normally contains water of the same concentration as the environment and is thus much more likely to act as a shock absorber, particularly in pelagic fish eggs subject to wave action. The embryo and yolk may regulate the body fluids from fertilisation (e.g., in pelagic plaice eggs: Holliday and Jones) by means of the vitelline membrane. In the demersal herring egg there is no regulation until the yolk is covered by cellular tissue at gastrulation. Many species such as pilchard *Sardina pilchardus*, mackerel *Scomber scombrus*, hake *Merluccius merluccius* and sole *Solea solea* have oil droplets, but these are not typical of pelagic eggs for they are absent in the sprat *Sprattus sprattus* and in *Gadus* and *Pleuronectes* species. Tendril-like outgrowths found, for example, in *Exocoetus* and *Scomberesox* eggs are as likely for attachment as for buoyancy.

There may be a mechanism to adjust specific gravity to the spawning medium. Solemdal found that variations in the osmotic pressure of the blood of female flounders, *Pleuronectes flesus*, at the spawning season depended on the salinity. This, in turn, caused changes in the specific gravity of the eggs when spawned. In water of low salinity the eggs were larger and their osmotic pressure lower. In the low saline water of the Baltic ($6.5\%_0$), fertilisation and development were possible but the eggs rested on the bottom. In the sprat there is a correlation between low salinity and larger eggs and also in herring and plaice.

In the larval stages large subdermal spaces as in the plaice and cod will reduce specific gravity as long as the body fluids are maintained at an osmotic pressure below that of the environment. Other species have long processes to the dorsal or lateral fins, e.g., the angler fish *Lophius* and deal fish *Trachypterus*. It may be significant that Orton quotes the latter species as having only moderate fin folds and subdermal spaces, but whether the increased surface area assists buoyancy is open to question; these may be organs to reduce predation. In freshwater, the larvae of salmonids can afford to retain dense yolk reserves as they rest on the bottom. Others like carp and bream have cement organs for attaching themselves to weed.

The first filling of the swim bladder is well known in freshwater fish. In the physostomatous salmonids like *S. trutta*, *S. gairdneri*, *Cristivomer namaycush*, and *Coregonus clupeaformis* the swim bladder is filled 1-3 months after hatching, depending on temperature or feeding. Air is taken in from the surface via the pneumatic duct and without access to the surface the swim bladder remains empty. The pneumatic duct remains open under such conditions and functions later. Lake trout, *Cristivomer namaycush*, were able to swim considerable vertical distances (280 m) with unfilled bladders suggesting that development in deep water presented no problems for filling the swim bladder. Physoclists like *Gasterosteus*, *Lebistes* and *Hippocampus* also require access to air to fill the swim bladder but if deprived of this at the critical time the pneumatic duct closes and the swim bladder cannot be filled later.

Pressure effects on fish larvae were tested briefly by Bishai. Newly hatched herring survived compression to and decompression from 4 atm and young plaice 37-50 mm long to and from 2 atm. Young salmonids could live at 5 atm to the end of the yolk sac stage, sometimes showing increased activity during compression or decompression. Older fry, about 20 days after yolk resorption, could withstand compression to 2 atm but had difficulty in dealing with air bubbles inside the body and in the swim bladder during decompression. Older fry still, about 60-140 days after yolk resorption, could not even withstand compression to 1.2 atm. At later stages, the ability to withstand compression increased but decompression was lethal. Qasim *et al.* observed changes in the vertical distribution of larval teleosts subjected to pressure changes. Young plaice larvae tended to swim upwards when the pressure was increased from 1 atm to 2 atm and sank when the pressure was returned to normal. Metamorphosing plaice did not show this response. Larvae of the blenny *Centronotus gunnellus* responded in a similar way to pressure changes equivalent to 25 cm of seawater and larvae of another blenny *Blennius pholis* to only 5 cm of seawater. Only the last of these species has a swim bladder. It is possible in such experiments that the larvae could have been responding to water currents produced by the pressure changes.

If this is not so, one may ask how larvae without swim bladders can respond to pressure.

Locomotion and Schooling

Most species show signs of activity within the egg, and movement is often an important part of the hatching process. The relatively large yolk sac at hatching must be a hydrodynamic embarrassment to the larvae, but in salmonids, where it is especially large, there is little movement as the larvae grow in a rather passive way within the stream bed. The buoyant oil globule found in many marine pelagic larvae may also make equilibrium difficult. Two sorts of movement are found at yolk resorption-a serpentine eel-like mode of swimming in the long thin-bodied clupeoid type and a more maneuverable movement allowing "backingup" with the use of the pectoral fin and marginal fin folds in the shorter flatfish type of larvae.

Swimming performance is shown in Fig. 13 in terms of burst speeds maintained for a few seconds, where the theoretical 10 body lengths/ sec (shown as dashed line) seems to fit quite well. Cruising speeds, of the order of 2-3 body lengths/ sec (ideally the maximum sustainable speed without an oxygen debt accumulating). Improvements in performance can be correlated with development and upturning of the caudal fin.

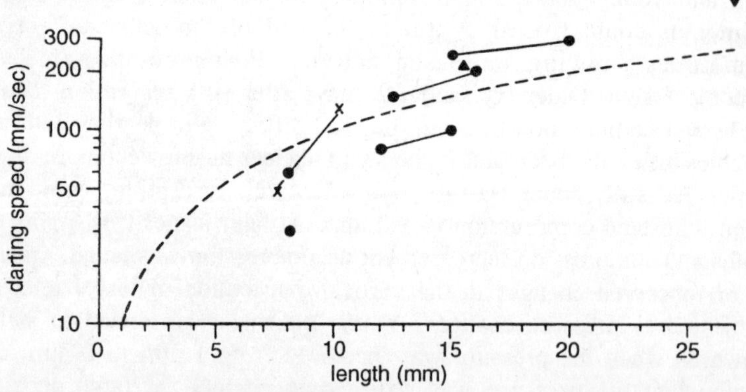

Figure 3.9 : Darting speeds of larvae of different length: (●–●) Clupea harengus 10°-15°C, (×—×) Pleuronectes platessa 6.5°-7.5°, (A) Carassius auratus, no temperature, and (V) Perca fluviatilis, no temperature.

If schooling occurs it usually starts at metamorphosis. Presumably the larvae, being transparent, do not provide very good mutual stimuli for keeping schools together visually. Shaw analysed the development of schooling in the silverside *Menidia* and found responses initiated between two fry at a body length of 8-9 mm with increasing participation as they grew. By 11-12 mm as many as 10 could be seen in parallel orientation. Fish reared in isolation in paraffin-coated bowls took a few hours to become integrated into existing schools. Generally speaking there was a considerable latency in schooling when "isolates" and "quasi-isolates" (isolated 5-7 days after hatching) were brought together at a size where schooling was normally well developed. Surprisingly, this initial latency was less among "isolates" than "quasi-isolates." In the newly emerged fry of *Oncorhynchus* species, schooling ceased in the dark and following subjection to light they needed 15 min or more to complete the school again, although in all species except coho this time became reduced with age. In herring schooling started about 2 months after hatching at a body length of 25-30 mm. Starving larvae tended to school more positively, but the school rapidly broke up when food was offered.

Rheotropic (optomotor) responses (orientation to currents or moving backgrounds) are often well developed in the larval stage as shown in some experiments on swimming performance and from the many observations on salmonid larvae by day and night. This response seems to be maintained by either visual or tactile cues, and the sign may change under shock treatment. In young salmonids a temperature rise of 12°C changed the rheotaxis from positive to negative, presumably a defensive response.

Searching Ability

The ability to search for prey or, in the case of visual feeders, "optically filter" the environment, depends on cruising speed and the distance at which food is perceived. Perception distances depend on the movement of the prey, its orientation in relation to the eye, contrast with the background, and illumination. Activity in terms of cruising speed may well be reduced after a period of feeding and is certainly influenced by feeding drives in older larvae. Starvation may also cause a reduction in activity. The volume searched per hour has been estimated in whitefish and herring larvae. When multiplied up on a daily basis the volume searched becomes very dependent on day length, that is, on latitudinal and seasonal effects.

MORTALITY, TOLERANCE AND OPTIMA

From hatchery, rearing, and experimental studies it is well known that fish eggs and larvae go through periods of mortality. Susceptibility to mechanical shock or nonoptimal temperatures and salinities may vary with age. Marine fish eggs seem especially delicate until the completion of gastrulation although this has not been systematically tested. Perhaps sensitive morphogenetic processes in the early stage or failure to osmoregulate are the cause. Similarly there are stages, such as the eyed stage in salmonids, where eggs

Table 3.10 : Volume Searched during Feeding

Species	*Size (mm)*	*Volume searched (liter/hr)*
Coregonus wartmanni (whitefish)	(?)10	14.6
Clupea harengus (herring)	8-16	0.3-2.0
Clupea harengus (herring)	10 13-14	1.5-2 6-8
Sardina pilchardus (pilchard)	5-7	0.1-0.2
Pleuronectes platessa (plaice)	6-10	0.1-1.8

may be particularly resistant to damage. After hatching, further stages of mortality may be experienced which can sometimes by connected with the further development of certain organs and with such physiological changes as the transition from cutaneous to gill respiration or the development of the swim bladder. Rearing studies on marine fish show a high mortality at the end of the yolk sac stage in species such as cod G. *morhua*, haddock *Melanogrammus aeglfinus*, pilchard *Sardina pilchardus*, herring *Clupea harengus*, and lemon sole *Microstomus kitt*. These "critical phases" are probably in part the result of unsuitable food being supplied, especially from the aspect of size. The question of whether critical phases at yolk resorption occur under natural conditions is a much vexed question, examined especially by fishery biologists looking for the factors controlling brood survival. J. C. Marr, in reviewing this hypothesis, concluded in two species (mackerel *Scomber scombrus*, and Pacific sardine *Sardinops caerulea*) that mortality was more likely to be constant or steadily decreasing without the catastrophic periods advocated by earlier workers. Undoubtedly some of the five species mentioned have small mouths and the finding of suitable food in the sea at the end of the yolk sac stage before the point of no return must be a serious problem.

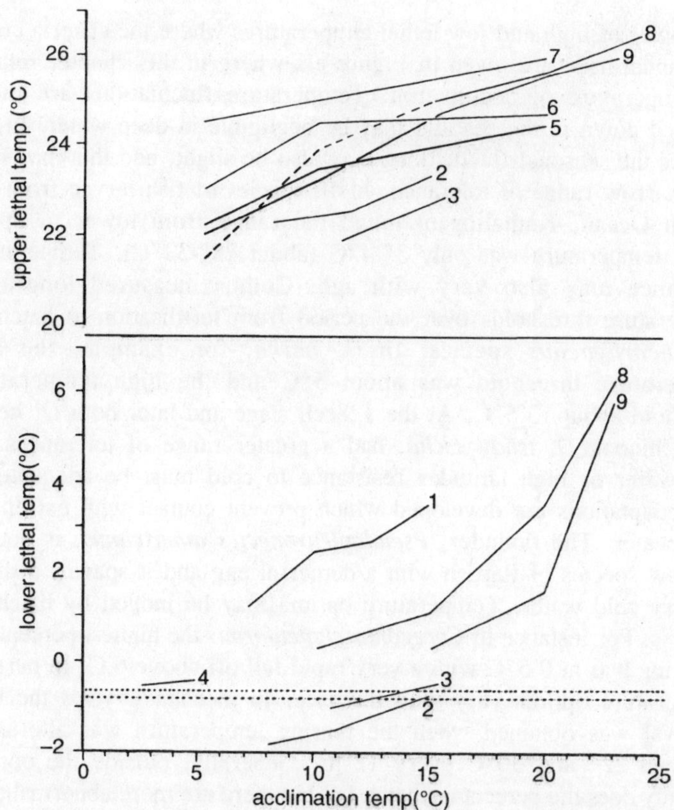

Figure 3.10 : Lower and upper lethal temperatures at different acclimation temperatures, criterion of lethal temperature being 50% survival after 24 hr; horizontal dashed lines give range of blood freezing point values. 1. Brevoortia tyrannus—in 24%o salinity, acclimated for 12+ hr. 2. Clupea harengus (spring spawned) and 3. C. harengus (autumn spawned)—in 34%$_0$ salinity, acclimated for long period 4. Oncorhynchus keta—in 28%$_o$ salinity, acclimated for 21 days. 5. Salmo salar, *6. S.* trutta, *and 7. S.* trutta (fario)-in *freshwater, acclimated for 5 days.* 8. Oncorhynchus tshawytscha, *and 9.* O. nerka.

The lethal levels of various environmental factors such as salinity, temperature, oxygen tension, pH, radiation, and mechanical stimuli have been assessed for fish eggs and larvae. Salinity tolerance is surprisingly wide, ranging in marine species which have been examined from about 5%$_o$ up to 45-50%$_0$ with even a wider range over short periods of exposure. Predictably, the lethal temperature varies with level of adaptation. A wide variety of criteria have, unfortunately, been used for defining lethal temperature, ranging from 50% mortality after one day (or longer) to the number of days to 50% mortality.

Examples of high and low lethal temperatures where the criteria could be standardised are given in Figure elsewhere in this chapter related to temperature of acclimation. Temperature fluctuations are much damped down in the sea and may be negligible in deep water. In the tropics the seasonal fluctuations may also be slight, and this may lead to a narrow range of tolerance. In 10 species of fish larvae from the Indian Ocean, Kuthalingam found the range from lower to upper lethal temperature was only 3°-4°C (about 28°-32°C). Temperature tolerance may also vary with age. Combs measured long-term temperature thresholds over the period from fertilisation to hatching in *Oncorhynchus* species. In *O. nerka*, for example, the low temperature threshold was about 5°C and the high temperature threshold about 13.5°C. At the 128cell stage and later both *O. nerka* and Chinook *O. tshawytscha*, had a greater range of tolerances. In freshwater or high latitudes resistance to cold must be adequate or else adaptations are developed which prevent contact with extremely cold water. The flounder, *Pseudopleuronectes americanus*, is one of the few species of flatfish with a demersal egg and it spawns inshore in very cold water. Temperature optima may be judged by hatching success. For instance in *Coregonus clupeaformis* the highest percentage hatching was at 0.5°C with a very rapid fall off above 6.C. In herring temperature optima vary with the race. In medaka *Oryzias* the best survival was obtained when the rearing temperature was alternated between 22° and 30°C every 12 hr. Generally, outside the optima not only does the percentage hatch fall but there are more abnormalities.

Oxygen is apparently only limiting at its low level. However, supersaturated solutions may be harmful if air bubbles are swallowed by the larvae. This "gas disease" can be fatal if the larvae cannot eliminate the bubbles from the gut as their buoyancy is disturbed. The eggs of the salmonids *S. salar*, rainbow trout *S.gairdneri*, Pacific salmon O. *tshawytscha* and *O. kisutch*, brook trout *Salvelinus fontinalis*, and lake trout *Salvelinus namaycush*, have been tested at low oxygen tensions. Most were able to resist levels of 2.5 ppm or even less (25% saturation) although development was retarded and the larvae smaller at hatching. Reductions in current flow had a similar effect. Alderdice *et al.* using chum salmon eggs, measured the retardation in hatching resulting from low dissolved oxygen levels at different developmental stages. The intermediate stages, about one-third of the way from fertilisation to hatching, were most susceptible. Hayes *et al.* defined the limiting tension not from its lethal aspect, but the tension at which normal oxygen consumption started to drop. In *S.*

salar eggs this was 3.0 ppm at 20 days, 7.5 ppm at 45 days, and 4.7 ppm at hatching. Other salmonid data on limiting tensions are summarised by Wicket. In the egg of chum salmon, *O. keta*, the calculated limiting tension varied from 0.72 to 3.7 ppm increasing with age (temperatures from 0.1° to 8.2°C). In the eggs of pike, *Esox lucius*, it varied from 1.2 to 4.1 ppm depending on age and temperature. This type of finding is of value in assessing the survival chances of fish eggs buried in gravel. Wickett, for example, has shown that the flow rates and oxygen content of the water may be quite inadequate for chum salmon eggs, being even as low as 2 mm/hr and 0.2 ppm.

Marine and other fish larvae were studied by Bishai. Herring larvae at hatching were "restricted" at 27-32% saturation and at 55-64% saturation a few days later. Larvae of the lumpsucker, *Cyclopterus lumpus*, just tolerated 43% saturation 3 weeks after hatching and young plaice 4-5 cm in length just tolerated 14% saturation. He found that the larvae of the salmonids, S. *salar* and *S.trutta,* could survive very low levels of saturation, 3-10%, after hatching, but this ability decreased with age, the minimum level tolerated being 16-28%. Again, criteria for defining lethal levels vary, and it is difficult to compare the results of different workers. To some extent acclimation to low oxygen increases resistance to hypoxia and there are also specific differences in tolerance.

Bishai has also summarised work on the effects of pH on larval fish. Compared with seawater (pH about 8.0) which is fairly well buffered, freshwater has a pH which is more variable and more susceptible to change by effluents or run-off. Bishai found that herring larvae could survive values between 6.5 and 8.5. Lumpsucker larvae had a low tolerance level of about 6.9, plaice (4-5 cm) 6.2-6.5, while salmonid larvae had a range of tolerance from about 5.8 to 6.2 up to 9.0, or even 10.0 in sea trout alevins. To some extent the levels of tolerance depend on the substances used to change the pH. It is possible that CO_2 for reducing pH has an additional toxic or respiratory effect compared with the use of HC1. Certainly Alderdice and Wickett found high levels of CO_2 inhibited oxygen consumption in the eggs of *O.keta.*

Radiation effects on salmonid eggs and larvae are reviewed by Hamdorf and Eisler. Strong visible light may cause early hatching, mortality, poor growth, and greater pigmentation. Exposure to light is less serious when the intensity is low or in the later stages, especially after the onset of pigmentation. It is, perhaps, not surprising that light is harmful to eggs and larvae which normally develop in complete or semidarkness among gravel or stones; increased activity probably contributes to its deleterious effect. Other species which develop in sand such as the grunion, *Leuresthes tenuis*, or near the sea bed such as herring may also show a poorer rate of hatching under lighted conditions. Careful controls are required in experiments since the light may cause other changes such as raising the temperature or increasing growth of phytoplankton in the water. The main effect on many other species is to accelerate hatching. Ultraviolet light has been tested on *Fundulus* and causes abnormalities of the skeleton, cylopia and twinning. In sockeye salmon, *O. nerka*, it causes premature hatching, vertebral abnormalities, high mortality, and delays pigmentation. Marinaro and Bernard exposed the eggs of pelagic fish such as pilchard *Sardina pilchardus*, mullet *Mullus*, horse mackerel *Trachurus* and "sargue" *Diplodus annularis* to sunlight with and without ultraviolet filters, finding a lower rate of hatching with ultraviolet. The transparency of many fish eggs and larvae may be an adaptation to prevent absorption of light as well as acting as a means of camouflage. X rays have been tested, for example, by Solberg on *Fundulus* embryos. They were most sensitive after fertilisation with a decrease of 10 times in sensitivity between early cleavage and 4-5 days later. Other work is reviewed by Eisler especially on salmonids. X rays have a number of effects such as higher mortality, decrease of growth, deformities, decrease in erythrocytes, and destruction of hematopoietic tissue and increased pigmentation. Generally, susceptibility decreases with age and in optimal conditions of temperature. Neyfakh used X rays for inactivating the nuclei in the developing eggs of the loach, *Misgurnus fossilis*. Irradiation during certain phases of development was followed by death at a very specific time, suggesting a periodical functioning of the nuclei.

Fish eggs in almost any environment are subject to mechanical stimuli which are potentially harmful—movement of gravel within a spawning redd, current borne objects on the sea bed, or wave action at the surface. The resistance of the egg may be measured by the maximum load it will take before the chorion bursts. They indicate an increase in the resistance of the chorion as it hardens subsequent to fertilisation and a decrease in the strength of the chorion prior to hatching, presumably resulting from hatching enzymes. Changes in the chorion may in part account for periods of varying resistance to external conditions.

Other harmful influences may be hatching enzymes, especially for embryos which may be somewhat unhealthy in other respects, or where eggs are very crowded and the enzymes are concentrated in the water after hatching. Stuart observed sediments attached to the chorion of loch trout. These might well reduce oxygen intake. The alevins, how ever, seemed to have the ability to disperse such sediments by means of respiratory currents passing over the body surface. With gill respiration sediments were aggregated by mucus and rendered less harmful.

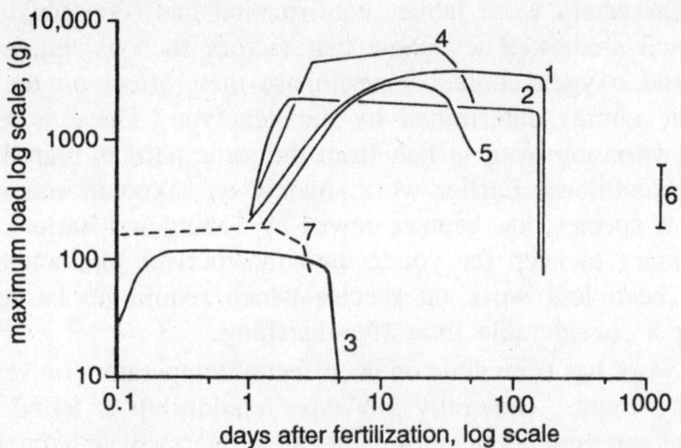

Figure 3.11 : The maximum load before bursting in eggs of different species during development. 1. Salmo salar, 2. Coregonus lavaretus, 3. Acipenser sp. ; 4. Salvelinus f ontinalis (?), 5. Salmo salar ; 6. Pleuronectes platessa (range, varying age) ; and 7. Engraulis anchoita.

A recent trend in research on tolerance is to use combinations of factors such as temperature, salinity, and oxygen concentration to see how they interact and to find out the optima. Kinne and Kinne used such combinations during the development of the desert minnow, *Cyprinodon macularius* (*i.e.*, temperature between 10° and 37°C, salinities from 0 to 85$\%_0$ and 70%, 100%, and 300% saturation of air in water). Mortality was least at near lethal temperatures when the salinity was held between one-half seawater and 35$\%_0$. The lethal temperature was lower in hypoxial conditions. It seemed that of the salinities used 35$\%_0$ was the optimum for incubation. Lewis, using menhaden larvae, *Brevoortia tyrannus*, found that temperature tolerance was greatest at salinities of 10-15$\%_0$, suggesting that isotonic conditions were optimum for survival. Forrester and Alderdice calculated from experimental data that the optimum salinity and temperature for hatching eggs of cod, *Gadus macrocephalus*, was 19.4% and 5.3°C.

MERISTIC CHARACTERS

Counts of vertebrae, myotomes, scales, gill rakers or fin rays have been used in racial studies. The variability in number and lability under different environmental conditions are striking examples of rather flexible raw material for evolutionary forces. Usually mean counts from samples of fish are required, the differences from race to race being inadequate for identification of individuals. While correlations between the counts in adults or young and the environmental conditions on the spawning ground first suggested that meristic characters were labile, confirmation has been obtained by experimental studies. They show that factors such as temperature, salinity, and oxygen content superimpose their effect on the range of meristic counts determined by the genotype. There is also an individual variation even in fish from the same parents reared under the same conditions. Earlier work, mainly on salmonid and inshore or estuarine species, has been reviewed by Taning and Barlow. Since it is necessary to keep the young until the meristic characters form there has been less work on species which require to be fed and reared for a considerable time after hatching.

Most work has been done on the effect of temperature on vertebral and fin ray counts, Generally a V-type relationship is found where the average vertebral count is minimum at an intermediate temperature. For example, in sea trout, *Salmo trutta*, Taning found this was at about 6°C; in plaice it is at 8°C ; at lower and higher temperatures the average counts were higher. With fin rays the V may tend to be asymmetrically inverted with counts highest at intermediate temperatures. In other instances the mean vertebral counts are inversely related to temperature, but the lack of an inflection may result from an insufficiently wide range of experimental temperatures being used.

Both intensity and duration of light and CO_2 may have an effect. Higher intensities in the grunion, *Leuresthes tenuis*, reduce the number of vertebrae. A 16-hr day caused lower caudal vertebrae and possibly anal fin ray counts in *Oncorhynchus nerka*, while in sea trout, *Salmo trutta*, increasing tension of CO_2 produced lower numbers of vertebrae.

The work of Heuts and Lindsey using different temperature-salinity combinations in *Gasterosteus aculeatus* were somewhat inconclusive resulting from the problem of getting sufficient numbers of survivors in an extensive series of experiments, some of which were in unfavorable conditions. Certainly it was shown that temperature effects vary with the salinity level. Using a freshwater and brackish water race, Heuts found the greatest effect of temperature on fin ray counts at the salinity of adaptation. Differential mortality and the unsolved problem of controlling, for example, oxygen and temperature independently have often led to difficulties in interpreting results.

It is the underlying reasons for meristic lability which is, perhaps, of most interest to the physiologist. Sensitive periods have been found during which lability of vertebral counts is at its greatest. In sea trout it is around gastrulation, with a further period when the last vertebrae are being preformed. It is also before hatching in herring and killifish. However, in plaice and paradise fish meristic characters are still susceptible to modification after hatching. Generally speaking, fin ray counts are determined later than those of vertebrae. Later studies on *Oryzias* showed that transfer to a higher temperature 2 days after fertilisation produced fewer vertebrae, whereas transfer after 6 days resulted in a higher vertebral count. Regular diurnal alterations of temperature between 22° and 30°C gave vertebral counts intermediate between those resulting from sustained temperatures of 22° and 30°C.

Of the various hypotheses put forward to explain meristic lability those of Gabriel and Barlow fit the facts best. Often, but not always, environmental factors which delay hatching produce higher counts. Alternatively, higher counts may be considered to come from nonoptimum conditions. It is likely that the environmental factors change the relationship between growth and differentiation. Hayes *et al.* in their studies of morphogenesis in salmon at temperatures from 1° to 15°C, certainly found that the relative appearance of certain anatomical characters varied with temperature. If differentiation is late more tissue is available to be differentiated, leading to a higher count. Thus one may consider an interaction of the Q_{10} for growth and the Q_{10} for differentiation of the character; as the temperature is varied so the relationship between growth and differentiation varies. Where V-shaped curves are obtained there are then points of inflection for the temperature coefficients. Garside and Fry, using normal and reciprocal hybrid fry of speckled trout, *Salvelinus fontinalis*, and lake trout, S. *namaycush*, found that the mean myomere count was lower where the fish developed on the speckled trout yolks, which were smaller. There was also an inverse relationship between myomere count and the degree of twinning in Siamese twins of speckled trout, the counts being lower where there was less shared tissue.

REARING AND FARMING

Techniques

Three lines of approach are of interest at the present time:

(1) Tropical and temperate fish culture for food
(2) Rearing of salmonids and sturgeon to maintain population size after intensive fishing or hydroelectric schemes
(3) Temperate marine fish rearing and farming

A good general review on fish farming by Iversen is available and on tropical species by Hickling. Some of the most popular species are the common carp *Cyprinus carpio*, mullets, *Tilapia*, milkfish *Chanos chanos*, the carp *Puntius javanicus*, and grass carp *Ctenopharyngodon idella.* Many of the species are reared in still water fertilised artificially or by sewage. The last four species mentioned are herbivorous which gives maximum efficiency of food utilisation. Some of the species, e.g., *Tilapia* and *Chanos*, can be kept in salt or brackish water. In the case of *Tilapia* and the Israeli carp, breeding takes place in the holding ponds; in the European carp special breeding ponds are used. On the other hand, the fry of *Puntius*, *Chanos*, *Ctenopharyngodon*, and mullet must be caught in the wild and transferred to rearing ponds. Generally in tropical culture there are no serious problems of feeding the young which are of sufficient size to take the food available.

The extensive breeding of salmonids presents no problem of food supply for the young, which are relatively large when the plentiful yolk has been resorbed. Rearing is usually in running freshwater although acclimation of the young of *Salmo* species to marine conditions is now being developed, with improved growth and freedom from disease. The question of dietetics is well advanced in the group (see papers in *The Progressive Fish Culturist*).

Marine fish rearing and farming, although attempted from time to time over the last hundred years, have only recently made significant advances. Starting with the liberation of millions of plaice and cod eggs or yolk sac larvae in coastal areas, only the later experiments by Dannevig seem convincingly exploitable. Very recently, *Mugil* and *Solea* have been established as self-perpetuating species in some of the Mediterranean coastal lagoons.

Following the earlier rearing in aquaria of some species to an advanced stage by highly empirical methods, a better experimental approach has now been adopted. Success in rearing *Pleuronectes platessa*, *Solea solea*, *Microstomus kitt*, *Clupea harengus* (Rosenthal, unpublished data), *Sardinops caerulea*, *Engraulis* and *Scomber* (unpublished data), and *Sardina pilchardus* has depended on some or all of the following factors:

(1) Food supply. The use of *Artemia* and *Balanus* nauplii, *Mytilus* trochophores, young *Tigriopus*, small nematodes, rotifers and oligochaetes and secondary foods like *Dunaliella* has improved survival rates, so also has the extensive collection and selection of natural plankton

(2) Antibiotics. Particularly with flatfish, dosages of mixed sodium penicillin (50 IU/ml) and streptomycin sulfate (0.05 mg/ml) have improved survival in the egg stage

(3) Black-walled tanks and uniform overhead illumination have helped feeding success by providing contrast of the food against the background and even distribution of food in the tank

(4) Tank hygiene

There are interesting physiological and behavioural aspects of rearing where the fish may be artificially crowded and conditions of stress set up. Size hierarchy effects prevail, but growth seems to be density dependent only from the beginning of metamorphosis. Bitten fins are found more often in the smaller members of a tank community but are not always dependent on density. Perhaps more serious is the incidence of abnormal (albino or semi-albino) pigmentation in flatfish. It is generally worse in smaller fish and in very crowded tanks but may be reduced by using *spawning* stock acclimated to hatchery conditions. Survival to metamorphosis is also better in larvae from acclimated spawning stock.

Sensory Deprivation

Hatchery-reared flatfish do not survive well when transferred to the sea, presumably because of high predation; hatchery-reared salmonids usually have a poorer swimming performance. In fact, the hatchery environment by preventing contact from predation, by often rather uniform physical and chemical conditions and lack of shelter may prevent the development of appropriate defensive responses and muscular systems. The need for high density may also encourage undesirable intraspecific relationships at the expense of the more desirable interspecific responses.

The more obvious shortcomings of the aquarium tank may be accompanied by sensory deprivation in a more general and less easily defined way. The uniformity of the aquarium environment may lead to insufficient sensory input. For instance, Qasim, when rearing the young of *Blennius pholis*, obtained better survival with a day-night regime than with continuous light. Blaxter, rearing herring, found a hint of better survival when the light conditions over the tank were continually changed and when air jets playing on the surface kept up a continuous series of ripples. Hoar found that young salmon would feed on chopped earthworm in darkness during the day, but not at night, perhaps because of an activity (or retinomotor) rhythm. Such rhythms might eventually be suppressed by continuous light. Activity rhythms with periods of rest might play as important a part in the development of the sense organs and nervous system as sufficient sensory input during the active phases.

CONCLUSIONS

The most substantial advances in recent years have been made in investigations of marine eggs and larvae. Further work is needed on the following:

(1) Factors determining spawning time and fecundity, especially of marine fish
(2) Permeability studies on the egg using isotope techniques, with associated studies on the fine structure of the membranes
(3) The order in which different substrates in the yolk are utilised and the interrelations between the organic materials in the yolk and embryo being used for growth and maintenance
(4) Differences in egg quality (of a subtle or less subtle nature) which may influence viability
(5) Feeding mechanisms and nutrition in larvae with special reference to the possible utilisation of small particles or dissolved matter
(6) The respiratory and circulatory system in the larval stage where no functional gills or hemoglobin are present
(7) Differences in the sensory systems of larvae and adults
(8) The endocrine system in larvae, especially at metamorphosis
(9) Metamorphosis itself, particularly changes in the blood, laying down of scales and pigment, and behaviour
(10) In rearing studies, so vital for investigations of larval stages and for techniques of fish culture, aspects of nutrition and stress need attention, especially the effect of tank conditions on growth, mortality, and the development of behaviour

4

METAMORPHOSIS

The term metamorphosis, when considered in the broadest sense in the animal kingdom, refers to any abrupt change in the form or structure of an organism during postembryonic development. The modification (transformation) of the organism, or of its tissues and organs, is usually in preparation for a change in environment, behaviour, or mode of feeding. When one applies this term to vertebrates, it is apparent that only some fishes and amphibians possess a metamorphosis during their postembryonic development. The fact that these two vertebrate groups are both mainly aquatic and share some common evolutionary history is probably highly relevant to their possession of metamorphosis. Every student of biology is aware of the dramatic metamorphosis that occurs between the tadpole and adult stages in anuran amphibians and of the immense value that is laid on the event for studies in developmental biology. However, the same cannot be said for fish metamorphosis. There has certainly been increased research activity in the ontogeny of fishes in recent years, but data on metamorphosis are sparse, despite the stress that was laid on the need for research by an earlier report in this series.

The lack of information on metamorphosis has proven to be an obstacle in many research areas of fish biology, particularly in systematics. In some cases fish formerly considered as distinct species are now known to be metamorphosing larvae of another species. Part of this deficiency in our knowledge can be explained by the absence of clearly defined parameters for assessing and comparing metamorphosis among the vast number of fish species. For example, it may seem practical to define metamorphosis in a flounder as the period of eye migration, but there is a need for a list of more general characteristics of fish metamorphosis so that this phase can be identified in species with less dramatic events. Furthermore, the existing literature is riddled with copious terms and synonyms for what might be considered as a fish metamorphosis, such as transformer, postlarval stage, transitional stage, transformation stage, and kasidoron stage. These terms reflect the fact that it is recognised that all fish species have an ontogenetic interval where some postembryonic change prepares them for their new role as juveniles or adults and that the degree of change is variable between species. However, it is not clear what postembryonic changes in fishes are true metamorphic events.

The objective of this chapter is to emphasize metamorphosis as a substantial and significant developmental strategy among fishes. This will be accomplished by providing criteria for identifying this ontogenetic interval which follow those used for other vertebrates, by placing metamorphosis in context with the entire ontogenetic process, and then dealing with metamorphic events and their morphological and physiological significance. No attempt is made to provide a compendium of all fish species that undergo true metamorphosis, for this will be left to those who may be stimulated to do so by this essay. Instead, an attempt will be made to provide some terms of reference for such an endeavor, which hopefully will involve a multidisciplinary approach.

METAMORPHOSIS AND FISH ONTOGENY

Definition

There have been some earlier attempts to place metamorphosis in fishes in the broad sense of all animal metamorphosis. Metamorphosis is usually defined for all vertebrates or all protochordates and chordates, and then fishes have been considered within these confines. For example, Barrington took exception to an earlier definition by Ahlstrom and Counts that fish metamorphosis or transitional stage is an interval during which marked changes occur in body proportions and structures without any marked increase in length. He felt that physiological change and change of habitat were ignored by this definition. Although these aspects and consequences of fish metamorphosis had not been overlooked, Barrington was particularly interested in including the physiological and behavioural changes of parr-smolt transformation (smoltification) which some call a second metamorphosis, a secondary metamorphosis, or a second type of metamorphosis. Just *et al.* did not feel that parrsmolt transformation met their criteria of chordate metamorphosis, while Wald suggested that first metamorphosis and second metamorphosis are necessary consequences of the expression of two separate genotypes, which often develop in parallel in the same organism. Many, but not necessarily all, larval genes are repressed at first metamorphosis when the adult genes are expressed progressively. Second metamorphosis occurs when the adult genes controlling sexual maturation and/or migratory behaviour and physiology are fully expressed. If one accepts this viewpoint, then first and second metamorphosis should not be considered as similar morphogenetic intervals. Norris called smoltification a second type of metamorphosis, which is found in species with a complex life cycle involving a migration between freshwater and marine habitats. Examples include postmetamorphic juveniles of the Salmoniformes, the Anguilliformes, and the Petromyzontiformes. It is clear that not all species possessing a second metamorphosis undergo a first metamorphosis, so that it should be understood that the term second metamorphosis implies only that it is a different type of metamorphosis from the first. Postlarval change or second metamorphosis is not a consideration of this present review, which will be concerned with postembryonic (larval) change or first *metamorphosis*. As outlined below, first metamorphosis in fishes is an example of *true vertebrate metamorphosis*.

A reiteration of the three criteria for general chordate first metamorphosis will be a useful beginning for an attempt at providing guidelines for assessing first metamorphosis in fishes. In abbreviated form these are:

1. A change in nonreproductive structures between embryonic life and sexual maturation but not embryonic development, sexual maturation, or aging.
2. The larva occupies an ecological niche different from the embryo and adult because of its form. This assures that late embryo and early adult (juvenile) periods are not considered.
3. Morphological change at the end of larval life (climax) is triggered by an external (environmental) and/or internal (e.g., hormonal) cue.

The above criteria when considered together are designed to exclude parr-smolt transformation and include all significant morphological change occurring during larval life and not just at climax. Although just *et al.* state that more than one of their criteria should be met before using the term metamorphosis, all three could be applied to metamorphosis in many larval fishes. However, it is questionable whether progressive morphological change that began in the embryo, such as ossification of the skeleton and continued development of the nervous system, are metamorphic events in fishes. In some cases these are found in young of fishes that follow a direct development from the embryo to a juvenile/adult with no intervening larval interval. The larval interval must be present in the life cycle to have a first metamorphosis. Furthermore, it is not absolutely clear that all cases of parr-smolt transformation would be excluded by these three criteria. Therefore, it seems that three additional standards or characters should be added to the above to more clearly define fish first metamorphosis. These are:

1. Generally characterised by a marked change in form that is not necessarily a rapid process (permits inclusion of transitional larval features).
2. The immediately postembryonic larva and the adult (or juvenile) do not look alike (eliminates all direct postembryonic development).
3. The process does not involve growth (e.g., increased snoutvent length) and may in fact feature a decrease in length (eliminates increase in length during larval life and direct development as a metamorphic event).

Note that transitional larval characters, as in criterion 1, are a difficult issue and will be given further consideration below. It is questionable whether their acquisition is an event of metamorphosis.

Place within the Ontogenetic Sequence

Historical Concepts

The task of placing first metamorphosis of fishes within the general context of true chordate metamorphosis has not been difficult. A more onerous task is to position first metamorphosis within the existing ontogenetic sequence of steps (intervals) of fishes. A problem of terminology and sequencing of life-cycle intervals has been with us for some time, but particularly with regard to those intervals that are postembryonic. Past descriptions indicate at least four developmenta' pathways (strategies) that various fish species take from the embryo to the adult.

Type 1. The young at hatching are a replica of the adult except in size and sexual maturity and proceed to adulthood by attaining these characters over variable (and often prolonged) periods of time.

Type 2. The young of type 1 may also proceed to a juvenile for a limited time before gaining through second metamorphosis the definitive adult characters.

Type 3. Exogenously feeding young or larvae arise from the embryo and acquire prejuvenile characters, which gradually transform or are lost during first metamorphosis before those of juvenile and/or adult stages are attained.

Type 4. After a postembryonic period of variable length in which larval form does not change to any appreciable extent, larvae undergo a marked, often instantaneous, first metamorphosis into a juvenile.

The juvenile interval in all larval pathways (types 3 and 4) is sometimes followed by a second phase of change less dramatic than the first called the second metamorphosis. Types 1 and 2 are direct development, type 3 is transitory or intermediate, and type 4 is indirect.

Type 4 (indirect) development in fishes is an example of a vertebrate first metamorphosis. There are only a few groups of fishes that have a first metamorphosis in their life cycle. Although well represented among Class Osteichthyes and all Petromyzontiformes, first metamorphosis is not present in either Myxiniformes or Class Chondrichthyes. Unquestionable examples of first metamorphosis among bony fishes are those illustrating dramatic transformation of the body form, such as seen in Anguilliformes (true eels), Notacanthi formes (spiny eels), Elopiformes (bonefish, ten-pounders, and tarpons), and Pleuronectiformes (flatfishes). In the past, the fishes of the type 3 (transitory) pathway have been difficult to categorise in a metamorphic grouping. In Dipneusti (lungfishes), Polypteriformes (bichir), Acipenseriformes (sturgeon), Lepisosteiformes (garpike), and Amiiformes (bowfin), larval organs for respiration, feeding, or locomotion develop to aid a planktonic existence and are lost before a metamorphic climax. These fishes all acquire and lose external gills sometime in early larval life and never undergo an abrupt change into a juvenile. These two features are sufficient evidence to permit the exclusion of these fishes from examples of Class Osteichthyes with a first metamorphosis. Furthermore, Ophidiiformes and Lophiiformes are examples of Euteleostei that also possess transitory larval structures that persist for some time but are not present in the definitive phenotype. Their transient features are cenogenetic adaptations. Some flatfish larvae also have an elongated second dorsal spine that disappears during metamorphosis. There is no question that the above species undergo morphological change during the larva period, but in many cases, such as in the kasidoron stage of gibberichthyids where a long pelvic appendage is acquired for larval life, the changes are not in preparation for adulthood. It is not known whether development of temporary larval structures is initiated during embryonic life and they grow in the larva to a position of prominence. If such is the case, it may be necessary in the future to exclude these from the examples of first metamorphosis. However, any of the above that involve abrupt change from the larval phenotype will continue to meet the criteria.

Recent Concepts

The most recent attempt at describing intervals of ontogeny in fishes is that of Balon. His saltatory model describes embryo, larva, juvenile, adult, and senescent periods, separated by major thresholds (i.e., a switch or rapid transition into a new stabilised state). Each period may be divided into phases. Steps are the shortest intervals of ontogeny, are separated by less dramatic thresholds, and are found within a phase. Metamorphosis is a major threshold separating the larva period from juvenile or adult periods. This model fits well with the criteria of metamorphosis as they are presented in this chapter. The postembryonic pathways of fishes presented earlier are reconsidered using the intervals of ontogeny of Balon. The early proposal of Balon did not adequately deal with metamorphosis; although the new scheme places this event in a better perspective with the rest of fish ontogeny, metamorphosis may occupy a considerable period of time and it is questionable as to whether metamorphosis should be relegated to a category such as a threshold rather than a phase of larval life.

Other recent schemes of intervals of the early life history of fishes call the larval growth period a premetamorphic interval and the change from larva to adult a transitional stage of larval life termed the "transformation stage". Metamorphosis is designated as an event of larval life. Therefore, to the present day, there is no term that has been universally accepted for the metamorphic interval of ontogeny, and this interval has not been always considered in the context of larval life. It is the proposal of the present writer that first metamorphosis in fishes be considered in the same context as in amphibian metamorphosis. That is, fish first metamorphosis is *a phase* in the larva period during which time postembryonic changes occur before the juvenile or adult periods are attained. First metamorphosis is obligatory during indirect development when a larva period is part of the life cycle-that is, the use of the term larva implies that there is a first metamorphosis in the life cycle. In some cases metamorphosis may be initiated early and extend over the entire larva period (for qualification see above comments on transitory larval structures). However, in a more classical sense, first metamorphosis is a second larval phase, which follows a first phase of larval growth (premetamorphic phase) and is marked by an abrupt transformation from the larval phenotype. Stages should be described during the metamorphic phase and should begin with the *initiation event* and end with the completion of the *climax event*, at which time the juvenile form and behaviour are present. The time at which these two events occur during metamorphosis should be a major consideration of any life-history study, even if they are, in some species, almost simultaneous events.

STAGING

There should be clearly delineated criteria or parameters that denote the intervals of the metamorphic phase of the larva period. In the past, external metamorphic characters have been used to describe the intervals of metamorphosis, and these intervals have been called *stages*. This author cannot provide any reason for discontinuing the use of this term. However, it should always be understood that ontogenetic processes are continuous and "stage" does not imply an instantaneous state. The Petromyzontiformes represent the only group of fishes where there are standard criteria for staging metamorphosis. The diversity of Osteichthyes and differences in their ontogeny will probably never permit a standardisation of stages across the various orders. However, a series of universal stages at the level of the genus is not beyond comprehension. This will not be difficult if the species within the genus are few and the events of metamorphosis are relatively similar. Cooperation and communication between lamprey biologists in the two hemispheres has resulted in the designation of seven clearly defined stages of metamorphosis, primarily based on changes in the mouth, eye, branchiopores, dorsal fins, and body pigmentation. As ultimately happened with amphibians once their metamorphosis was divided into a universally acceptable series of stages, lampreys during their metamorphosis have been introduced as a universal and important experimental tool for studies in developmental biology.

Staging of metamorphosis in osteichthians is only rarely provided with descriptions of the early life history. With few exceptions, these have been provided for those species undergoing little larval change prior to a dramatic remodeling of body form at climax. The bestknown examples of first metamorphosis in bony fishes are those of Anguilliformes, Elopiformes, Notacanthiformes, and Pleuronectiformes, but even in these cases, data on staging are sparse. In individual species of these orders, first metamorphosis is often classified within a continuum of postembryonic developmental stages that extend to the sexually mature adult. Thus, for example, early metamorphosis and late metamorphosis are stages 4a and 4b, respectively, in the postembryonic development of plaice, and prometamorphic, midmetamorphic, and postmetamorphic are stages 2 to 4, respectively, in the series of six postembryonic stages of a gonostomatid. Although Fukuhara and Seikai *et al.* use stages F, G, and H to describe early, middle, and late stages, respectively, in metamorphosis of Japanese flounder, Miwa and Inui refer to these three stages as prometamorphic, climax, and post-climax. Recently, first metamorphosis in bonefish *Albula* was given as the second phase of a two-phase larval developmental period; however, there was no attempt at subdivision into stages. Six stages are described for Anguilliformes, but the specific criteria are not well defined. One of the most detailed descriptions of first metamorphosis in a bony fish is that available for *Megalops atlanticus*. Although metamorphosis is stage 2 of a three-stage life cycle (stage 1, larva; stage 3, adult), this interval is subdivided into four phases, each characterised by a specific body length and both internal and external features. In other orders, definitive staging criteria have not been provided because of the paucity of specimens available during metamorphosis. In the case of Pleuronectiformes, the small numbers of metamorphosing individuals in field samples are explained by the fact that the phase is transitory, avoidance is increased as the animals become larger, and metamorphosing individuals may change habitat. However, there is now excellent potential for rearing flatfish larvae in the laboratory. As the metamorphosing individuals are becoming increasingly available for analysis, it is now important that suitable staging criteria be utilised.

The use of external criteria for staging metamorphosis in most bony fishes has largely been ignored, despite the early encouragement of the value of such features as myotome numbers and fin position in assessing advancing biological age. Recent evidence indicates that there may be progressive external changes that can be used in some species to permit staging of metamorphosis. For example, in Congridae, the forward advancement of the subterminal anus and its accompanying anal fin orgins and the developing pterygiophores and actinotrichia are key characters. Similar morphogenetic events characterise metamorphosis in other eels. In *Siganus lineatus*, it would seem that progressive changes in pigmentation could be followed more closely to provide stages. In this species a brown-head omnivorous stage is the only feature that presently separates metamorphosis from a dark-head carnivorous larval stage and a herbivorous juvenile stage. Development of nostrils and olfactory lamellae can be used for staging in some flounders. To ensure that the earliest stages are determined, there should be a correlation of the time of initiation of both internal and external changes. Murr and Sklower took a similar approach in an early attempt at staging metamorphosis in eel leptocephali.

TIMING

One of the least understood aspects of metamorphosis in fishes is the factors that determine the time at which the phase begins. If the entire'postembryonic interval in species with larvae is considered as a first metamorphosis (e.g., species with transitory larval structures), then a determination of the time of the onset of this phase is not a problem. However, if the ontogeny of a fish species is characterised by a climax during which a major remodeling takes place after a long, uneventful larva period, the timing of the initiation of the phase is undoubtedly of great significance to the ultimate survival of the postmetamorphic individual. Thus, in most cases, climax is a highly synchronized event that is directly correlated to the state of physiological preparation of the organism, to environmental conditions, and to availability of food for the recently metamorphosed juvenile.

Body Length, Growth Rate, and Age

The ontogenetic pathway (i.e., direct or indirect) followed by fish species is related to the amount of yolk in the egg. The availability of nutrients to the embryo and larva probably has a bearing also on the length of larval life and the time of the onset of metamorphosis. There is also evidence that some marine fishes may delay metamorphosis in order to maximise dispersal. Most fish species must reach "metamorphosing size" before they can undergo a climax change into a juvenile. The body length at which metamorphosis occurs is species-specific and is related to the duration of the larva period, that is, age of the individuals. Data from lampreys serve as a good illustration of this point. The duration of larval life of lampreys can be ascertained by examining length-frequency data in larval (ammocoete) populations in a site which is stable and restricted. This method of age determination indicates that species of lampreys reach metamorphosing size within a range of 2½ years (*Mordacia mordax*) to 6¼ years (*Lampetra planeri*). However, it is quite likely that ammocoetes of some species may remain within the final year class for at least another 1-2 additional years, and maybe up to a total of 18 years, without a substantial increase in length. During this "arrested growth phase", ammocoetes prepare themselves for the nontrophic phase of metamorphosis by building up their reserves of lipid. Length-frequency data for aging of ammocoetes of the sea lamprey (*Petromyzon marinus*) have been supported by measurements of the yearly growth pattern of the larval opisthonephric kidney. Growth patterns in calcareous otic elements could also prove to be an important determinant of the duration of larval life in lampreys.

In lampreys there appears to be no correlation between the size at metamorphosis and the length of sexually mature adults. Thus, ammocoetes of the anadromous form of *P. marinus* (whose adults may eventually reach 83 cm) enter metamorphosis at a smaller size than their landlocked counterparts, whose spawning adults rarely exceed 50 cm in length. Furthermore, there is evidence indicating that in paired species (i.e., a parasitic species and a closely related nonparasitic species), ammocoetes of the nonparasitic species enter metamorphosis at a mean length greater than that of parasitic species. There is no increase in length in nonparasitic species following metamorphosis, for the entire adult period is nontrophic and is characterised by maturation of the gonads and depletion of energy stores.

For any given population, the time (age) at which an ammocoete beings initial metamorphosis is likely to be dependent on larval growth rate. Because growth rates are highly variable, there is a wide range in the time at which metamorphosis can occur within a species. Environmental factors such as population density and temperature are an important influence in this regard. Thus, 119-130 mm is the metamorphosing size of lampreys in the small Dennis Stream in New Brunswick, whereas in other larger watersheds of that province, where animal density is less, the metamorphosing individuals are generally larger and show a narrower size range (J. H. Youson and G. M. Wright, unpublished data). Experimental evidence also indicates higher growth rates and earlier metamorphosis in low-density populations.

The length at which osteichthians enter metamorphosis has also

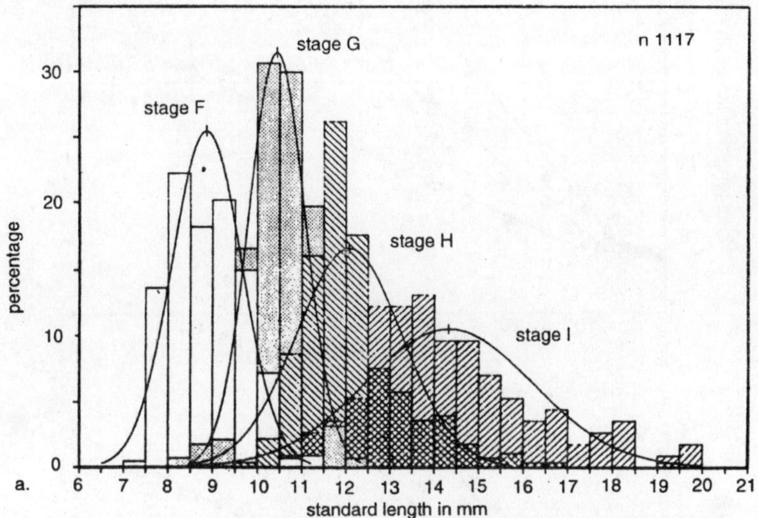

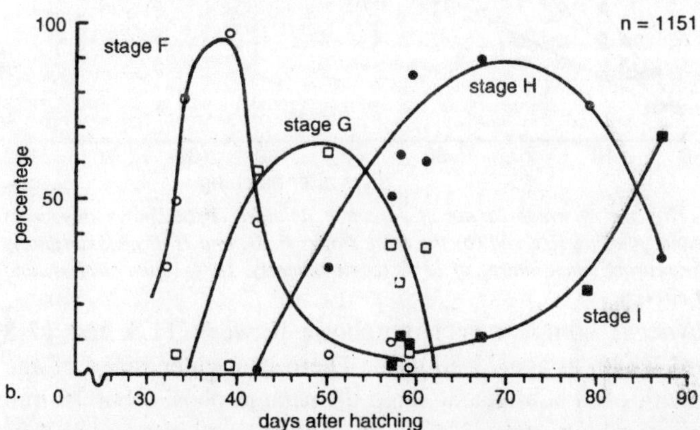

Figure 4.1 : A broader range of (a) sizes and (b) ages is noted when larvae of the Japanese flounder Paralichthys olivaceus advance through early (F), middle (G), and late (H) stages of metamorphosis to the juvenile period (I).

been an important consideration and will likely continue as an important criterion for determining the potential timing of the. climax event in individual species. However, as in lampreys, there is considerable individual variation in the age and length at which bony fishes enter metamorphosis. For example, laboratoryreared Japanese flounder *Para-*

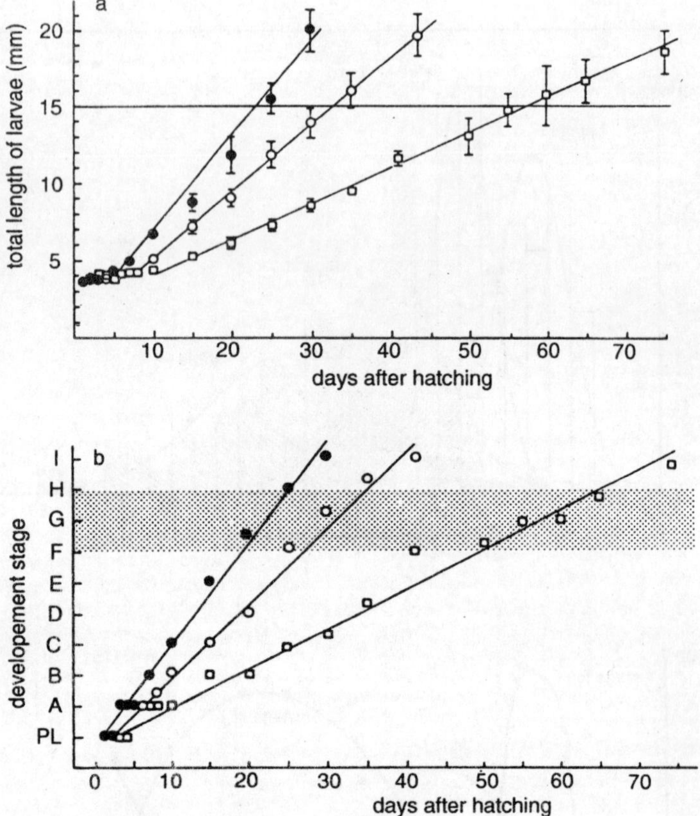

Figure 4.2 : The age at which larvae of Japanese flounder (Paralichthys olivaceus) *reach* (*a*) *metamorphosing size and* (*b*) *the three stages F, G, and H of metamorphosis varies with the rearing temperatures of 13°C* (*open squares*), *16°C* (*open circles*), *and 19°C* (*closed circles*).

lichthys olivaceus complete metamorphosis between 11.4 and 17.5 mm standard length at age 25-50 days. There is a wider range of age and length with each subsequent stage of metamorphosis, but 15 mm is the size at which most larvae metamorphose into juveniles. Policansky demonstrated that size (length), and not age, is the single most important determinant of the timing of metamorphosis in starry flounder *Platichthys stellatus*. He viewed this as a character of first metamorphosis that is expressed throughout the animal kingdom.

In fishes, length at metamorphosis has been directly correlated with the initiation of internal morphological change, with age at metamorphosis as determined with otoliths, and with changes in behaviour and physiology. The wide range of sizes at which eel and flatfish species enter metamorphosis has some significance for discussions of evolution, systematics, and adaptation and will be considered later. The basic difference between the dramatic metamorphoses that occur in eels and flatfishes is length of time to climax after hatching. Whereas at least the European-bound 22-year-old *Anguilla* leptocephali enter climax at about 75 mm length, larval flatfishes of 15-25 mm climax only several weeks after hatching. However, in both Anguilliformes and Pleuronectiformes there are exceptions. For example, the moringuid and muraenescoid eels metamorphose at 20 cm and 2-5 months after hatching and flatfishes metamorphosing at 12 cm have been reported. There seems to be a wide variation in age at metamorphosis of flatfish larvae in laboratory crosses.

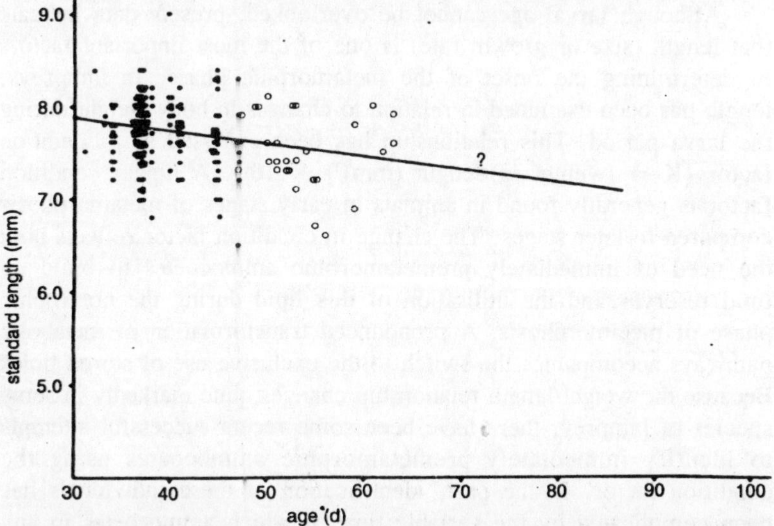

Figure 4.3 : Plot of standard length versus age (days) at metamorphosis of the starry flounder Platichthys stellatus. *Closed circles represent animals reared at 12.06 ± 1.35°C and open circles those reared at 9.79 ± 0.71°C.*

The relationship between age and length of larvae at metamorphosis is an important parameter for consideration in both lampreys and bony fishes. However, the studies of Policansky on the laboratory-reared starry flounder indicate that, although age and length influence the timing of metamorphoric climax, the influence of length may be strong in bony fishes. This study appears to contradict an earlier investigation on plaice, which concluded that age and temperature are more important than length in controlling the onset of metamorphosis.

A more recent report on Japanese flounder indicates that higher temperature results in reduced time (age) to metamorphosis and smaller metamorphosing individuals. These data imply that factors other than age, body length, and growth rate influence metamorphosis in flounder. In contrast, carefully controlled field experiments on sea lamprey *Petromyzon marinus* in the Big Garlic River, Michigan, showed that slow growth can extend larval life to up to 18 years from the expected 5-8 years. Also, low population density results in higher mean length at metamorphosis and shorter larval life in this lamprey species. These studies may provide us with examples of the controlling effect of growth rate (i.e., the time it takes to reach the critical length) on the time of onset of fish metamorphosis.

Physiological Preparation

Although larval age cannot be overlooked, present data indicate that length (size or growth rate) is one of the most important factors in determining the onset of the metamorphic phase. In lampreys, length has been examined in relation to changes in body weight during the larva period. This relationship has been analysed as a condition factor $\{K = [\text{weight (g)}/\text{length (mm)}^3] \times 10^6\}$. A higher condition factor is generally found in animals in early stages of metamorphosis compared to later stages. The change in condition factor reflects both the need of immediately premetamorphic ammocoetes to build up lipid reserves and the utilisation of this lipid during the nontrophic phase of metamorphosis. A pronounced transformation of metabolic pathways accompanies the switch to the exclusive use of stored lipid. Because the weight/length relationship changes quite markedly in some species of lamprey, there have been some recent successful attempts to identify immediately premetamorphic ammocoetes using the condition factor. In the past, identification of these individuals has been complicated by the variable time at which ammocoetes in any population enter metamorphosis.

Metamorphic climax is a nontrophic phase in many bony fishes, and major reorganisation of metabolism occurs at or near climax. It is becoming a well-documented fact that development of specific metabolic pathways and the storage of energy are important features of the larva period in bony fishes. It would be of interest to investigate whether a parameter such as condition factor can be assessed in order to denote immediately premetamorphic climax in other fishes. It is well established that the bonefish *Albula sp.* utilises its stored energy during metamorphosis, and the same may be true for other members

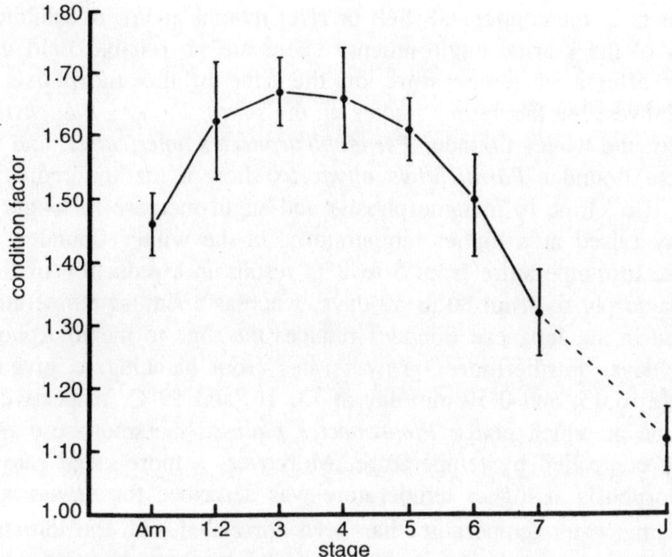

Figure 4.4 : The mean condition factor (and 95% confidence limits) from anadromous Petromyzon marinus. Am, ammocoete of metamorphosing length; stages 1-7 of metamorphosis; J, juvenile 6 months after stage 7.

of the Superorder Elopomorpha. Furthermore, condition factor is presently being used to assess larval growth rates in fishes and therefore could be applied through metamorphosis. In addition, RNA-DNA ratios and both nucleic acid and protein changes in metamorphosing flatfishes have been useful indicators of alterations in nutritional status and morphogenetic events. Buckley indicated that RNA-DNA ratios are particularly valuable since they are not affected by either age or size during larval life but increase during metamorphosis. Fukuda *et al*. used the proteinDNA ratio to follow increased cell size at metamorphosis following a rapid decline during larval growth.

Temperature

Tesch suggested that the time at which late-metamorphosing leptocephali of *Anguilla sp.* arrive at the mouths of rivers in northern Europe and North America from the Mediterrean can be correlated with temperature. Tucker reinterpreted the original data of Schmidt and claimed that temperature conditions encountered by leptocephali resulted in "eco-phenotypes" of *Anguilla anguilla*. However, a recent alternative suggestion is that the varying time of metamorphosis in *Anguilla* sp. may reflect the influence either of stimulating factors in the waters of the continental shelf or river mouths and/or of inhibitory factors of the marine environment. There are no reliable field data on the effects of temperature on the time of metamorphosis in Osteichthyes, but laboratory studies on the starry flounder *Platichythys stellatus*, the winter flounder *Pseudopleuronectes americanus*, and the Japanese flounder *Paralichthys olivaceus* show a marked reduction in age (i.e., time to metamorphosis) and slight increase in length of progeny raised at a higher temperature. In the winter flounder, an increase in temperature from 5 to 8°C results in a reduction of time to metamorphosis from 80 to 49 days, whereas a similar temperature increase in the Japanese flounder reduces the time to metamorphosis by 31 days. Furthermore, growth rates from hatching to juvenile are 0.22, 0.43, and 0.59 mm/day at 13, 16, and 19°C, respectively. The time at which plaice *Pleuronectes platessa* metamorphose may also be controlled by temperature. Moreover, a more rapid rate of metamorphosis at higher temperature was described for *Albula* sp..

Rising water temperature has been correlated with the initiation event of metamorphosis in lampreys of metamorphosing size. Laboratory and field experiments indicate that most lampreys metamorphose at 20-25°C, and a drop in temperature between 7 and 10°C can delay the onset by 4-5 weeks.

Behaviour

Fish behaviour during metamorphosis has received little attention, but with increased observations of organisms in laboratory culture, this will likely be a subject of future study. Information that can be gained from such observations may be similar to the following in *Siganus lineatus*. The larvae of this species are carnivores, but when metamorphosis commences they spend less time swimming in search of live food and stop to gaze at algae. Later in metamorphosis they also begin to ingest the algae. This behaviour is linked to a lengthening of the intestine and a temporary omnivorous diet during metamorphosis. Adults of S. *lineatus* are herbivores and have developed a long intestine during metamorphosis to aid in assimilation of their new source of nutrition. Increased swimming speed has been noted as an indicator of the onset of metamorphosis in the Japanese flounder and is correlated with morphological change in the finfold. Other behaviour related to metamorphosis has been noted in lampreys, where metamorphosing individuals seek shelter in a coarser substrate and faster flowing water than premetamorphic animals. Also, changes in swimming behaviour accompany the initiation of metamorphosis in those species of bony fish which switch from a pelagic larva to a demersal juvenile.

CONTROL

There have been numerous studies on the factors involved in initiation of the second metamorphosis in fishes, such as parr-smolt or yellow-silver eel transformation, but data on the control of first (true) metamorphosis (larval metamorphosis) are sparse. There is little doubt that environmental cues such as water temperature and, perhaps, photoperiod are important but, as in amphibians, hormones probably play a significant role. Early studies on both lampreys and bony fish were ambivalent in their assessment of the role of hormones. Most recent studies suggest that both environmental and hormonal cues interact to trigger and maintain fish species during the climax of metamorphosis.

Environmental

Environmental temperature not only affects the duration of larval life but is also a trigger for metamorphosis. Purvis subjected ammocoetes of *Petromyzon marinus* of similar metamorphosing size and at the same time of the year to either 20-21°C (aquarium), to 14-16°C (Big Garlic River) or 7-11°C. (Lake Superior) and found the incidence of metamorphosis to be 75-100%, 46-76%, and 5-10%, respectively. Increased incidence of metamorphosis at higher temperature had been previously shown in the Southern Hemisphere lamprey *M. mordax*. It is now common practice to hold ammocoetes in the laboratory at 20-25°C during the time that they would normally initiate metamorphosis in their natural environment. This dependence on temperature to trigger or to maintain metamorphic climax in lampreys explains why this phase of larval life is synchronized among individual populations of the same species. There is a reasonable amount of evidence from the studies of Policansky and Sekai *et al.* that larval life in starry flounders can be shortened by triggering of metamorphic climax with higher temperature. In the Japanese flounder, 24, 32, and 55 days are the times to metamorphosis at temperatures of 19, 16, and 13°C, respectively.

Eddy and Cole and Youson have shown that pinealectomy (removal of both the pineal and parapineal in this case) will prevent metamorphosis in lampreys. These two organs possess cells that have all the potential for photoreception. Although photoperiod may have some involvement in regulating the onset of metamorphosis, carefully controlled experiments suggest that the pineal complex (pineal and parapineal) may monitor other environmental cues, such as temperature, which in turn stimulate internal mechanisms to initiate the event. Survival of larvae of the gilthead sea bream *Sparus aurata* past their metamorphosis is directly correlated with the duration of the photoperiod. This may be related to growth rates, for there is an increased chance of encountering prey within a longer photoperiod and, under such conditions, metamorphosing size can be attained within a shorter time period.

Hormonal

In any experiments concerning the stimulation, retardation, or prevention of fish metamorphosis, it is important to be certain that immediately premetamorphic animals are used. This problem is particularly significant in studies of lamprey metamorphosis, where the duration of larval life varies among individuals in a given population. As noted earlier, condition factor is valuable for identifying immediately premetamorphic ammocoetes. We now know that the pineal complex and pituitary have some involvement in metamorphosis. Pinealectomy and hypophysectomy prevent metamorphosis in ammocoetes which have reached a certain condition factor. Removal of the rostral pars distalis inhibits metamorphosis, while extirpation of only the caudal pars distalis will retard the process but permit its completion. The rostral pars distalis is important in initiating metamorphosis, but the caudal portion must be present to complete the phase. According to Joss, these experimental results imply that more than one hormone is involved in lamprey metamorphosis and that they are likely thyrotrophin, somatotrophin, and gonadotrophin. However, cells of the caudal pars distalis, but not the rostral portion, show increase in activity (cell division and granulation) during the early stages of metamorphosis in *P. marinus* (*G. M.* Wright, personal communication). Furthermore, present assumptions of Joss of hypophyseal hormone activity during lamprey metamorphosis are complicated by the fact that growth of the caudal pars distalis during metamorphosis is much greater in nonparasitic *Lampetra planeri* than in parasitic *L. fluviatilis* (*M. W.* Hardisty, personal communication). This would suggest implication of this region of the pituitary in regulating gonadal growth and sexual maturation, both of which only occur in the nonparasitic species at this time. No doubt future studies on lampreys will be directed toward a definitive identification of the hypophyseal hormones and the cells involved in this stimulus to development. In the meantime, we are only left with an impression

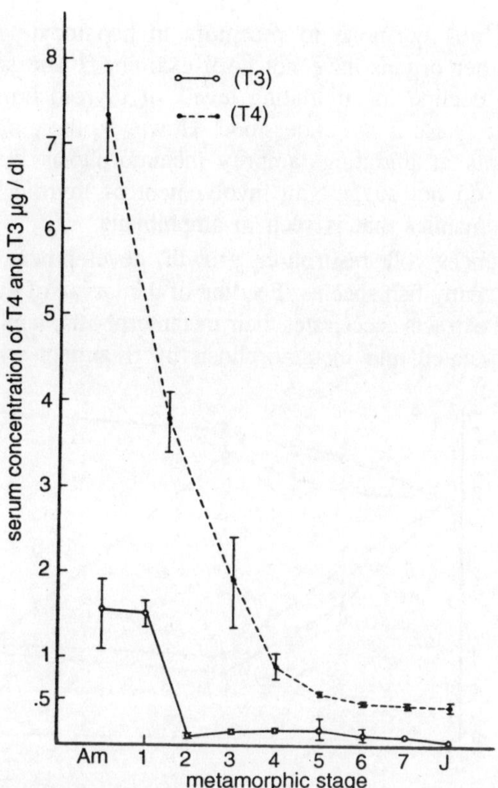

Figure 4.5 : Sera concentrations (µg dl^{-1} ± 2SE) of triiodothyronine (T3) and thyroxine (T4) in ammocoete (Am), metamorphosing stages (1-7), and juveniles (J) of Petromyzon marinus.

that both rostral and caudal pars distalis are evidently involved in lamprey metamorphosis.

Assuming that there is some hormonal control from the pituitary, the mode of action of these hormones may be to act directly on the tissues undergoing the developmental change or, as in amphibian metamorphosis, to stimulate other endocrine glands to release hormones that act upon the tissues. In light of the importance of thyroid hormones to amphibian metamorphosis, the only hormones to be examined during fish first metamorphosis are thyroxine (T4) and triiodothyronine (T3). In lampreys, transformation of the larval endostyle to a thyroid gland with follicles is initiated very early in metamorphosis and is accompanied by a dramatic decline in serum levels of both T4 and T3. The drop in T3 levels is not correlated with increase in binding capacity of this hormone to receptors in hepatocyte nuclei of the liver, but other organs have not been examined. The significance of the marked decline in circulating levels of thyroid hormones to the metamorphic phase is not understood. However, these data, and those from attempts at initiating lamprey metamorphosis by external T4 stimulation, do not suggest an involvement of thyroid hormones, at least in the manner that is seen in amphibians.

T4 enhances yolk resorption, growth, development, and survival of larvae of many fish species. Feeding of the larvae of the mudskipper with thyroid extracts accelerates their metamorphosis, while leptocephali may be stimulated into metamorphosis by treatment with T4.

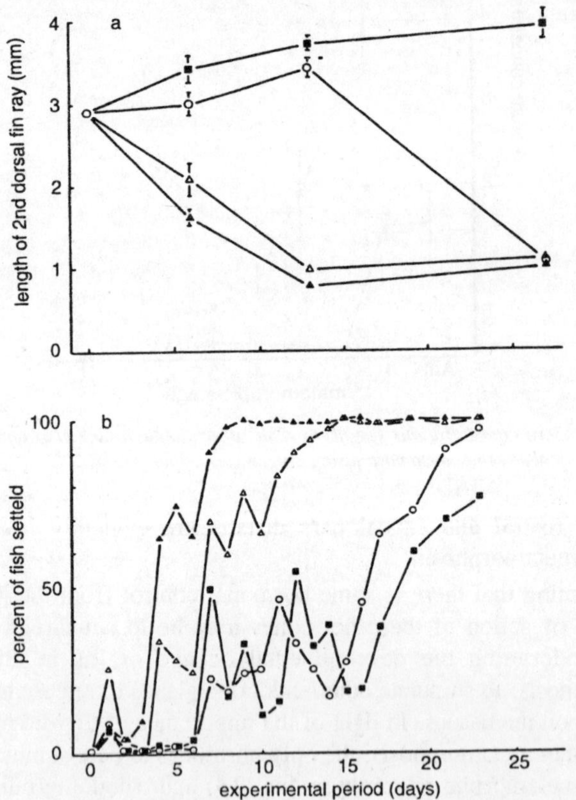

Figure 4.6 : (a) Effects of thyroxine (T4) and thiourea (TU) on the length of the second elongated dorsal fin ray (O, control; ❑ 30 ppm TU; L, 0.05 ppm T4; Δ, 9.1 ppm T4. (b) Effects of T4 and TU on settling behaviour of flounder larvae (0, control; ❑, 30 ppm TU; L., 0.05 ppm T4; Δ, 0.1 ppm T4. T4 treatment terminated after 16 days.

Recent studies on flounder larvae have provided definitive evidence that the thyroid plays a major role in initiating first metamorphosis in bony fish. T4 induces metamorphosis, while antithyroid agents arrest the process. The *effect* of T4 is dose dependent with 100 ppb and 10 ppb (but not 1 ppb) in sea water resulting in metamorphosis. T3 is several times more potent than T4 in inducing metamorphosis. These experimental studies confirm earlier histological investigations that showed an increase in activity of the thyroid gland during eel and plaice metamorphoses. Histological and immunohistochemical studies with antisera to thyroxine (T4) has revealed that the thyroid follicular cells show high activity during metamorphic climax in Japanese flounder and appear inactive at postclimax. These authors also observed that cells in the proximal pars distalis that are immunoreactive to antithyrotropin showed an activity during metamorphosis that reflects the control of the anterior pituitary on thyroid activity. Furthermore, the administration of T4 and thiourea reveals that negative feedback regulation exists in flounder. These studies are a clear indication that a thyroid-pituitary axis is involved in metamorphosis in flounder.

Sterba was able to produce partial metamorphosis in ammocoetes by corticortrophin injections and there seems to be an increase in activity of the essential enzyme Δ^5-3β-hydroxysteriod dehydrogenate in the presumptive adrenocortical (interrenal) tissue at this time in the life cycle. Whether adrenocorticosteriods have any significance in fish first metamorphosis, as they have recently been shown to have in amphibian metamorphosis, should be a future consideration.

DURATION

The length of time between the initiation of metamorphosis and the appearance of a definitive phenotype (juvenile, smolt) is generally species-specific. In those species that initiate metamorphosis soon after hatching of the embryo in order to remodel transitory embryonic larval structures, this phase takes up most of larval life. In contrast, in species in which metamorphosis is initiated and completed as a single dramatic climax, the event may last only a few days to a few weeks, such as in Pleuronectiformes and *Albula sp.*, or be as long as to 3-4 months, as in most lampreys, and 9-10 months, as in *Auguilla sp.* and a few lampreys. As metamorphosis is usually an interval of ontogeny in many fishes when there is an alteration in feeding or digestive mechanisms, the termination of the metamorphic phase may be determined by the time at which the animal is able to resume, begin, or alter feeding. Thus, metamorphosis is completed when

Siganus lineatus becomes a carnivore, when *P.marinus* starts feeding parasitically, or when Pleuronectiformes begin to inhabit the bottom. That the acquisition of the external characters of the juvenile is not sufficient evidence to assume the end of metamorphosis is also illustrated in the ability of the animals to osmoregulate in their new niche. The ability of *Anguilla sp.* to osmoregulate in fresh water and the concomitant change in behaviour are important criteria for assessing the termination of the metamorphosis of glass eels from leptocephali. Metamorphosis in anadromous lampreys involves a transformation of organs of osmoregulation, and the completion of this development, such as in the esophagus and kidneys, dictates the time at which migration into saltwater occurs and the end of the larva period. The duration of the metamorphic phase of larval life in all lamprey species, whether parasitic or nonparasitic, is long and probably is a reflection of the complexity of change that takes place, particularly in those tissues and organs that are (were) involved in osmoregulation in present and ancestral individuals. Thus, all extant nonparasitic species of lamprey undergo the same metamorphic change over a protracted phase common to freshwater and anadromous parasitic species.

EVENTS

The metamorphic interval of fish ontogeny involves two major events: initiation and climax. These events may be separated by a considerable time period or be almost simultaneous. They are characterised by a number of morphological and physiological changes that ultimately determine the metabolic and behavioural patterns of the juvenile. This section describes the changes occurring during these events.

Changes in Body Length and Proximate Composition

The larva period is usually typified by some growth as the animal acquires nutrition through an exogenous feeding habit or, perhaps, through integumentary absorption of dissolved organic matter. In species that have a significant and prolonged climax event, growth usually stops at metamorphosis, and this has been corroborated through examination of otoliths. In many species there is actually a decrease in body length, and this can be attributed to the cessation of feeding. The extent of decrease in body length is particularly variable among bony fishes and may be related to the degree of ossification of the skeleton that took place in earlier larval life: that is, it is more difficult for an animal to shrink in the presence of a hard-tissue endoskeleton. Although both lampreys and most elopomorphs show a reduction in length of at least a few millimeters during metamorphosis, a most spectacular reduction is seen in the bonefish *Albula sp.*, where larva of 60-70 mm reduce by over 50% of their length. This process takes only 8-12 days and is due to the resorption of a transparent, extracellular gelatinous matrix. A similar resorption of a gelatinous mass occurs during the metamorphosis of other elopomorphs (orders Anguilliformes, Elopiformes, and Notacanthiformes), but the decline in length is not always as marked as in *Albula*. However, *Megalops atlanticus* shows a characteristic decline in length from 28 to 13 mm during its "negative growth phase".

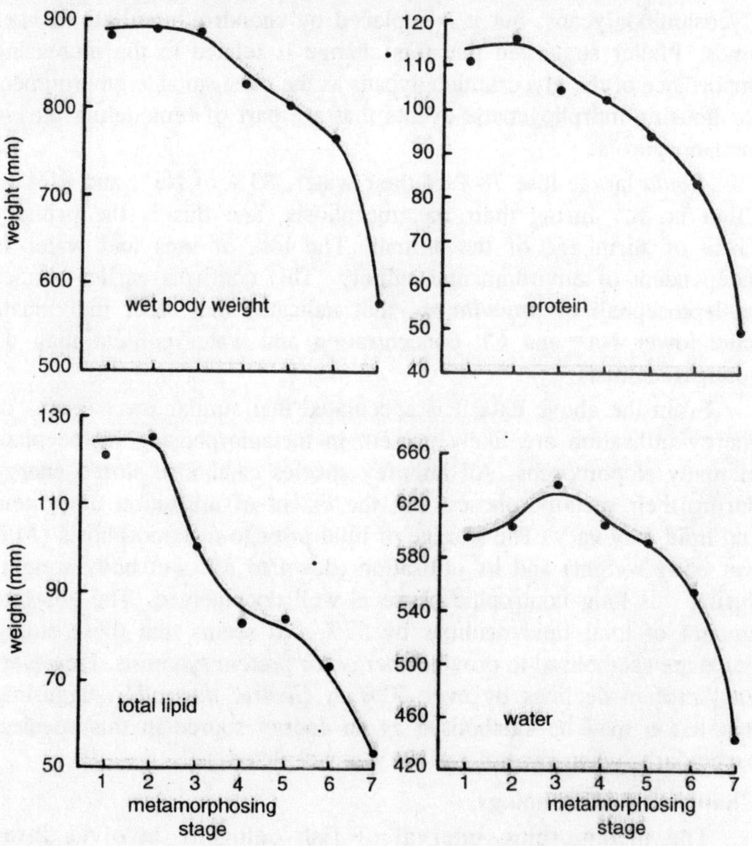

Figure 4.7 : The wet body weight and the weight of water, total lipid, and protein calculated for standard animals representing stages 1-7 in the metamorphosis of Geotria australis.

According to Pfeiler, the leptocephali of all elopomorphs utilise the stored carbohydrates and lipids of the gelatinous matrix during their nontrophic phase of metamorphosis. However, definitive evidence has only been obtained during metamorphosis of *Albula sp.* In this bonefish there is a 50% loss of total lipid, 83% loss of carbohydrate content, and 52% decline in ash. Although protein content shows no significant change, nonprotein nitrogen declines from 70 to 58% of total nitrogen from the beginning to the end of metamorphosis. Most of the proteins and carbohydrates that are lost are associated with the glycosaminoglycan fraction of the gelatinous matrix. In early metamorphosis keratan sulfate is the principal component of the glycosaminoglycans, but it is replaced by chondroitin sulfate at later times. Pfeiler suggested that this change is related to the increasing importance of the glycosaminoglycans as the most suitable environment for housing morphogenetic events that are part of remodeling during metamorphosis.

Albula larvae lose 78% of their water, 83% of Na^+, and 91% of Clbut no K^+ during their metamorphosis, and this is the primary cause of shrinkage of the animal. The loss of ions and water is independent of environmental salinity. This confirms earlier studies on leptocephali of *Anguilla sp.* that indicated that older individuals have lower Na^+ and Cl^- concentration and water content than do younger animals.

From the above data it is speculated that similar mechanisms of energy utilisation are likely present in metamorphosing leptocephali of many elopomorphs. All lamprey species catabolise stored energy during their metamorphoses, but the extent of utilisation of protein and lipid may vary. The storage of lipid prior to metamorphosis (14% wet body weight) and its utilisation (down to 8% wet body weight) during this long nontrophic phase is well documented. The absolute amount of total lipid declines by 57% . It seems that these stored lipids are catabolised to provide energy for protein synthesis. However, total protein declines by over 50% in *Geotria australis*, suggesting that it too may be catabolised as an energy source in this species, particularly when neutral fat has been depleted.

Changes in Morphology

The metamorphic interval of fish ontogeny involves three morphogenetic processes: transformation of larval tissues and organs into those of the adult, regression and eventual loss of larval structures, and the formation of new adult tissues and organs from anlagen which may have persisted since embryonic life. The complexity and breadth of this phase of the larva period in fishes are illustrated by example in the following discussion.

Transformation

Presently the term transformation is used in the literature to describe most metamorphic events, when in fact it is only one of three morphogenetic processes taking place. Transformation involves both tissue regression and proliferation in functioning larval structures and is responsible for such spectacular features as eye migration in

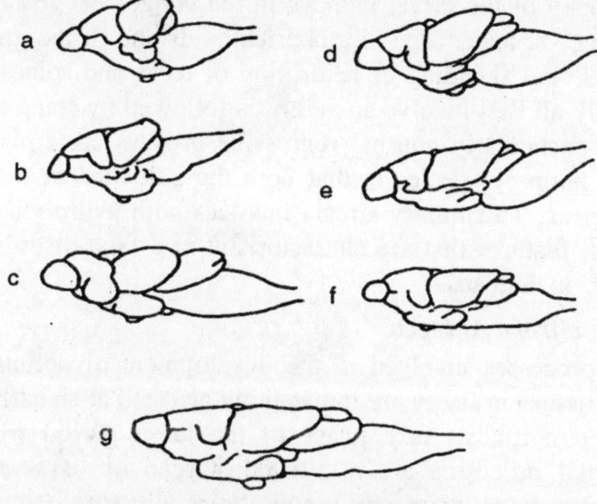

Figure 4.8 : Morphological change of shape and proportion of the regions of the brain of the conger eel conger myriaster during metamorphosis (a-g).

flounders, changes in the shape of the willow leaf-like leptocephalus to a glass eel, modification of the buccal funnel of ammocoetes to the suction-like disc of juvenile lampreys, and lengthening of the tail of the ribbonfish. Examples of more subtle internal changes involving tissue transformation are elongation of the intestine in logperch, *S.lineatus*, and turbot, remodeling of the blood supply to the pelvic and pectoral fins in *Auguilla spp.*, transformation of the brain in conger eel, forward movement of dorsal and anal fins in many species, creation of the asymmetry in the projections of the right and left olfactory bulbs, 90° relative displacement of the vestibular and ocular coordinates, and development of the lateral line system in flatfishes. Transformation of ammocoete tissues and organs has received considerable attention. A few such transformations are development of intestinal folds and chloride cells of the gills, modification of the ammocoete endostyle to a thyroid with follicles and the bile duct into an endocrine pancreas, and displacement of the larval hemopoietic sites (typhlosole and nephric fold) to the fat column.

Regression

Regressive processes require that tissue components be either degraded by hydrolytic enzymes or catabolised as an energy source. The utilisation of stored lipids by ammocoetes or of the gelatinous (glycosaminoglycans) matrix of eels illustrates regression by catabolism. In the case of ammocoetes, as lipid of the nephric fold is gradually used, the site becomes progressively occupied by newly formed kidney tissue. Atrophy of the larval kidneys in the conger eel and lampreys, loss of the pelvic appendage in gibberichthyids and of the gas bladder of flounders, and shedding or resorption of teeth and spines in many species likely all first involve an autolysis followed by some resorptive process. A particularly unusual regressive process takes place in the liver of all lamprey species in that both the gall bladder and all bile ducts disappear. This biliary atresia involves both hydrolytic enzymes and fibrosis, features that are characteristic of a fatal disorder, of the same name, in humans.

Development from Anlagen

Many processes involved in the development of definitive adult organs and tissues in fishes are initiated but arrested at an early embryo period. Metamorphosis is a phase of the larva period when these developmental processes are reinitiated. Organ or tissue rudiments (termed anlagen or primordia) and their cells are stimulated to differentiate during metamorphosis. They either may have been arrested in an advanced form of development so that they resemble, but do not function like, adult structures, or are still in an unrecognisable form. In the latter case, the anlagen may be a clump or a cord of undifferentiated cells or be represented by cells diffusely scattered among differentiated cells. Thus, the retina of ammocoetes does not complete its development until metamorphic climax, while the definitive kidneys and adult esophagus, buccal glands, and olfactory sacs all originate from separate anlagen, which exist in an inactive but recognisable form throughout earlier larval life. This latter form of development is seen in the formation of the intermuscular bones of flounder, the definitive kidney of eels, and the definitive teeth of many fishes. Some of the cartilaginous skeleton associated with the new feeding apparatus of adult lampreys develops from scattered undifferentiated cells that congregate to form a blastema in early metamorphosis.

Physiological Change

Blood

There is little direct information on physiological or biochemical change in bony fishes during metamorphosis. Most present data are implied from observations of premetamorphic larvae and juvenile or mature adults. Blood has received the greatest attention. Hemoglobin and erythrocytes are absent in leptocephali of many eel species, and the second hemoglobin of adult tilapia is not present in larvae. Near the middle of metamorphosis in the lamprey *P.marinus* there is a mixed pattern of larval and adult hemoglobin, but the typical adult pattern is present in the juvenile. The initiation of this phase in lampreys is marked by a decrease in the hematocrit. There is usually a concomitant drop in hemoglobin concentration as erythrophagocytosis takes place in the liver. The nature and quantity of serum proteins also undergo modification during lamprey metamorphosis. Crossed immunoelectrophoresis has shown a larval protein, AS, remains until early metamorphosis but subsequently disappears. A gradual increase in concentration of two serum proteins, SDS-1 (a glycoprotein) and CB-III (a lipoprotein), which eventually will make up 85% of the total serum protein in mature adults, marks the beginning of metamorphosis. The functions of these larval and adult proteins are as yet unknown. A female-specific serum protein (FSSP), likely vitellogenin, is first detectable in metamorphosing stage 4 of the nonparasitic species *Lampetra reissneri*. This event coincides with the beginning of vitellogenesis in the liver and with the onset of sexual maturation in females of this species. It would be of interest to establish whether the prolonged juvenile period of parasitic species results in a delay in the time of appearance of FSSP in the serum.

Lampreys undergo a dramatic alteration in iron metabolism at metamorphosis. The plasma of larval *G. australis* has 19,000 μg Fe/dl and only 34 μg Fe/dl in mature adults. The major iron-binding protein of larvae is ferritin, while that of adults is more like transferrin. During metamorphosis the liver gradually accumulates iron and eventually becomes iron-loaded by the juvenile period, while at the same time plasma nonheme iron becomes much lower. A mechanism present in the intestinal epithelium of ammocoetes for elimination of excess iron is likely lost at metamorphosis. The effects of the loss of this mechanism on body iron concentration remains to be investigated.

Oxygen Consumption

There is a gradual increase in oxygen consumption during lamprey metamorphosis (29.3 to 60.4 μl g^{-1} h^{-1} at 10°C), and this accompanies extensive modification to the branchial chambers. The animal develops a high affinity for oxygen, which can be linked to changes in the hemoglobins at metamorphosis but also might be somewhat related to the large number of catecholamine-secreting cells developing at this time. A change in activity is correlated with a high rate of hemoglobin synthesis and subsequent oxygen-carrying capacity during metamorphosis in herring and plaice. Laurence has provided a comprehensive comparison of oxygen utilisation of a larval bony fish in premetamorphic and metamorphic phases. In the larva period of winter flounder *Pseudopleuronectes americanus*, the absolute values of oxygen consumption (microliters per hour) increase until metamorphosis, at which time they decline. Oxygen consumption increases after the completion of metamorphosis. Metabolic rate [(μl μg^{-1} h^{-1}) $\times$ 10^3] decreases with increasing size from hatching through metamorphosis. Both of these parameters increase during the larva period when the temperature is raised. The change in values during the larva period are believed to reflect behavioural and morphological changes as a result of a switch from cutaneous to gill respiration.

Electrolyte Balance

There is a limited amount of evidence that metamorphosis in eels may involve an alteration in electrolyte balance. It seems that *Anguilla* sp. is able to modify its osmotic permeability so that net water movements are minimised in either fresh water or seawater. Such a mechanism exists in *Albula* sp. throughout metamorphosis. Although larvae of this species lose most of their water and NaCl content during metamorphosis, this net loss is not effected by subjecting animals to dilute (896), normal (35%o), and concentrated (48%) seawater. The losses are due to absorption of a larval gelatinous matrix. Serum osmolality rises during early stages of lamprey metamorphosis and gradually attains the adult values (—300 mosmol kg^{-1}) by the end of the phase. An attempt to acclimate metamorphosing landlocked *P. marinus* to even 10%o at stages 5 and 6 results in serum osmolality of up to 340 mosmol kg^{-1} but only 300 at later stages.

Hormones

Morphological evidence indicates that there is a change in activity of the thyroid gland during first metamorphosis in all fishes. However, to date, this has only been documented in lampreys through analysis of sera concentration of the major thyroid hormones. Both sera T3 and T4 concentrations decline abruptly at the very onset of metamorphosis and remain close to these levels throughout the rest of the life cycle. There is little doubt that this drop in circulating levels of thyroid hormones is a significant physiological change, but it must be kept in mind in any interpretation of significance that sera levels of T3 and T4 in ammocoetes are as much as 10 times those of most vertebrates and that juvenile levels are within the normal vertebrate range. Future studies must be directed toward analysing the relationship of sera T3 and T4 concentrations with changing metabolic activity.

Metamorphosis in eels and that in lampreys share the common feature of being the phase of larval life when a specific peptide hormone is first synthesized by the endocrine pancreas. Whereas insulin appears in eel metamorphosis to accompany the somatostatin that has been present since early larval life, somatostatin appears during lamprey metamorphosis to join existing insulin. There is some suggestion from morphological and histochemical observations that adrenocortical hormones may be significant during lamprey metamorphosis; however, no definitive evidence exists.

Metabolism

In metamorphosis of the whitefish *Coregonus sp.*, climax is marked by an increase in activity of enzymes of the glycolytic pathway (phosphofructokinase, pyruvate kinase, and lactic dehydrogenase) and decline in activity of enzymes of the oxidative and citric acid cycle. Therefore, during metamorphosis of whitefish there may be a switch from a total utilisation of amino or fatty acids to increased use of glycogenolysis. The biochemical pathways have not been examined in *Albula sp.* during metamorphosis, but there is strong evidence to indicate that endogenous lipid and carbohydrate stores are utilised as a source of energy. Similarly, during the metamorphosis of lampreys, both lipids and carbohydrate stores are mobilised. A rise in serum glucose is correlated with a rapid decline in liver glycogen but a slower decline in muscle. An eventual cessation of the liver lipogenic process occurs as the activity of the relevant enzymes decline. The digestive enzyme complement changes at metamorphosis in the lake surgeon *Acipenser fulvescens* and is genetically predetermined. Metamorphosis of this species represents a transition period when lipase, amylase, and chymotrypsin concentrations decrease and pepsin and trypsin increase.

Free amino acid composition has been examined for *Albula sp.* during their metamorphosis. Although there are no changes in ninhydrin-positive substances, there is a significant decrease in a number

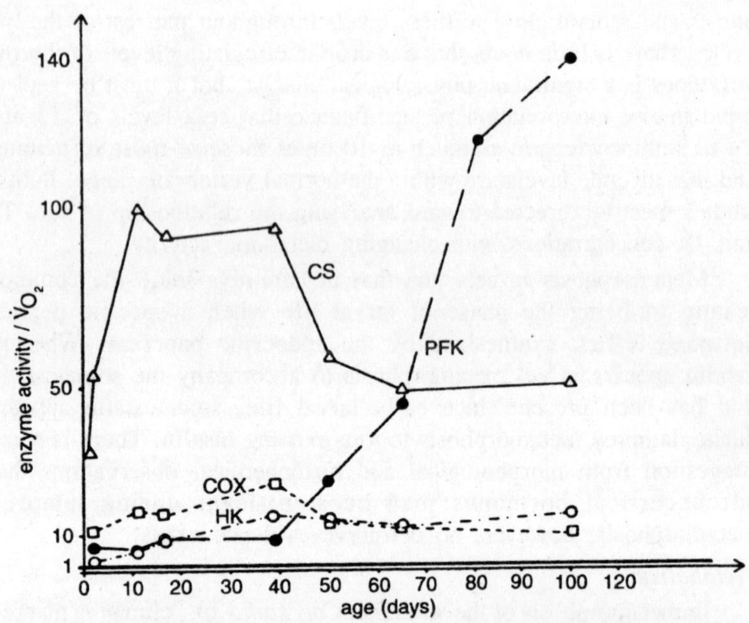

Figure 4.9 : Molar ratios of enzyme activity and oxygen consumption development of Coregonus sp. *from citrate synthase (CS), phosphofructokinase (PFK), cytochrome oxidase (COX), and hexokinase (HK). The ratios were calculated on the bases of stoichiometric oxygen requirements for each enzyme-catalyzed reaction.*

of essential amino acids. Leucine, isoleucine, phenylalanine, histidine, valine, methionine, lysine, and arginine represent 47% of the total amino acids in early metamorphosis but are reduced to only 23% in advanced stages. Taurine is the most abundant free amino acid in whole-body extracts and accounts for 36% and 59% of the total, by weight, in early and late metamorphosing individuals, respectively. This amino acid may be important as an organic osmolyte in order to maintain intracellular volumes at a time when there is a high extracellular loss of water and salt during resorption of the gelatinous matrix. However, taurine may also increase as a necessary compensation for the decrease in total essential amino acids. The nonessential amino acids as a group generally show only small change during metamorphosis of the bonefish, but certain components show marked increases (glycine and glutamic acid) while others are reduced (tyrosine and serine).

Behavioural Changes

The activity levels of metamorphosing fishes usually are related to whether this phase is trophic or nontrophic and to the degree of metamorphic change. In species that continue to feed, such as *Siganus lineatus*, the carnivorous larvae are active, but with early metamorphosis and increased coiling of the gut they become omnivorous, periodically active, and stop to examine benthic algae and detritus. Late metamorphosis is marked by further coiling of the gut, less activity, and a completely herbivorous diet.

Changes in feeding behaviour (pelagic to benthic) have been noted for the logperch *Percina caprodes* during its subtle and inconspicuous metamorphosis, and are related to development of the stomach, elaboration of the pancreas, and the transition from a physostomous to a physoclistous swimbladder. Blaxter and Staines noted a decreased ability of plaice and sole to search for food during metamorphosis, that is, a drop in distance covered per minute and a decrease in time spent in feeding activity. They found a species variation in the ability to take food in the dark during eye migration. On the other hand, in many flatfishes there is a decline in ingestion rates and rate of food requirement at metamorphosis due to their high efficiency of converting food to growth at this time in the life cycle. T4 stimulation will shorten the period of time required for metamorphosing flounder to become benthic. It is believed that *Taenioconger sp*. form colonies during their metamorphosis, which is a prelude to their adult fossorial behaviour. The increase in density of late larvae or metamorphosing individuals may be due to transport by tidal currents during their pelagic life. Changes in swimming performance accompany the onset of metamorphosis and often coincide with the development of pink and red muscle fibers and/or alterations to the finfold.

Many behavioural changes are related to the development or alterations of the nervous system. The presence of the lateral line system during the extension of the tail of the ribbonfish provides an antenna for the reception of water displacement and lowfrequency sound. It may have acquired a head up-tail down position in order to sense predators that are below the field of view of the eyes. The flatfishes have received a considerable amount of attention because of the potential interruption to their sense of balance during eye migration and when there is no compensatory change of the vestibular system as they lose their symmetry. There are slight differences in behaviour between plaice and turbot during metamorphosis, but by the end of metamorphosis the influence of light on the sense of balance is almost lost. As metamorphosis is approached in these two species, behavioural visual acuity reaches a plateau and remains unchanged. Nocturnal activity and the switch from a pelagic to benthic habitat in flounders coincides with the development of twin cones in the retina.

SIGNIFICANCE

Ontogeny and Phylogeny

Development is of two forms, ontogenetic and phylogenetic, and it is clear that, as in amphibian metamorphosis, first metamorphosis in fishes often reflects on both of these. According to Gould, there has been a change during evolution in the relative time of appearance and rate of development of characters that were present in the ancestor (heterochrony). This is one of the ways in which ontogeny has been involved in evolution and would explain the presence of a number of developmental strategies during the ontogeny of fishes. Metamorphosis is one developmental strategy in fishes that has been selected to permit a delay in development of definitive characters. A comparison of individuals within some groups of similar fishes suggests that variation in the duration of their larval life (age or length at metamorphosis) may reflect this evolutionary trend. For example, *Megalops atlantica* metamorphoses at an earlier age (2-3 months) and a smaller size (30 mm) than many other species of eels. Does this mean that this species is representative of a more "primitive" condition among the eels? This may not necessarily be the case, for the variation in length of larval life between North American and European varieties of *Anguilla sp.* may be essential to their distribution and be a reflection of the delay in the trigger to metamorphosis rather than differences in their evolutionary history. The tendency for lengthening of larval life during the phylogenetic development of a species has also been suggested both for flatfishes and lampreys. In Pleuronectiformes, larvae metamorphose over a range of 5-120 mm, but most are between 10 and 25 mm. Metamorphosis at the smaller and larger sizes are examples of "derived states," and species have either shortened larval life because of limited food or habitat or have prolonged it in order to ensure dispersal. However, at least in bony fishes, the developmental strategy of delayed metamorphosis does not always imply a period of increased larval growth.

The concept of "paired species" among lampreys, with a nonparasitic form derived from a closely related parasitic species, has introduced many questions about the evolutionary history of lampreys. For example, there have been suggestions of the introduction of pedomorphosis among extant lampreys where sexual maturation may accompany or precede metamorphosis. There is no definitive evidence to dispute a claim that fossil *Mayomyzon* may have been pedo morphic. Its size of 3-6 cm is approximately one-third the length of most extant lampreys at metamorphosis or as juveniles, implying that, if larval life was present in ancestral lampreys, it must have been of short duration. Zanandrea also described neotenous ammocoetes in relatively primitive nonparasitic *L. zanandreai*. However, we cannot assume that *Mayomyzon* in its small adult size was less specialised than extant adult lampreys, for the fossils may be of juveniles that had just recently metamorphosed from slowgrowing ammocoetes of advanced age. Recently, Vladykov gave his assessment of previous reports of neoteny in lampreys and concluded that it does not exist. However, discovery on the west coast of Canada of newly metamorphosed individuals that are capable of feeding in the laboratory while sexually mature may open a further chapter to this discussion. A second major question is, are the extant nonparasitic lampreys, which undergo a prolonged larval period and a rapid sexual maturation following metamorphosis with no distinct juvenile period, representative of more recent trends in lamprey evolution? In other words, is there a repetitive trend toward the elimination of feeding in postmetamorphic lampreys? Hardisty believes that lamprey evolution has been characterised by a resistance to a change in the time to sexual maturity, but the duration of larval life, and hence the time when metamorphosis is initiated, have been subjected to a great deal of adaptive modification. In the case of nonparasitic species, circumstances favored an extension of larval life and a corresponding reduction of adult life, while they favored a lengthened larval life but a retention of an extended juvenile period in parasitic species.

Szarski claims that a larval form is present in the life history of those fish species that show primitive characteristics and a larva period must be concluded by metamorphosis. It is his belief that a larva period was present in all vertebrate ancestors. A contrasting view has been provided for amphibians. As labyrinthodonts possessed large amounts of yolk in their eggs and development may have been direct, larval life and metamorphosis in extant amphibians may be a derived state. In this context, it is noteworthy that another extant agnathan and the most primitive living vertebrate, the hagfish, lacks a larva period and has large yolky eggs-that is, it has direct development. If lampreys and hagfishes share a common ancestry, then metamorphosis in lampreys may also be a derived state. However, differences in ontogeny between the two extant agnathans can probably be explained by the fact that they followed quite different evolutionary histories. Taking the opposite stand, if retention of a larva period and metamorphosis are primitive and less specialised ontogenetic features among fishes, there must be selective advantage for their persistence in some species. The maintenance of a complex life cycle (including metamorphosis) in many anuran amphibians is related to the fact that tadpoles are highly specialised suspension feeders. This method of feeding is adopted in order to take advantage of rapid rises in primary production. Similarly, following detailed observations of metamorphosis in *Albula* sp., Pfeiler makes a claim that retention of the leptocephalous strategy in all orders of the "primitive" teleost Superorder Elopomorpha has been selected because the premetamorphic larvae live in a nutrient bath that permits absorption of dissolved nutrients across the integumentary epithelium. With this simple and efficient method of obtaining nutrients, there was likely selective pressure for extension of the larval growth phase, which was followed by a metamorphic phase. However, there is no concrete evidence that the leptocephalous strategy is primitive. As in the case of the lamprey ammocoete, it is unwise to place too great a phylogenetic burden on the leptocephalus. Both the ammocoete and the leptocephalus are highly specialised for their particular mode of life and, during the course of their respective evolutionary histories, the adult and larval forms have increasingly diverged in their morphology and habits of life. This divergence in both groups has resulted in specialised, and far from primitive, larvae and adults and in a radical metamorphosis.

Divergence in larvae (tadpoles) and adults also seems to be a feature of the evolutionary history of anuran amphibians. In fact, there is evidence to support the view that the metamorphic process has evolved, and some extant anuran species with tadpoles may have evolved from species that lacked a larva period. In summary, it cannot be assumed from our present evidence that the larval developmental strategy with its terminal phase of metamorphosis is a primitive form of ontogeny in fishes.

Szarski states that the greater the differences between the larva and the adult in both fishes and amphibians, the more striking is the metamorphosis. Notable among these differences are the habitat, the behaviour, and the manner in which nutrients are acquired. Therefore, metamorphosis in fishes is characterised by variable need to break down larval structures and often to construct adult structures from embryonic anlagen. Thus, the almost sedentary, blind, suspension-feeding ammocoete, which resides in a burrow of silt in a freshwater stream, undergoes a dramatic first metamorphosis to a juvenile with suctoral mouth and rasping tongue, a salivary gland secreting an anticoagulant, a well-developed eye, a new form of ventilation (tidal), and innumerable changes to the digestive, excretory, and repiratory systems, which permit, in many species, a migration to saltwater. The major changes in the body form of eel leptocephali and larval flatfishes during metamorphosis also emphasize the great differences between these larvae and their definitive forms and the specialisation of individuals in both periods of the life cycle.

Dispersal

Northcutt and Gans and Jollie propose that ancestral larval fish were suspension feeders like modern-day lamprey ammocoetes, while Mallatt models them as raptorial feeders on phytoplankton and zooplankton much like most larvae of extant bony fishes. Regardless of the evolutionary history of feeding mechanisms in larval fishes, the premetamorphic phase of the larva period of extant fishes has been selected by many species, for it provides an opportunity for storage of energy, for growth, and for dispersal. These features of the larva period are related to pelagic feeding, a feature most agree is characteristic of the larvae of ancestral fishes.

Balon considers metamorphosis to be a very costly interval, in terms of energy loss, during the ontogeny of fishes. The regression, remodeling of temporary larval organs, and development of new organs consume valuable energy. Furthermore, larvae of small size are more susceptible to predation. Therefore, small larvae have to be produced in large numbers or small numbers of large larvae have to be especially equipped to ensure survival of the species. Thus, leptocephali or teleplonic larvae are equipped by their shape for dispersal by flotation over long distances where they will not compete for food with their definitive form. This passive, but selective, transport is not a high energy consumer and is also practiced by flatfishes. In such cases, the larva period is probably protracted for the above reasons. One may then take the view that the larval interval is a specialised period of ontogeny that initially evolved to take the organism through the "critical period", that is, the shift from yolk to exogenous energy source. Success in a pelagic life resulted in a prolongation of the period. Size disparity in metamorphosing individuals of several species suggests that the larger forms may represent highly successful larval adaptation to planktonic feeding and retardation of metamorphosis. This has ultimately resulted in greater dispersal, such as in *Anguilla sp*. and in the landlocked form of the sea lamprey, *P.marinus*. The success of the sea lamprey in the Great Lakes of North America is not so much a result of the ability of the adult to feed and to breed but instead is a reflection of the tremendous success of the ammocoete to colonise a new habitat. Colonisation is one of the objectives of the larva period in vertebrates, and the ammocoete has utilised its efficient, suspension-feeding mechanism to advantage in order to grow to metamorphosing size. Another successful case of larval dispersal is seen in eels. According to Pfeiler, the thin epidermis and the thin, leaflike body of high surface to volume ratio are responsible for the successful dispersal of the eel leptocephali. Circumstantial evidence is provided suggesting that the shape of the leptocephalus is not just related to passive locomotion. Instead, the high surface to volume ratio is important for efficient epithelial uptake of dissolved organic matter from the surrounding medium. This method of acquiring nutrition during the premetamorphic phase is important to meet the energy demands of both growing leptocephali and metamorphosing larvae. The latter obtain their nutrient requirements through breakdown of a the gelatinous matrix that was formed prior to the commencement of metamorphosis.

Physiological and Behavioural Adaptation

There are many examples among the fishes that illustrate the significance of metamorphic change to the survival and existence of their definitive phenotype. Many of the lesser known examples, such as the lengthening of the tail in ribbonfish and of the intestine of *Siganus lineatus*, and the change in the digestive enzymes of the lake sturgeon, have been mentioned in preceding sections. Blaxter has recently summarised the development of sensory structures in fishes and how these relate to behavioural modifications. Flatfishes, are constantly referred to in any discussion of adaptive change in vertebrates.

Many have found that metamorphosis in Pleuronectiformes provides an opportunity to study adaptive changes in the vestibulo-ocular reflex system. Concomitant with a 90° relative displacement of the vestibular and ocular coordinate systems during metamorphosis is the likelihood of development of pathways that permit secondary vestibular neurons of the horizontal semicircular canal to make contact with vertical eye-muscle motoneuron pools on both sides of the brain. This is a necessary response to permit suitable eye movements in the new benthic environment of fully metamorphosed individuals. Similarly, the development at metamorphosis of a duplex (rods and cones) retina from a retina in pelagic leptocephali with only rods is an adaptation to a need for enhanced visual acuity in elvers and juvenile flounders in a new environment.

Multiple hemoglobins seem to be the rule for adult fishes, but these hemoglobins are not all present in larvae. The addition of at least one more hemoglobin at metamorphosis allows the adult to exploit warmer, more saline environments, with less oxygen than in larval niches. Pelagic behaviour may account for the complete absence or low concentration of hemoglogin in some larvae. In lampreys, in addition to the change in hemoglogin, metamorphosis in anadromous species is characterised by the development of a new kidney and ion-transporting epithelia in the gills and intestine to permit adaptation to a marine environment.

The adaptive value of many metamorphic alterations is not fully understood. For example, there is a loss of the gallbladder and the bile ducts in all lamprey species. Although water conservation for a marine existence seems plausible, an aductular liver is not seen in any other marine vertebrate. In parasitic lamprey species that feed on the blood and body fluids, the ingested material may be in a form that can be readily absorbed across the intestinal wall without the emulsification properties of bile. However, lamprey species that ingest large quantities of muscle and even organs also lack a gallbladder and bile ducts. As iron is a normal component of bile, iron-loading of tissues of postmetamorphic lampreys may be a consequence of their biliary atresia, but there also seems to be little adaptive value to this manifestation of metamorphosis. A definitive explanation for the presence of the transitory larval structures in a number of bony fishes also awaits further study.

SUMMARY AND CONCLUSIONS

Metamorphosis in fishes is found at two intervals of their ontogeny. A first metamorphosis occurs in all species having a larva period and is the phase that terminates this period and that leads into a juvenile period. Bony fishes with a classical first metamorphosis during their ontogeny are Anguilliformes, Elopiformes, Notacanthiformes, and Pleuronectiformes. First metamorphosis is also present in Petromyzontiformes but not in Class Chondrichthyes or in Myxiniformes. A second metamorphosis may occur in many fish species when the juvenile undergoes sexual maturation to the definitive adult, but it is not a true metamorphosis. First metamorphosis is a true vertebrate metamorphosis: it is a phase of the larva period in which change takes place in nonreproductive structures, it results in a change of niche, and it is characterised by a dramatic and abrupt change in form that is likely triggered by internal and environmental cues. In fishes, first metamorphosis does not usually involve an increase in length, and the marked change in form assures that the larva and adult do not look alike. First metamorphosis is emphasized as a phase of the larva period with initiation and climax events, but in many species these two may occur almost simultaneously. Stages of metamorphosis can be delineated by careful evaluation of changes in external characters.

The time of first metamorphosis varies among the orders of fishes but is usually related more to body length than to age. Growth rate within a population may also be an important determinant. In species with a protracted and nontrophic metamorphic phase, physiological preparation is essential to the completion of metamorphosis. The highly synchronised timing of metamorphosis of various populations of the same species suggest environmental influence such as temperature and photoperiod.

There is increasing evidence to indicate that first metamorphosis in fishes is controlled by a contribution of both environmental (particularly temperature) and hormonal factors. Thyroid hormones are likely involved in initiating and maintaining metamorphosis in bony fishes, and regulation of secretion of the thyroid gland is controlled by thyrotropin from the pituitary. Lamprey metamorphosis may involve the interaction between several hormones and also seems to require coordination from the pituitary gland.

The duration of the metamorphic phase often reflects the magnitude of the morphological and physiological change or the importance of synchronising the termination of change with the availability of food or with environmental conditions. During the initiation and climax events, developmental processes include resorption of stored material for energy, regression (loss) of larval structures, transformation of larval structures into adult tissues and organs, and the development of new tissues and organs from embryonic anlagen or primordia. Physiological change includes alteration of blood chemistry, electrolyte balance, hormones, and metabolism. Behavioural modifications are concomitant with morphological and physiological change.

First metamorphosis in fishes is significant in discussions of the relationship between ontogenetic and phylogenetic development, yet it is uncertain whether it reflects a primitive condition or is a more recently derived state in vertebrate evolution. In some cases the larva period provides an opportunity during ontogeny for the wide dispersal of the species at low energy expenditure. Although it is a highenergy-consuming phase and often is marked by extensive depletion of stored energy, metamorphosis has a climax event whose completion is synchronised to a time when energy can be obtained through feeding or when conditions are ideal for continuing through to a sexually mature adult. A larva period that terminates with a metamorphosis has been a highly successful developmental strategy among fishes.

It is proposed that first metamorphosis in fishes be given more attention during studies of fish ontogeny, that standard criteria for defining this phase be adopted, and that universal staging be used as much as possible at the genus or species level. If such recommendations are followed, it will ensure the successful application of this large and specialised group of vertebrates to aquaculture, to furthering our knowledge of animal diversity, and for use in many other areas of biology. Fishes are presently utilised as essential research tools in neurophysiology and pathology. The fields of developmental biology, nutrition, and taxonomy will be immediate benefactors of any new information on fish first metamorphosis.

5

Fishing Twines

The Synthetic Polymer Fibres and Filaments ; their general characteristics, chemical and biological properties, with special reference to their use in Fishing Gear

High polymer chemistry, called also macromolecular chemistry, is a very young science, dating back only 35 years. The term "macromolecule" was first introduced in 1922 by Staudinger, who used it to describe the high molecule hydrocarbon obtained by the hydrogenation of natural rubber. Since then a great deal of research has been carried out and eventually, W. H. Carothers, discovered a fibreforming synthetic condensation polymer, for which the name "nylon" was coined.

The researches of W. H. Carothers and his associates led, within a relatively short time, to further important developments in the field of polymers, and to the discovery, by Whjnfield and Dickson, of polyester fibres from polyethylene therephthalate.

During the period 1939-1940 a great expansion took place in high polymer research. Some old polymers were carefully reinvestigated and many new ones discovered. But the effect of the discoveries and researches made during the 1930s and the 1940s have not yet been fully investigated, and their impact is not yet fully realised.

Besides polyamides and polyesters, the production of many new synthetic fibres-polyethylene, polyacrylonitrile, polyvinyl chloride, polyvinylidene chloride, polyurethanes and various vinyl copolymers-was also started and reached the commercial stage.

The chemical industry was thus given the opportunity of producing a wide variety of new fibres, the characteristics of which could be, to a point, modified in order to make them suitable for specific needs and end-uses. It has been rightly stated by Harold DeWitt Smith' that "mankind has initiated a second great textile project, namely, the creation of textile fibres". This,implied, however, extensive research for improving, on one hand, the characteristics of the fibres for the various applications and developing, on the other, new uses or expanding already accepted ones.

At the same time the vast industrial expansion of the post-war years called for ever increasing quantities of those specialised products known as "industrial textiles" and imposed more and more exacting requirements. The new synthetic fibres have enabled the textile industry to keep abreast of the increased demand and to meet successfully the particular needs of processes and trades using textiles.

The industrial use of man-made fibres has increased very rapidly. In the United States, industrial uses took a substantial proportion of man-made fibres in the early 1950s; in England, 34 per cent. of all continuous filament man-made fibre yarn went to industrial uses in 1953 and 1954.

Industrial textiles comprise, however, a very wide range of products, from cloth for filter presses to conveyor belts, from tyre cords to tarpaulins, from ropes and cordages to boat sails, from cigarette filter tips to fishing nets. Each of these applications requires efficient and lasting performance, under the respective conditions of use, and the textile technologist has to meet very different needs and stringent specifications. In the case of industrial textiles, quality is of paramount importance and where these products are part of a process, a breakdown in performance will at least cost money and may even cost lives. It is imperative, therefore, that quality and performance characteristics be of the highest possible standard.

Not so many years ago, we could hardly speak of a "fishing industry". For thousands of years, fishing had been a handicraft and down the ages the methods and equipment showed little progress. Now the picture has changed, and is changing, rapidly. To keep pace with development, scientific and application research is essential; the more the properties of the various fibres are known, the better we shall be able to employ them in the most useful and appropriate way.

PROPERTIES

The following main properties of fibres are to be specially considered for use in the fishing industry:

Density
Tenacity
Tensile strength
Knot strength
Loop strength
Elastic properties
Toughness
Stiffness
Water absorption
Effect of: heat
age
sunlight
chemicals
sea-water
Resistance to bacteria, mildew and insects

Density

Density According to J. T. Marsh "the density of a substance is the quantity of matter contained in unit volume". The relative density is the ratio of comparison between a given substance and a standard one (usually water).

When the C.G.S. System is used (unit of volume : c.c., unit of mass : gram-standard substance : water at 4 deg. C.) the values of absolute and relative densities are identical. Specific gravity and specific volume are reciprocally proportional, so that the lower the density, the higher the bulk and vice versa.

Strength

Strength is a general name for defining both tenacity and tensile strength of fibres. The difference between these two terms is clearly stated by Marsha as follows: "Tenacity is the breaking force in terms of the fibre or yarn denier, whereas tensile strength is the breaking force in terms of the unit area." The first is expressed in terms of grams per denier, the latter in terms of grams per square millimetre. In this respect, it must be pointed out that values for tenacity are not directly comparable, as they do not take into account the density of the material; values for tensile strength, on the other hand, are directly comparable.

The breaking force is often expressed as "breaking length" which according to "Textile Terms and Definitions" is: "the length of a specimen whose weight is equal to the breaking load"

The conversion formulae are:

—breaking length (in Km.) = 9 × tenacity (in g/den.)

—tensile strength (kg./mm.^{2}) = breaking length (in Km.) × specific gravity.

Knot strength

Knot strength is the tenacity of a fibre or a yarn in which a plain knot has been tied. As Prof. Viviani points out "the knotting test may give an idea of the transverse strength of the fibre". The ratio of the knot strength to the single fibre or yarn strength gives the effect of bending due to the tying of the knot.

Loop strength

Loop strength. To show to what extent a fibre or a yarn is affected by bending, the "loop strength ratio" is a very useful figure. To determine this value the free ends of linked loops of fibre or yarn are secured in the grips of the testing machine, and the load required to break them is found.

Elastic properties

Elastic properties, i.e. extensibility and elastic recovery of a fibre are as important as strength. Too much stress has been placed on tenacity as the most valuable property of textile materials and, although high breaking load is of significance, extensibility is at least of equal importance. An inextensible fibre, or a fibre possessing very poor extensibility—even when combined with high tenacity—is of little actual value, if the extension at break is considered'. For the evaluation of a fibre, however, the extension at break is not the only point to be considered; from the practical point of view, the behaviour of the fibre when stressed or extended to a degree not reaching the breaking point is of great value.

The elastic behaviour of a fibre is referred to as Young's Modulus, which is the relationship between stress and strain at loads below the elastic limit. It is expressed in grams of stretching force per denier of fibre (g/den.) and corresponds to the tension required to produce an extension of 1 per cent. The values thus expressed are based on the denier of the fibre and not on. the crosssectional area and may be converted into kilograms per square millimetre by multiplying by 9 times the specific gravity in grams per cubic centimetre.

It must be borne in mind, in this respect, that textile fibres do not obey the physical law (Hooke's Law), according to which strain is proportional to stress or, to be more accurate, they obey this law up to a point, called "yield point" beyond which the fibre exhibits a "viscous or plastic flow". Consequently Young's Moduli are useful for comparison, but only in the region where Hooke's Law applies, when they are "constants".

Elasticity is the power of recovery from strain or deformation. The total extension of a fibre is formed by two components: an elastic extension, which is recoverable on release of stress, and a permanent (or plastic) elongation which is not recoverable. In its turn elastic extension comprises partly "immediate recovery" and partly "delayed recovery", so that the time factor has to be taken into account in determining it.

According to Kornreich "the elastic limit is reached when stress results in a permanent extension, known as `elongation "'. The ability of a fibre to recover from strain is of importance for several end-uses.

Toughness

Toughness. Of practical importance also is the "toughness", "work of rupture", or "energy absorption" of a fibre. According to Kaswell' "the load is a force, the extension is a distance and the area under the loadextension (stress-strain or stress-elongation) diagram is the product of force and distance, or work (energy)". The area therefore depicts the fibre's ability to have work done upon itself, i.e. to absorb energy. The "toughness index" may be just the same for a fibre of high tenacity and low extensibility as for fibre of low tenacity and high extensibility. In general, the higher the "toughness index", the better is the fibre from the standpoint of use, as work of rupture shows the ability of a fibre not only to absorb energy, but also to withstand a sudden load. (i.e. a "live-load").

Stiffness

Stiffness is defined by Harold DeWitt Smiths as resistance to deformation. Average stiffness is the ratio of breaking stress to breaking strain; elastic stiffness the ratio of stress to strain at the yield point.

Moisture content

Moisture content affects the physical properties of fibres and, in particular, their tenacity, extensibility, rigidity and swelling. In this respect the hygroscopic nature of fibres, i.e. their power to absorb and desorb water, is of importance, as water acts as a plasticizer to a higher degree for hydrophilic fibres and to a lesser degree, or scarcely at all, for hydrophobic ones.

Water influences the fibres in various forms, e.g. as relative humidity (in the atmosphere), as liquid water (water or imbibition), and as steam; furthermore, the action of cold water is different from that of hot water.

The amount of atmospheric moisture that a fibre is able to absorb when exposed, fully dried, to surrounding air is referred to as "regain" and is expressed as a percentage of the dry weight of the fibres. To ensure that the regain of different fibres be comparative, tests are conducted in a standard atmosphere. "Standard regain" is the amount of moisture a fibre absorbs at 65 per cent. relative humidity at a temperature of 20 deg. C. (international standards). The standard testing atmosphere allows a tolerance of ± 2 per cent. R.H. and ± 2 deg. C.

The moisture content of the fibres affects their tenacity in different degrees. Natural cellulosic fibres (cotton, ramie, linen) show an increase in tenacity from the dry to the wet state; regenerated cellulosic fibres are, on the contrary, stronger in the oven-dry state than moist; synthetic fibres show no significant variation in strength from the dry to the wet state.

Liquid water, entering the fibre structure, causes swelling according to the water imbibition properties which are related inversely to wet-strength. Hot or boiling water affects the plastic properties of the fibres. For most of the synthetic fibres it causes shrinkage and, for some types, even degradation.

The action of steam is dependent upon its temperature and saturation. It has a great influence on the plasticity of the fibres, and it is accompanied by shrinkage, more pronounced with synthetic than with other fibres.

The property of synthetic fibres to shrink under the action of heat (wet or dry) is utilised by a special treatment (thermo-setting) to impart to fabrics and articles a dimensional stability, which is permanent as long as they are not treated at higher temperature than that of setting.

Brief definitions of mechanical properties

Brief definitions of mechanical properties. It might be useful to sum up briefly the basic mechanical properties of textile fibre materials to be evaluated under the influence of tensile forces as defined by Harold De Witt Smith:

strength — the ability to support a load

elongation — the deformation produced by a load, along the

		line of action of the load
stiffness	—	resistance to deformation
toughness	—	the ability to absorb work
elasticity	—	the ability to return to original shape and dimensions upon cessation of deforming force
resilience	—	the ability to absorb work without suffering permanent deformation.

SYNTHETIC POLYMER FIBERS

The physical and chemical properties of some synthetic polymer fibres, all satisfactorily resistant to sea water and fresh water, are given below:

Polyamides

Polyamides: The principal commercial polyamide fibres are:

polyamide 66 (Nylon) made from polyhexamethylene adipamide

polyamide 6 (Perlon) made from polycaprolactam

polyamide 11 (Rilsan) obtained by auto-condensation of ω-aminoudecanoic acid (based on castor oil)

They are produced both in the form of continuous filament yarn and staple fibre, in a wide range of deniers.

As already mentioned, the discovery of polyamide 66 is due to Dr. Wallace Hume Carothers, who since 1928 had been entrusted with research in the laboratories of E.I. Du Pont de Nemours Co.

It should be mentioned, incidentally, that many suppositions have been made and several explanations (often fantastic) given about the origin of the name "nylon". In a publication of E.I. Du Pont de Nemours Co. it is stated, however, that "the name given to the new product was 'nylon'-a term just as synthetic as nylon itself. It was chosen because it was short, catchy, pleasant to the ear and not easily mispronounced".

Polyamide 66 (nylon) and polyamide 6 (Perlon) yarns and fibres are available in normal and high-tenacity types. The characteristics of each quality and type are as follows:

Polyamide 66 (Nylon)

	Normal tenacity	*High tenacity*
(1) Density	1.14	
(2) Tenacity dry (g/den.)	4.5—6	6.5—8.5
breaking length: dry (Km.)	40—54	58—76
wet (%of dry)	85—90	90

(3) Tensile strength (Kg./mm²)	46—62	67—87
(4) Knot strength (% of tenacity)	85—90	
(5) Loop strength (% of tenacity)	84—87	
(6) Extension at Break % : dry	26—32	15—20
wet	30—37	18—28
(7) Elastic recovery	100%	100%
up to 8%		at 4%
(8) Toughness	1.08	0.77
(9) Stiffness (g.p.d.)	18	32
(10) Water absorption at 65% R.H.	4.5%	
at 100% R.H.	7—9%	
(11) Effect of heat melting point	250°C	
softening point	235°C	

Effect of age. Virtually none.

Effect of sunlight. Loss of strength on prolonged exposure.

Effect of chemicals

Concentrated hydrochloric acid, concentrated nitric acid, sulphuric acid cause rapid disintegration of the fibre. Phosphoric acid degrades the fibre, at elevated temperatures. Formic acid in concentrations above *80* per cent. will dissolve the fibre.

Acetic acid, oxalic acid, glycollic acid, lactic acid in high concentration have similar effect to that of formic acid. Other organic acids have little or no effect on the fibre.

The resistance to alkalis is outstanding.

Bleaching agents containing available chlorine will cause degradation on the fibre; hydrogen peroxide too causes degration though less rapid.

Solution of neutral inorganic salts have no effect on the fibre, at room temperature. At 100 deg. C concentrated solutions of magnesium perchlorate, lithium bromide or lithium iodide will dissolve fibre. The fibre is inert towards liquid ammonia, sulphur dioxide and cuprammonium solutions.

Petrol, mineral oils, benzene, xylene, ethers, esters, ketones, alkyl halides and mercaptans have no effect, even at elevated temperatures.

Alcohols (except benzyl alcohol), glycols, and aldehydes (except formaldehyde, glyoxal, and chloral under certain conditions) have only a little effect.

Benzyl alcohol, nitrobenzene, chlorhydrins, chloral hydrate, nitroalcohols, and adiponitrile are solvents for the fibre at high temperatures.

Phenols, particularly phenol itself (i.e. carbolic acid), meta-cresol, and cresylic acid are solvents for the fibre.

Resistance to bacteria and fungi

Untreated fibres are immune to attack from all micro-organisms, whether fungi or bacteria (i.e. air borne). Yarns and cords show extreme resistance to attack by marine algae.

Resistance to insects

There is no known instance of insects or moth larvae deriving sustenance from polyamide 66 fibres.

Resistance to abrasion

In the whole field of textile fibres, polyamide fibres have the highest resistance to abrasion.

Polyamide 66 fibres are also produced in the form of staple fibres; the characteristics of these are the same as for continuous filament, except for the following items:

Tenacity: dry (g/den.) . .	4.1—5
wet (% of dry) . .	85—90
Breaking length : (Km.) . .	37—45
Tensile strength: (Kg/mm^2) . .	42—51
Total extension: % dry . .	37—40
wet . .	42—46
Stiffness (g.p.d.) . .	11

Polyamide 6 (Perlon)

	Normal tenacity		*High tenacity*
(1) Density		1.14	
(2) Tenacity dry (g/den.)	4.1—5.8		6.5—8
breaking length: dry (Km.)	37—52		58—72
wet (% of dry)		85—90	
(3) Tensile strength (Kg./mm^2)	42—60		67—82
(4) Knot strength (% of tenacity)		85—90	
(5) Loop strength (% of tenacity)		85—90	
(6) Extension at Break : dry	24—30		16—24
wet	27—34		19—23
(7) Elastic recovery	100% up to 8%		100% at 4%

(8) Toughness	0.67	0.68
(9) Water absorption at 65% R.H.	4.8%	
at 100% R.H.	7-9%	
(10) Effect of heat melting point	217°C	
softening point	170-180°C	

Effect of age.

Virtually none.

Effect of sunlight

Loss of strength on prolonged exposure

Effect of chemicals.

Resistance to bacteria and fungi, Resistance to insects, Resistance to abrasion — Similar to polyamide 66 fibre.

Polyamide 6 fibres are available also in the form of staple, which have as in the case of polyamide 66, a lower tenacity and a higher extension than continuous filament.

Polyamide 11 (Rilsan)

Polyamide 11 (Rilsan) fibres are relatively new, as their production was started only a few years ago. For this reason, the data about the properties and characteristics of these fibres are not yet complete.

(1) Density . .	1.04
(2) Tenacity dry (g/den.) . .	4.7—5.5
breaking length: dry (Km.). .	42—50
wet (% of dry) . .	100%
(3) Tensile strength (Kg./mm^2) . .	44 - 51
(4) Extension at Break%: dry and wet .	25 - 40
(5) Elastic recovery . .	100% up to 8%
(6) Water absorption at 65% R.H. .	1.2%
at 100% R.H. . .	3%
(7) Effect of heat melting point .	189°C
softening point .	178°C

(8) *Effect of chemicals.* The resistance to chemicals (acids) alkalis, organic solvents, mineral salts and oxidising agents, is slightly superior to that for polyamide 66 and 6.

In respect to ageing, effect of sunlight, resistance to bacteria, fungi, insects and to abrasion there is no substantial difference between Rilsan, nylon and Perlon.

Yarns, cords and ropes made from polyamides (66 and 6) are already well established in the field of fishing gear, because of their special characteristics, namely high strength combined with extensibility, low absorption of moisture and water, quick drying, outstanding resistance to abrasion, bacteria and fungi and seawater and algae.

POLYESTER FIBRES

The principal polyester fibre is made from polyethylene terephthalate, which was first synthesised by Whinfield and Dickson in the laboratories of the Calico Printers Association, in Great Britain.

Mr. Whinfield designated the fibre "Terylene" for convenience in writing and was surprised that the name has survived.

We report below the data regarding polyester filament yarn ;

	Normal tenacity	*High tenacity*
(1) Density	1.38	
(2) Tenacity dry (g/den.)	4—5.5	6.0—7
breaking length (Km.)	36—50	54—63
wet (% of dry)	100%	
(3) Tensile strength (Kg./mm^2)	50—68	74—87
(4) Loop strength (% of tenacity)	90	80
(5) Knot strength (% of tenacity)	70	70
(6) Extension at Break dry and wet	27—17	15—7
(7) Elastic recovery	97% at 2% 80% at 8%	100% at 2% 90% at 8%
(8) Toughness	0.78	0.5
(9) Stiffness (g.p.d.)	23	51
(10) Water absorption at 65% R.H.	0.4%	
at 100% R.H.	0.5%	
(11) Effect of heat melting point	260°C	
softening point	230-240°C	

Effect of age. Virtually none.

Effect of sunlight. Only slight loss of strength on prolonged exposure.

Effect of chemicals. Resistance of a high order to chemical attack; outstanding resistance to mineral and organic acids (expecially hydrofluoric acid). Only fair resistance to alkalis but adequate for most purposes. Resistant to oxidising agents and to organic solvents (except at the boiling point). Most phenols dissolve the fibre.

Resistant to bacteria, fungi and insects.

Resistance to abrasion: ranks second to polyamide fibres only.

Polyester staple fibre is also commercially available; the characteristics are the same as those of continuous filament, except for tenacity and extension, which are respectively lower and higher than for filament yarn.

Vinyl fibres

The first vinyl fibre was described in the literature nearly fifty years ago. Polymer chemistry had first of all to prepare polymers possessing suitable properties for spinning and as a result of a great deal of research and experiments, fibres having satisfactory physical and chemical properties were obtained. From the standpoint of utilisation in the fishing industry, the following vinyl fibres are to be mentioned: polyvinyl chloride, vinylidene chloride (Saran), polyethylene (Courlene and Courlene × 3) and polyvinyl alcohol (Vinylon).

We summarise below the main properties of these fibres:

Polyvinyl chloride fibres

(1) *Density* . .	1.39
(2) *Tenacity dry (g/den.)* . .	2.7—3.7
breaking length (Km.) . .	24—33
wet (% of dry) . .	100%
(3) *Tensile strength* (*Kg./mm*^{2}) . .	33 - 46
(4) *Loop strength* (% of tenacity) .	70%
(5) *Knot strength* (% of tenacity) .	70%
(6) *Extension at Break dry and wet* .	13 - 30
(7) *Elastic recovery* . .	80 to 85% at 3%
(8) *Water absorption* . .	0.1 at 95% R.H.
(9) *Effect of heat*	softens at 110—120°C shrinkage starts at 60° to 70°C.

Effect of age. Virtually none.

Effect of sunlight. Substantially unaffected after prolonged exposure.

Effect of chemicals. Resistant to acids, including aqua regia and to concentrated (caustic) alkalis. Dissolves or swells in some aromatics, chlorinated hydrocarbons, ketones, esters. Generally good resistance to other chemicals.

Resistant to bacteria, fungi, insects and algae and sea water.

Vinylidene chloride fibres

(1) *Density* . .	1.70
(2) *Tenacity dry* (g/den.) . .	1.5—2.6
breaking length: dry (Km.) . .	13.5 - 23
wet (% of dry) . .	100%
(3) *Tensile strength* (Kg./mm^2) . .	23—40
(4) *Loop strength* (% of tenacity) .	70—80
(5) *Knot strength* (% of tenacity) .	70—80
(6) Extension at Break (%) dry and wet . .	18—33
(7) *Elastic recovery* . .	98 to 100% at 5%
(8) *Water absorption* . .	0'1% at 95% R.H.
(9) *Effect of heat melting point*. .	150—160°C

Effect of age. Virtually none.

Effect of sunlight. Slight discolouration after prolonged exposure.

Effect of chemicals. Unaffected by most acids, including aqua regia. Affected by some alkalis, including concentrated ammonium hydroxide, and sodium hydroxide. Substantially inert to organic solvents. Generally good resistance to other chemicals.

Resistant to bacteria, fungi, insects, moth larvae, algae and sea water.

Polyethylene fibres

(The data refer to polyethylene yarns produced by Courtaulds Ltd. under the registered trade-names of "Courlene" and "Courlene × 3".)

	Courlene	*Courlene × 3*
(1) *Density*	0.93	0.96
	(i.e. the lowest density of all the synthetic polymer fibres-giving good flotation properties).	
(2) *Tenacity dry* (g/den.)	1—1.5	4.0—6.0
breaking length: dry (Km.)	9—13.5	36—54
wet (% of dry)	100%	
(3) *Tensile strength* (Kg./mm^2)	8.5—12.5	34.5—52
(4) Extension at Break (%) dry and wet	25—50	20—40
(5) *Elastic recovery*	90 to 95% at 5%	
(6) *Water absorption*	practically none	

(7) *Effect of heat* — Courlene softens above 90°C; melts between 110° and 120°C. Courlene × 3 (High tenacity) softens at 120°C; melts at 135°C.

Effect of age. Virtually none.

Effect of sunlight. Loss of strength after prolonged exposure.

Effect of chemicals. Remarkable resistance to attack by alkali, acids, solvents, organic salts, unequalled in this respect by other fibres.

Resistant to bacteria, fungi, insects, moth larvae, algae and sea water.

The abrasion and electrical resistance of Courlene × 3 is very good and both Courlene and Courlene × 3 are flexible at very low temperatures such as 70 deg. C.

Polyvinyl Alcohol Fibres

Vinylon is the generic term for polyvinyl alcohol fibres: these are produced in staple and iin filament.

	Staple		Filament	
	Normal Tenacity	High Tenacity	Normal Tenacity	High Tenacity
(1) *Density*	1.26—1.30		1.26—1.30	
(2) *Tenacity dry* (g/den)	4.2—6.0	6.7—8.0	3.5—4.5	7.7—9.2
breaking length:				
dry (Km.)	38—54	60—72	32—41	69—83
wet (% of dry)	77—85%	80—85%	80—90%	80—90
(3) *Tensile strength* (Kg./mm)	50—70	78—94	41—53	90—110
(4) *Loop strength* (% of tenacity)	35—43%	35—40%	90—95%	58—65
(5) *Knot strength* (% of tenacity)	60—67%	64—70%	75—80%	39—52
(6) *Extension at break*				
(%) dry	17—26	13—16	14—19	9—20
Wet	19—30	14—17	14—22	12—22
(7) *Elastic recovery*	75—80% at 3 %	78—82 at 3 %	70—90% at 3 %	85—98 at 3
(8) *Water absorption*				

(at 65 % R.H.) 4.5—5'0% 4.5—5'0%3'5—4.0% 3'0—5'0
(at 95 % R.H.) 10—12.0

(9) *Effect of heat*
melting point 220—225°C
shrinkage starts at abt. 200°C

Effect of age. Virtually none.

Effect of sunlight. Loss of strength after prolonged exposure.

Effect of Chemicals. Concentrated sulphuric, hydrochloric, formic acids cause decompositions or swelling. Strong alkalis cause yellowing but do not affect strength. Good resistance to organic solvents. Soluble in hot pyridine, phenol, cresol. Resistant to oils.

Resistance to bacteria, fungi, insects: not affected.

Polyvinyl chloride, polyvinyl alcohol and polyvinylidene chloride fibres are being used in making set nets, trawl nets and seine nets. Polyethylene is the only textile fibre that has a specific gravity lower than that of water. Owing to this low density, the diameters of Courlene filaments are greater than those of other textile fibres of the same denier. In the case of 125 denier monofilaments, for instance, the diameters in 1/1000ths/inch are Courlene 5.4, nylon 4.9, acetate 4.5, viscose 4.2.

Courlene and Courlene × 3 are used in making ropes of all kinds for use on board ship. The tensile strength is sufficient to make it suitable for most applications, while improvements being made may increase the tensile strength from 4 to 6 g/den. to 8 g/den. The yarn is spundyed, which makes it easy to produce coloured ropes; it does not absorb water and therefore does not swell; ice does not form inside it, and snow, etc. can easily be brushed or shaken off. Its specific gravity is low and ropes made from it will float on water. The yarns, being inert, will not support fungal or weed growth.

The fibre is useful where tensile strength and buoyancy are important, as in trawl nets where the upper net will tend to rise of its own volition; it is claimed that consequently fewer floats are required. The fibre has been used both for the main-lines and snoods in longline fishing, and for weaving canvases which do not become hard and are relatively easy to handle in wet, cold conditions. Insoles made from Courlene provide thermal insulation in the fisherman's boots.

MAN-MADE FIBRES AND FINISHES IN PROTECTIVE CLOTHING FOR FISHERMEN

The fishermen are exposed to wet, cold climatic conditions and must, to work efficiently, receive adequate protection. In the past they wore oilskins, made of a heavy cotton, flax or similar fabric, coated with many layers of linseed oil. Today many of the basic fabrics are made from 100 per cent. viscose rayon staple or blends with cotton. The fabrics are produced with good tear and tensile strength and a surface smooth enough to allow even coatings, now generally of p.v.c. All types of garments, such as coats, sou'westers and leggings, are made from them.

One of nylon's chief contributions is as a lightweight base for waterproof clothing. Fabric of 1½ to 2½ ozs./sq. yd. can be successfully proofed with p.v.c. neoprene and polyurethane resins. These cloths have the advantages of lightness, flexibility and a smooth p.v.c. coating which is of special advantage in collar fittings where the fabric may be flat against the wearer's face.

Another innovation is the double textured rainproof jackets, coats and leggings. These are made from two similar fabrics laminated with rubber. The textile fabric provides resistance to abrasion and the rubber lamination provides waterproofing.

The traditional seamen's jersey can be made in a blend of 50 per cent. viscose staple, 50 per cent. wool or 15 per cent. nylon, 35 per cent. viscose rayon staple, 50 per cent. wool. The latter blend has good resistence to abrasion and can be made to resist shrinkage. The same yarns have also been used in heavyweight half-hose and seamen's stockings, comfortable in use and hard wearing.

Yarns made from 100 per cent. viscose rayon staple and viscose rayon/staple cotton blends can be knitted on interlock machines, and lend themselves to coating with vinyl compounds, rubber or p.v.c. Gloves are now made of this type of fabric and also rubber leggings and waders. The fabrics can be raised to give a fleecy lining.

SYNTHETIC TWINES

Fishing gear is dependent on several factors, such as the fishing techniques to be employed, the fish to be caught, the material for nets and ropes, and so on. With regard to the latter, there are requirements which largely influence characteristics and efficiency of the gear. They are:

the thickness or diameter of the netting twine;

the weight of the twine in air and in water;

sinking speed;

strength and extension up to the breaking point;

strength and extension knotted; friction resistance;

resistance to sunlight and seawater;

knot fastness;

resistance to factors such as shock, heat, chemicals and fatigue;

elasticity;

susceptibility to dye and dye fastness;

stiffness or handiness;

plying and fabricating capacity.

In evaluating these properties different methods and means would give different values. Therefore when selecting one kind of material out of a group each attribute must be evaluated with a uniform method which should approximate the actual fishing conditions as far as possible.

The efficiency of a fishing net is partly predictable from the quality of the twine, therefore the results of twine testing will be first described in regard to two of the factors (1) the nature of the fibre, and (2) the number of twists given to the yarn as well as the strand.

The Structure and Size of Net Twines

One group of synthetic fibres made of spun yarn just as with cotton or hemp includes Kuralon (Manryo) and Mulon yarns. As in cotton the strands made by twisting the yarn are then again plied to form twine. Most of the synthetic yarns used for fishing nets are made equivalent to English 20's of cotton yarn in size, and are usually of 2 or 3 plies. The size is indicated by the number of yarns used for a strand with the number of strands (e.g. No. 5 with 3 plies), or by the total number of yarns contained in a twine (e.g. 15 yarns in 3 plies). The latter is more often used.

In another group the continuous filaments are plied into strands or threads. Amilan (nylon), Saran, Krehalon, Teviron, Envilon, Kyokurin and Kuralon No. 5 (Manryo No. 5) generally belong to this group. In this case, however, the strand may be formed by a single filament (monofilament) or by yarns consisting of several filaments (multifilament). Amilan, Teviron, and Kyokurin, and small diameter twines of Saran, Krehalon have strands made of multifilament yarns. Kuralon No. 5 and other large diameter twines, such as those of Saran and Krehalon, have strands made of monofilament. Table elsewhere in the chapter illustrates how these yarns, strands and twines are constructed.

Table 5.1 : Construction of Some Kinds of Netting Twines.

Material	*Chemical name*	*Construction of yarn and/or strand and twine*
Amilan	Polyamide	$(250D/15F_1 = Y) \times n_1 = S,\ S \times n_2 = T$
"	,,	$(210D/15F_1 = Y) \times n_1 = S,\ S \times n_2 = T$
"	,,	$(11OD/30F_1 = Y) \times n_1 = S,\ S \times n_2 = T$
,,	,,	$(\ 60D/20F_1 = Y) \times n_1 = S,\ S \times n_2 = T$
Kuralon No. 5 (Manryo No. 5)	Polyvinyl alcohol	$(500D = F_2) \times n_2 = S,\ S \times n_2 = T$
Saran	Polyvinylidene	$720D/6F_1 \times n_2 = S,\ S \times n_2 = T$
,,	Polyvinyl chloride	$1080D/9F\ xn_4 = S,\ S \times n_2 = T$
,,	"	$(360D = F2) \times n_3 = S,\ S \times n_2 = T$
"	"	$(1000D = F_2) \times n_3 = S,\ S \times n_2 = T$
Krehalon	"	$720D/4F_1 \times n_4 = S,\ S \times n_2 = T$
"	,,	$1080D/6F_1 \times n_4 = S,\ S \times n_2 = T$
,,	,,	$(360D = F_2) \times n_3 = S,\ S \times n_2 = T$
,,	"	$(1000D = F_2) \times n_3 = S,\ S \times n_2 = T$
Teviron	,,	$(300D/30F_1 = Y)xn_1 = S,\ S \times n_2 = T$
Envilon	"	$(450D = F_2) \times n_3 = S,\ S \times n_2 = T$
Cotton	not applicable	$(20\text{'s} = Y) \times n_1 = S,\ S \times n_2 = T$
Kuralon	Polyvinyl alcohol	$(20\text{'s} = Y) \times n_1 = S,\ S \times n_2 = T$

D : denier.

F_1 : multifilaments. F_2: monofilaments.

VY : yarn.

n_1 : number of yarns contained in a strand. Each yarn of this type comprises multifilaments.

S : strand.

n_2 : number of strands (usually 3 but sometimes 2 or 4) constructing a twine.

T : twine.

n_3 : number of monofilament contained in a strand. The thickness of a strand varies depending on n_3.

n_4 : number of multifilaments, usually one or two bundles.

Table 5.2 : Number of Twist per 30 cm. of Twines.

Number of yarn		4	6	9	12	15	18	21	27	36	45	54	60	75	84
Amilan	A	132	123	111	—	72	69	—	64	57	51	—	43	35	—
	B	198	189	167	—	112	124	—	131	117	104	—	78	70	—
Kuralon (Manryo)	A	—	96	81	—	59	50	—	44	38	36	—	33	29	—
	B	—	241	189	—	115	92	—	78	62	57	—	52	49	—
Saran	A	—	76	68	64	59	57	—	53	45	33	—	25(66)	23	—
	B	—	135	123	110	98	94	—	83	75	60	—	45	41	—
Krehalon	A	—	68	63	—	56	51	—	48	46	44	—	39	36	33
	B	—	104	100	—	93	90	—	87	85	81	—	75	66	60
Teviron	A	138	90	96	84	—	60	51 (24)	—	48	—	—	42	—	—
	B	246	202	158	148	—	123	114	—	102	—	—	84	—	—
Cotton	A	—	102	95	—	75	66	—	60	51	42	—	38	34	—
	B	—	405	336	—	264	212	—	171	135	90	—	60	53	—
Kyokurin	A	81	68	57	51	—	—	42	—	31	—	—	—	—	—
	B	126	114	95	81	—	—	65	—	52	—	—	—	—	—

A: twist for twine. B: twist for strand.

Table 5.3 : Diameter of Wet Netting Twines

(Unit: mm.)

Ply/No. of yarn of C20-equivalent	3/6	3/12	3/18	3/24	3/30	3/36	3/42	3/48	3/54	3/60	3/66	3/72
Amilan	0.68	0.92	1.15	1.37	1.56	1.76	1.94	2.11	2.26	2.41	2.54	2.78
Kuralon (Manryo)	0.83	1.10	1.36	1.60	1.83	2.08	2.23	2.42	2.59	2.75	2.90	3.04
Saran	0.66	0.86	1.03	1.22	1.37	1.52	1.66	1.77	1.88	2.00	2.09	2.18
Krehalon	0.66	0.88	1.05	1.24	1.40	1.55	1.69	1.83	1.95	2.06	2.16	2.25
Teviron	0.70	0.96	1.21	1.43	1.64	1.84	2.03	2.21	2.37	2.03	2.66	2.79
Kyokurin	0.67	0.90	1.08	1.28	1.46	1.63	1.78	1.93	2.06	2.18	2.28	2.40
Cotton	0.78	1.06	1.32	1.56	1.77	1.98	2.18	2.36	2.53	2.70	2.83	2.97

In the Table, 250/15F × n_1 of (nylon) Amilan, for instance, means that a strand is made by twisting a number of yarns of 250 total denier, each of which is constructed in 15 filaments. The thickness of a filament in that yarn is 16.6 denier approximately. As a (nylon) Amilan yarn of 210 denier corresponds in thickness to cotton-20-counts (20's), the number of yarn of any other denier in this twine, equivalent to cotton 20's (hereinafter called C20-equivalent), is converted by dividing the total denier of the twine by 210 denier. The twines consist of 3 and sometimes 4 strands, e.g. for some salmon gillnets. Similar calculations can be made for other synthetic fibres produced in Japan.

In all these synthetic products, the numbers of yarn for C20-equivalent twine are obtained by dividing the total denier of the twines with the thickness of the C20-equivalent. It should be remembered, however, that the thickness of C20-equivalent determined for these twines does not necessarily represent, in a strictly physical sense, the true thickness of cotton 20's.

Thickness of Netting Twines

No matter how similar may be the indicated thickness of different kinds of twines, the indication does not warrant that they are all alike in diameter, circumference or weight per unit length. Naturally, the real thickness of a twine, viz. the whole area of the horizontal section of all the filaments of the twine, varies according to the kind of twine, making it impossible to assess the characteristics of twines on a comparable basis. As, however, no other standard than the conversion into C20-equivalent has become available for the purpose, characteristics of both synthetic and natural twines will be compared with the help of this conventional method.

The thickness of net twines has been measured in a wet state. The twines were immersed in water for 24 hours and hung until the water stopped dripping from them. The twines used are generally considered to be of medium twist. Roughly speaking, the twist below that range has about 3 to 5 per cent. greater thickness.

Weight of Net Twines in the Air and in Water

Weight in the air

In the air, the weight of fishing nets has a close relation, particularly when they are wet, with the loading capacity of small boats as well as with the working conditions for fishermen. Although weight is largely governed by the thickness of the twine, mesh size and type of knot, a table of weights prepared by actually weighing nets in use is useful reference. In preparing such a table, the weight per metre of dry twine and the rate of weight increase for wet twine may be taken as the basis for estimating the wet weight of various types of nets.

The wet weight of a net in air can be obtained by multiplying the known dry weight of the net with the ratio W_w/W_d.

The situation is somewhat different with nets processed with resins. The amount of water that sticks to the nets differs according to the kind of fibre and thickness of the resin coating as well as the water absorption or repellency of the resin, so that it is difficult to estimate the wet weight of a net. However, an experiment has proved that resin-processed nets gain in weight over those not processed by 30 to 80 per cent. which includes the weight of the resin adhering to the material.

Coal tar treated nets in dry state are 70 to 150 per cent. heavier than non-treated nets. If the coal tar is diluted with creosote or gasoline, the weight gain is approximately from 50 to 100 per cent. Use of a centrifugal machine when dyeing can reduce the amount of tar on the net and bring the weight gain down to 60 to 80 per cent.

Underwater Weight

Increase in underwater weight of twines implies a quicker sinking capacity and in some cases a better shape of the net. This is an important quality for fixed nets, purse seines and stickheld dip nets.

Table V presents the underwater weight of various kinds of netting twines on the basis of the unit length of one metre measured in air. In the case of webbing, the same relation may be expected between synthetic twines and cotton (Wo/Woc) and between tarred synthetic and tarred twines (Wt/Wtc.)

Sinking Speed

Sinking velocity, an important factor especially for such types of nets as purse seines, has a close connection with the thickness and specific gravity of the raw material, is affected by the degree of twist given to the yarn, strand and twine, and by the smoothness of the twine surface. In a sinking speed test, a piece of twine, 2 cm. long with each end glued, was immersed in water for 24 hours. The air bubbles were then removed from the surface of the twine. The terminal velocity shown by the test piece in sinking straight down in a salt solution (specific gravity 1.020) was then determined. The solid line indicates the sinking speed of non-treated twine, and the dotted line the sinking speed of the tarred twine. It is obvious that the velocity differs even between the same kind and quality of twines depending upon the thickness of the twines as well as on the nature of the raw material, the number of twists, twisting technique, dyeing and/or heating.

Table given in the chapter elsewhere shows comparative sinking speeds for various twines. The coal tar treatment, which was first used for cotton twines to increase the sinking speed by preventing water absorption, has been found still more effective when applied to some synthetics. For this reason, coal tar and similar products are widely used in Japan.

About 80 per cent. tar adhesion results in the quickest sinking speed while, according to our experiments, twines over or undercharged with tar sank slowly.

Tensile Strength of Net Twine, and the Knot Strength

Differences in Strength

Tensile strength of twine depends upon that of the raw material used, whether the twine consists of short or long fibres, and on the number of twists given to the strand and twine. The balance between the primary twist for making the strands from the yarns and the secondary twist to make the twine from strands is also very important because, while an increased twist augments the tensile strength up to an optimal point, an excessive increase in the twist produces the opposite effect. This also applies to the strength of a fishing net itself.

Table 5.4 : Comparison of Sinking Speed between Non-treated Twines and Tarred Twines.

Material	*Items*	*3 plies, 15 yarns*	*3 plies, 30 yarns*	*3 plies, 60 yarns*
	Vo/Vc	0.57	0.54	0.51
Amilan	Vt/Vo	1.68	1.65	1.75
	Vt/Vtc	0.55	0.60	0.66
	Vo/Vc	0.74	0.73	0.66
Kuralon	Vt/Vo	1.88	1.55	1.53
(Manryo)	Vt/Vtc	0.75	0.76	0.77
	Vo/Vc	1.77	1.65	1.50
Saran	Vt/Vo	1.09	1.08	1.07
	Vt/Vtc	1.20	1.18	1.17
	Vo/Vc	1.65	1.56	1.43
Krehalon	Vt/Vo	1.14	1.10	1.10
	Vt/Vtc	1.17	1.15	1.14
	Vo/Vc	1.43	1.21	1.17
Teviron	Vt/Vo	1.20	1.14	1.10
	Vt/Vtc	0.93	0.92	0.92
	Vo/Vc	1.37	1.27	1.21
Kyokurin	Vt/Vo	—	—	—
	Vt/Vtc	—	—	—
	Vo/Vc	1.00	1.00	1.00
Cotton	Vt/Vo	1.63	1.50	1.38
	Vt/Vtc	1.00	1.00	1.00

The number of yarn is equivalent in the thickness to cotton 20's.

Vo : sinking speed (cm./sec.) of non-treated twine.

Vc : ,, ,, ,, non-treated cotton twine.

Vt : ,, ,, ,, tarred twine.

Vtc : ,, ,, ,, tarred cotton twine.

Among twines with various twists used for fishing in Japan there is a difference of about 15 to 20 per cent in tensile strength, that is, an overtwisted and weaker twine may be found side by side with a correctly twisted stronger one, as shown in fig. 2. Kondo and Koizumi pointed out the same in regard to both cotton and synthetic twines. Even with identical twines, and whether the balance is good or not, there are differences of 10 to 15 per cent. in the tension resistance

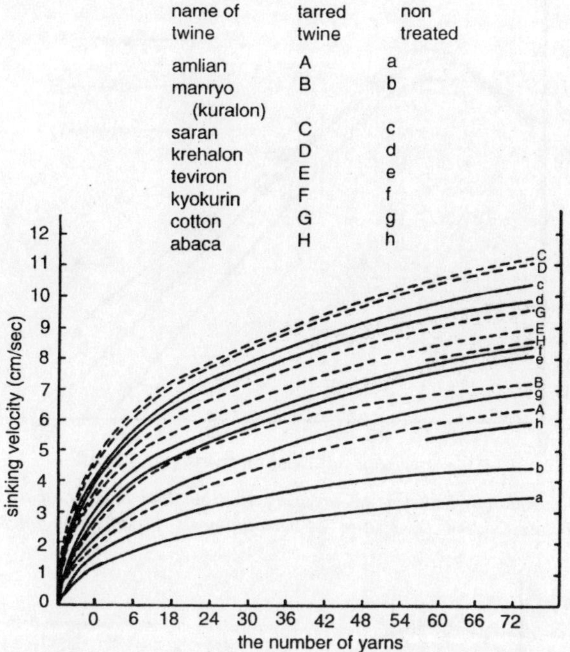

Figure 5.1 : Relation between the sinking velocity and thickness of netting twines of various kinds. Dotted lines show the sinking velocity of tarred twines, and solid lines non-treated twines.

because of the different numbers of twists. That is why careful attention must be paid to the amount of the twist in the twine when buying fishing nets and ropes.

Table 5.5 : Coefficients a_1" and a_2 for Breaking Strength of Wet or Dry Twines

	Wet a_1	*Dry* a_2
Amilan	0.89	0.94
Kuralon (Manryo)	0.78	0.96
Saran	0.70	0.68
Krehalon	0.60	0.56
Teviron	0.72	0.68
Kyokurin	0.76	0.81
Cotton	0.59	0.47
Envilon	0.76	0.74

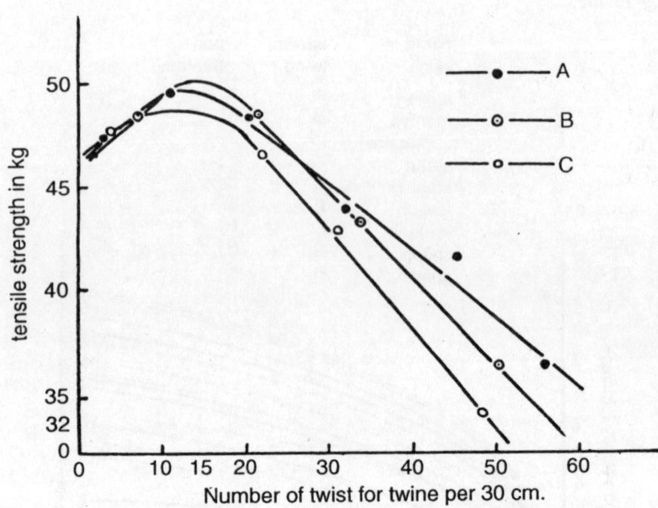

Figure 5.2-a. Relation between the breaking strength and number of twists given to twines, A, B and C, 1000 D, 3 plies, 24 monofilaments, produced by different makers.

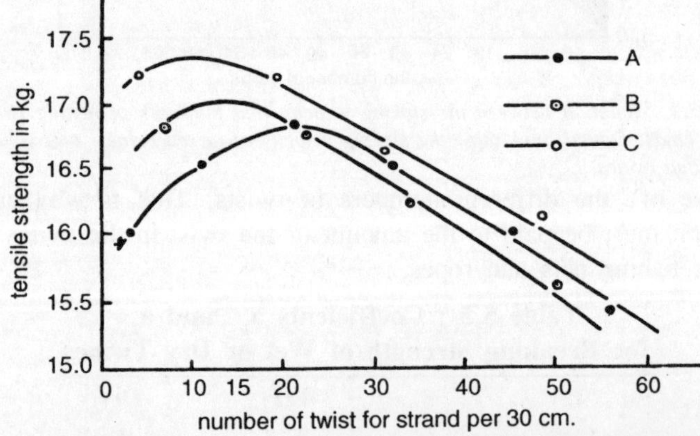

Figure 5.2-b. Relation between the breaking strength and the number of twists given to a strand of the twines.

In regard to the difference in the wet and dry strengths of netting twines, an experiment proved that natural fibre twine is about 10 to 20 per cent. stronger wet than dry, but the contrary is true with Amilan and Kuralon. With Teviron, Envilon, Saran and Krehalon, the wet twines are slightly (3.5 per cent.) stronger than dry ones. Similar results applied to the tensile strength of knots of fishing nets.

Temperature is another influencing factor, the tensile strength of nets decreasing by 10 to 20 per cent. in temperatures between 30 deg. C. and 0 deg. C., and the strength of twines by about 5 to 10 per cent. according to the kind of twine. The effect of temperature on the tensile strength of synthetic twines and knots is greater than on natural twines.

There is a disparity in the breaking strength between twines of less than 50 cm.; the longer the stronger they are, and *vice versa,* whereas between twines of a same material and thinner than 60 yarns of C20-equivalent, there is little difference in the breaking strength" An approximate breaking strain for each twine specified obtained by multiplying a_1 or a_2 with number of C20-equivalent yarn.

Strength at Knot

A series of tests have been conducted at temperature 18.0 deg. C. + 1.5 deg. C. for tensile strength at knot with 30 cm. long wet pieces of twine. In one test, two pairs of the legs *(AB* and CD) of an English knot were pulled asunder.

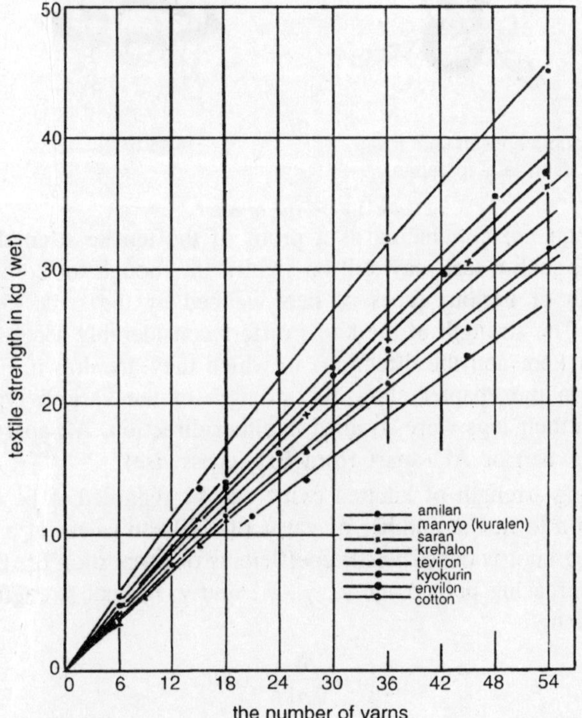

Figure 5.3 : Wet tensile strength of twines of various kinds, each with the number of yarns equivalent in thickness to cotton 20's.

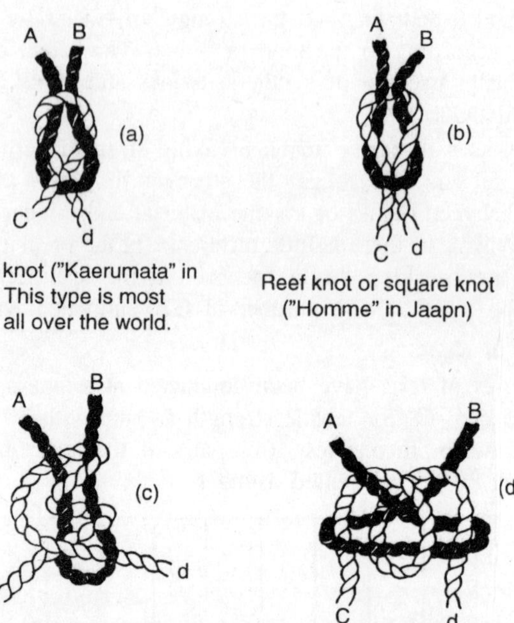

Figure 5.4 : Type of Knot.

Although not convincing as a proof of the tensile strength of a fishing net, such a test may still be significant enough for comparing the strength of various types of nets webbed by the same knotting technique. The strength of the knots differs considerably according to the type of knot and the directions in which they are drawn. Figures elsewhere in the chapter show the strength of the various types of knots when their legs were strained in either direction, AB apart from CD (lengthwise) or AC apart from BD (crosswise).

Breaking strength of knotted twine can be regarded to be nearly in proportion to n, the number of yarns of C20-equivalent of a twine in which the knot is constructed; coefficients of proportion β_1, β_2, β_3, and β_4. Decreasing percentage γ_1, γ_2, γ_3, and γ_4 in knot strengths are represented by

$$\gamma i = \left(1 - \frac{\beta i}{2\alpha i}\right) \times 100$$

where α_1 is derived from Table, and β_i represents any one of β_1 to β_4.

In figure elsewhere in the chapter, knots c and d are the types required for repairing broken nets and making gillnets, and nets in which the knots should never work loose. Knot c is called the double English knot, and Knot d the Lock knot. However, the tests showed no noticeable differences in strength between the English knot and those stated above. A study is now under way in connection with the strength of a knotless net.

Extension of twine and knots

The breaking extension of various kinds of twine and their knots is affected by the heating and stretching procedures during twisting. Most of the twines tested were not elongated, except Amilan which was extended 10 per cent.

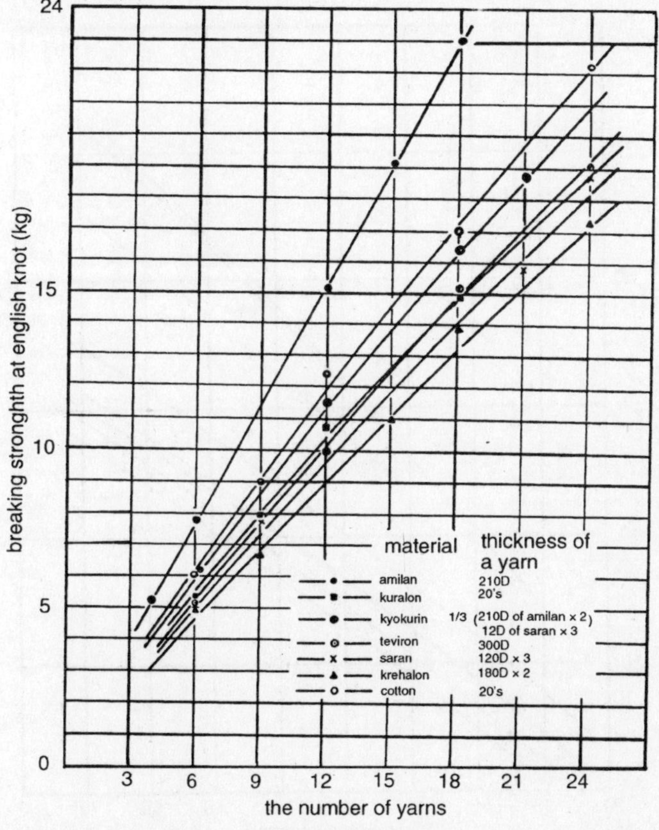

Figure 5.5 : Relation between wet strength of an English knot and the number of yarns of C20-equivalent less than 24. In the test, two pairs of the legs of the knot, AB and CD, as in fig. 4 were pulled asunder.

Abrasive Resistance

The relation of twist to frictional pressure and strength

Generally speaking, there are two forms of friction which wear down fishing nets; one between net twines at the seam and the knots, and the other against comparatively hard substances, such as the sea bottom, hull of the boat, and net or line haulers. The friction between the net twines and hard objects was examined by use of a device shown in figure elsewhere in the chapter. In order to keep the temperature constant, water was made to drip on the sample of twine under test. The twine stretched across an oil-stone C fixed on a block. The block was kept moving back and forth between E and F at 80 oscillations per minute, and the number of rubs counted.

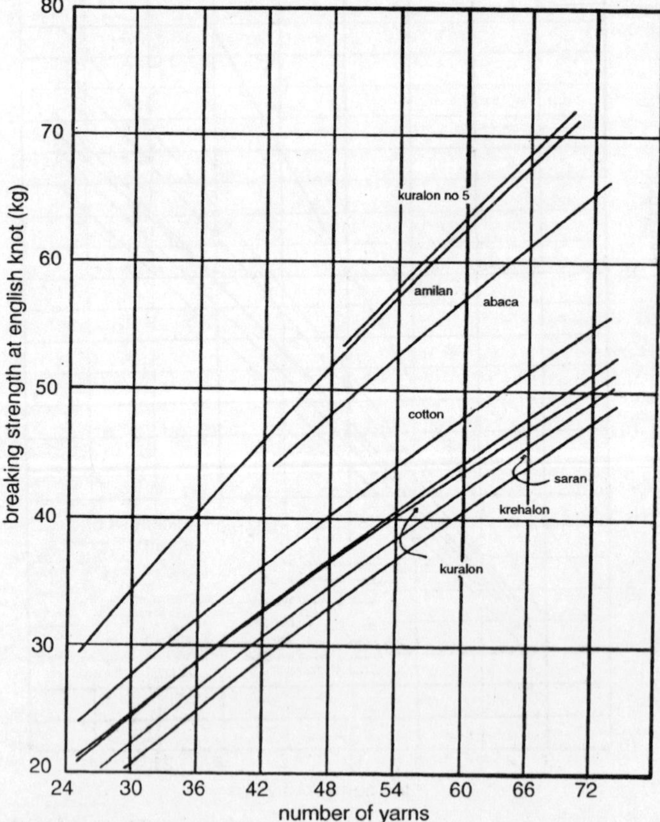

Figure 5.6 : Relation between wet strength of an English knot and the number of yarns of C20-equivalent more than 24.

The results of these tests with Amilan twines of various twists. The number of twists in the twines. Of the twists, those of the strand affect the abrasive resistance most, as they seem to play the greatest role in increasing the rigidity of the twine. Judging from the results, it seems that hard twisted twine should be used in that part of the net which comes into friction with hard objects. The utmost care should, however, be taken to ensure that the number of twists in the strand should not be too much to the detriment of tensile strength for the sake of friction resistance.

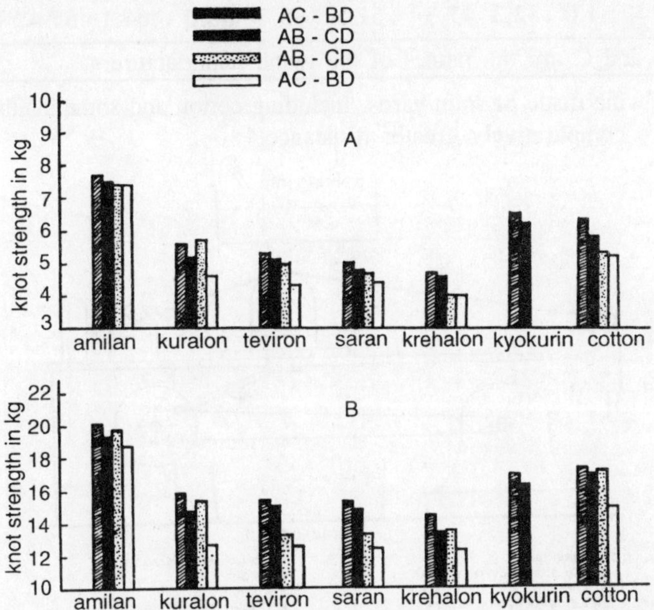

Figure 5.7 : A-B. Relation between the knot strength and types of knot of various twines. In test A, twines consisting of 6 yarns, and in B 12 yarns, of C20-equivalent were used.

The size of net twines is not always in conformity even when they have the same number of yarns, consequently, it is not feasible to compare the abrasive resistance of one type of fishing net with another that has different characteristics. However, for general purposes, comparison between various net twines made of C20-equivalent yarns. The method and the device employed for the test were identical with those described in the preceding paragraph. For this test the twines were of medium twist, having the same twist range.

Table 5.6 : Number of Twist of Amilan Netting Twine used for the Wear Test
(Indicated on the basis of length 30 cm.)

Kind of products	*Low Twist (L)*			*Medium Twist (M)*			*Hard Twist (H)*		
	For twine	*For strand*	*For yarn*	*For twine*	*For strand*	*For yarn*	*For twine*	*For strand*	*For yarn*
A	56.4	39.2	56.3	67.7	65.7	89.7	77.1	68.0	87.0
B	70.0	46.3	66-.0	65.8	50.4	65.6	78.5	50.8	69.9
C	54.0	32.3	47.3	51.5	35.7	53.0	64.1	55.4	71.7

A, B and C are the names of the twine manufacturers.

Twine made of spun yarns, including cotton and some synthetics, have a comparatively greater resistance.

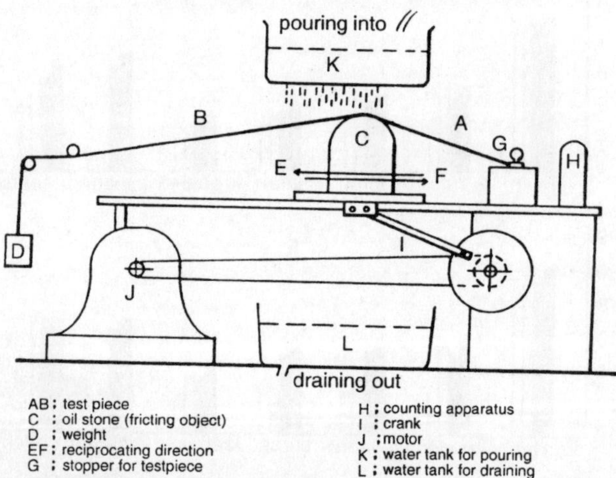

Figure 5.8 : A device used for testing the wear resistance of twines. The velocity of the reciprocative abrasion is 80 per minute both back and forth.

Numerals in parentheses show the percentage of contraction the length of twine heated before test.

The solid lines show the abrasivesive resistance of tarred twine, and one may see that tar treatment enhances the abrasive resistance of net twine. As a result of tarring long fibre twines resist friction better than others which have the same number of C20-equivalent yarn.

With resin treatment the increase in abrasive resistance depends upon the thickness of the coating. A 3 to 5 percent. adhesion may double or quadruple the strength.

Table 5.7 : Contraction in Length of Netting Twine by Heating in Water (Unit %)

Heating temperature	40°C	60°C	70°C	80°C	90°C	100°C
Amilan	0	0	0	-1	-1	-1
Kuralon (Manryo)	5	7	8	11	12	15
Saran	0	6	1	3	5	6
Krehalon	0	1	3	5	7	11
Teviron	5	15	25	30	45	56
	(3)	(8)	(15)	(20)	(30)	(35)
Cotton	6	7	8	8	8	9

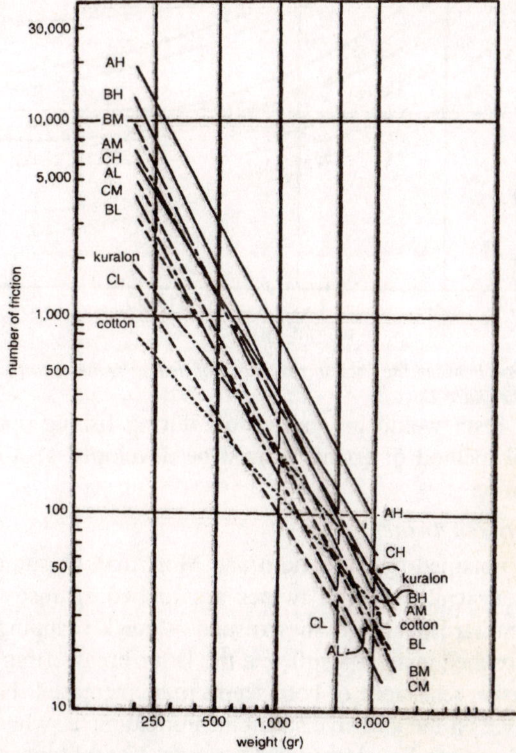

Figure 5.9 : Friction resistance of Amilan netting twines, A, B and C, each produced by different makers with different twists. The test temperature was kept at 18°C+ 1°C. The notation AH, for instance, indicates the twine made by A maker with hard twist.

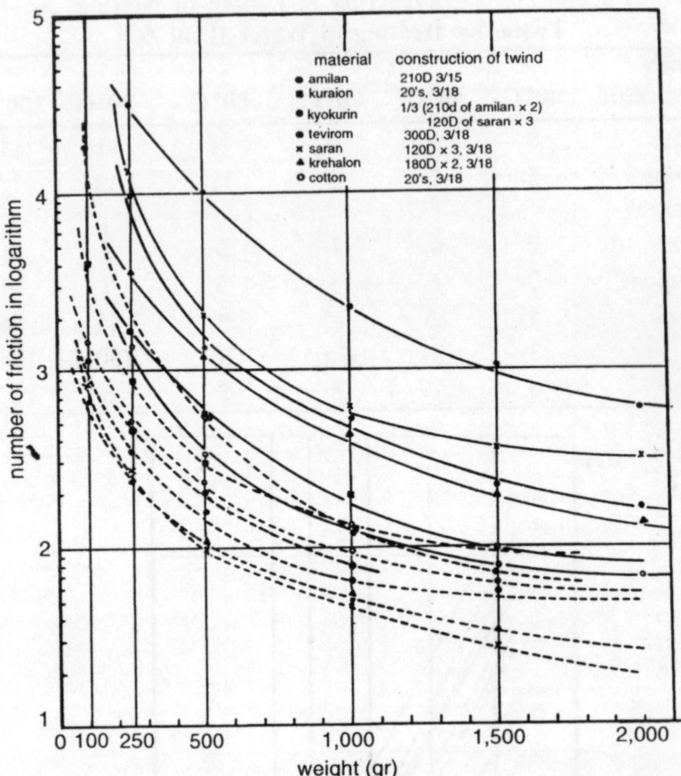

Figure 5.10 : Relation between the friction resistance of netting twines and the fricative load increased at the place D.

But, as the resin would be washed off during fishing operations a more practical method of fixing it must be developed as a counter measure to dilution.

Interfriction between twines

The results obtained by Miyamoto and Mori may be summarised as follows: When synthetic fibre twines are rubbed against another, abrasion between hard quality twines results in quick snapping; when a hard twine is rubbed against a soft one the latter breaks first; if both are soft the abrasive resistance of both seems to be increased. Hardness has a big influence on the abrasive strength. For example, when cotton twine is rubbed against Kuralon twine it needs 260 rubbings to snap the latter, yet Kuralon can withstand up to 530 rubbings against the cotton. When Kuralon is rubbed against Saran it snaps after only 60 rubbings and the same applies when cotton is rubbed against Teviron and Kyokurin. Between twines of the same kind, the number of rubs needed to break them is mostly as many as, but sometimes a little more than, the number required to break cotton.

Other Characteristics

Knot fastness

With natural fibre twines, knot slippage hardly constitutes a problem as compared to synthetic fibre twines. In Japan, the synthetic twines which lend themselves best to knotting are Kuralon, Teviron, Envilon, Krehalon, Saran, Kuralon No. 5, Kyokurin and Amilan in the order named. For fixing the knots, various types of knots used by Japanese fishermen. Most synthetic fibre nets except Kuralon and Mulon need heat treatment to fix the knots.

Heating with hot water, gas, or air, is widely applied to synthetic fibre nets for knot setting, as the treatment can reduce the amount of complicated knotting, which means that the finished product weighs less. Amilan nets may be heated at the temperature 100 deg. C. or higher, while the others, except Kuralon, can be treated at lower than 72 deg. C.

Resistance to sunlight and seawater

One group of tarred net twines immersed in the sea and another exposed to the sunlight, both for one year. From the table it appears best to use coal tar with low isolated acid content or to neutralise the tar with alkali before use.

Characteristics of ropes

For comparison, the breaking strain of manila and synthetic ropes is reproduced in Table elsewhere in the chapter.

NYLON FISHING NETS

Specification For Ideal FIsh Net Fibre

(1) The basic cost should be low.

(2) Processing into net form should be cheap, easy and efficient. For example, shrinkage on setting should be low.

(3) In both wet and dry states it should have high strength, by which is meant:

(a) high tensile strength;

(b) good ability to withstand repeated shocks;

(c) good flex strength or fatigue resistance;

(d) high knotting efficiency and

(e) good resistance to abrasion which leads to fine, long-lasting nets, capable of holding large catches.

(4) It should maintain its strength in use.

(5) It should have good dimensional stability and should not distort in size or shape during use.

(6) It should have low moisture absorption so that the increase in weight is small when a net becomes wet and handling is consequently easier.

(7) The fibre should have a low specific gravity, since this allows a greater length of netting for a given weight of yarn, and may permit lighter fittings and savings in power and manhandling. On the other hand, a low specific gravity may not be desirable where quick sinking of the net is required.

(8) It should be resistant to damage and attack by chemicals, oils, moulds, bacteria, insects and vermin in order that treatments and routine maintenance can be kept to a minimum.

(9) The performance of the fibre should remain constant at extremes of temperature.

(10) The net should hold the fish firmly when caught, yet not damage them.

(11) Some types of fishing nets may demand other requirements, such as translucency for gillnets or a, lower initial elastic modulus as is required for salmon gillnets.

Table 5.8 : Specific Gravity of Fibres.

Nylon 66 and nylon 6	1.14
Polyvinyl alcohol	1.30
Polyester fibre 1.38 Hemp	1.48
Flax	1.50
Cotton	1.52
Polyvinylidene chloride	1.72

The initial modulus (calculated by measuring the load to produce a 5 per cent. extension under conditions of 70 deg. ± 4 deg. F. and 67 ± 2 per cent. R.H.) for nylon yarns lies in the range 20 to 40 grams/ denier, with nylon 66 high tenacity yarns at the top end of the bracket, while, for comparison, the initial modulus of high tenacity polyester fibre lies in the range 90 to 100 grams/denier.

The basis of the specification can be extended to methods of fishing which do not employ nets. It could include the principles

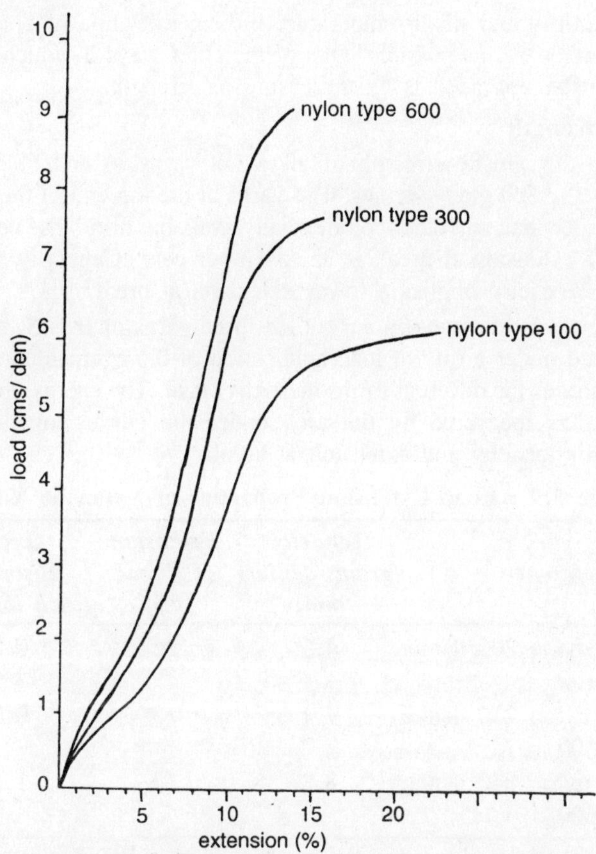

Figure 5.11 : Load/Extension Curves of BNS nylon 66 yarns.

involved in the choice of fibre for long-lines or whaling foregoers, etc. There is now a wide range of fibres available to satisfy the diverse needs of the fishing industry. None meets, nor could meet, the specification completely.

Even when one fibre is isolated and examined alone, it can be seen that there is ample variety of choice. Thus, nylon can be divided into classes according to the type of polymer used, and whether the yarn is assembled from continuous filament or spun fibre; within each of these classes there is a range of yarns with varying properties.

The task facing a supplier of continuous filament nylon for the production of fishing nets is to provide a yarn which measures well in comparison with the specification. Some of the conditions are met automatically (e.g. low specific gravity, rot resistance, high strength, good stability and small temperature influence), while others have to be met a. .early as possible. One of the chief ways in which a nylon yarn can be upgraded is by increasing its strength.

Yarn Strength

The dry tensile strength of all nylon yarns-66 and 6lies in the range 4.5 to 9.0 grams/denier. The yarns at the top end of the bracket are stronger than any other commercially available fibre. The equivalent range of extension at break is 25 to 12 per cent. Generally speaking a higher tenacity implies a lower extension at break.

Typical load-extension curves for three yarns of B.N.S. nylon 66 (measured under a rate of load application of 0.5 grams/denier/sec.), and illustrates the different forms these may take. The energy absorption of each, as measured by the area under the curve, together with values for tenacity and extension at break.

Table 5.9 : Load/Extension Properties of Nylon 66 Yarns.

Yarn Reference	*Tenacity (grams/denier) min.*	*Extension at (break (%)*	*Energy Absorption (inch lbs./inch)*
B.N.S. nylon 205 denier type 100	4.5	22	0.27
B.N.S. nylon 210 denier type 300	7.4	15	0.31
B.N.S. nylon 840 denier type 600	8.8	13	1.11

It will be seen that the energy absorption of 210 denier type 300 on a weight for weight basis is higher than the other two yarns. This very high energy absorption has been a factor in its ready acceptance for fishing nets.

Twine Strength

As in the case of all fibres, nylon twines are made with twist in order to bind the yarns together. The degree of twist and the construction control the feel of the twine. Increasing the degree of twist in a nylon yarn lowers the tensile strength. The ratio of twine strength to aggregate yarn strength is known as the doubling efficiency. Most nylon twines are made with a doubling efficiency of 95 per cent. or more. This can be affected, not only by twist, but also by conditions of setting and heat stretching.

Twine strength is commonly expressed in lbs. or other units of weight. Another method is to refer to specific strength (or breaking length), this being defined as the greatest length of twine which can be supported by a single piece of that twine without breaking it. The supporting piece may be dry, wet or knotted.

Table 5.10 : Average Specific Strength of Fibres.

	Twine		*Mesh*			
			Single Knot		*Double Knot*	
	Dry (yd.)	*Wet (yd.)*	*Dry (yd.)*	*Wet (yd.)*	*Dry (yd.)*	*Wet (yd).*
Nylon 66	63700	56000	30800	28100	34300	29900
Nylon 6	39100	35200	24400	22000	26600	23100
Polyester fibre	48600	50200	20000	20700	24100	23500
Linen	44000	58800			19900	3160

Specific strength (or breaking length) (in yards)= Twine strength (lbs) x twine weight (yds./lb.)

This unit is of value in comparing different fibres. Results on nets tested in Canada appear in Table elsewhere in the chapter, which lists the specific strength of dry arid wet twine and knots. Nylon 66 will be similar to type 300.

Knot Strength

When a knot is tied in a twine, it constitutes a place of weakness, and reduces the effective strength of the twine. The term "knotting efficiency" can be used to describe the ratio of dry knot strength to dry twine strength (or alternatively wet knot to wet twine strength). The knotting efficiency, measured either way, of nets made from nylon 66 yarn is of the order of 40 to 50 per cent. for single knots and 50 to 60 per cent. for double knots.

An investigation has been carried out into the factors influencing the loss of strength on knotting. This has resulted in a clearer appreciation of the mechanism of knotting and the effect of yarn properties on knotting efficiency.

Practical tests have shown that every type of knot has a different knotting efficiency, and that their order of efficiency approximates to the same for all fibres examined (nylon 66, nylon 6, polyester fibre, and polyvinylidene chloride). The differing configurations of various knots is responsible for the variations; it has been shown that a decrease in the angle through which the loop is formed decreased the loop strength; increasing the number of loops or hitches in a knot increased

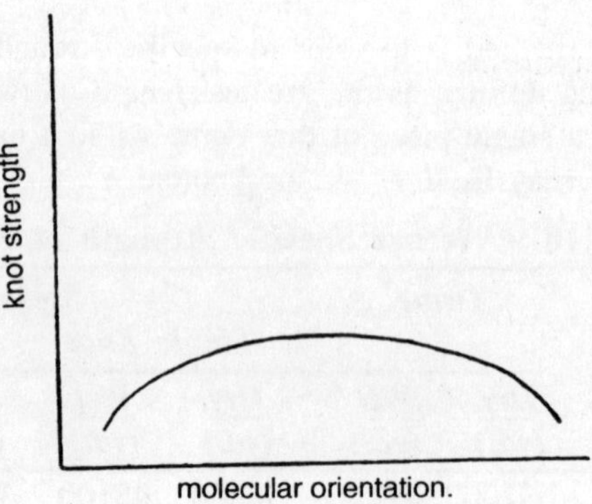

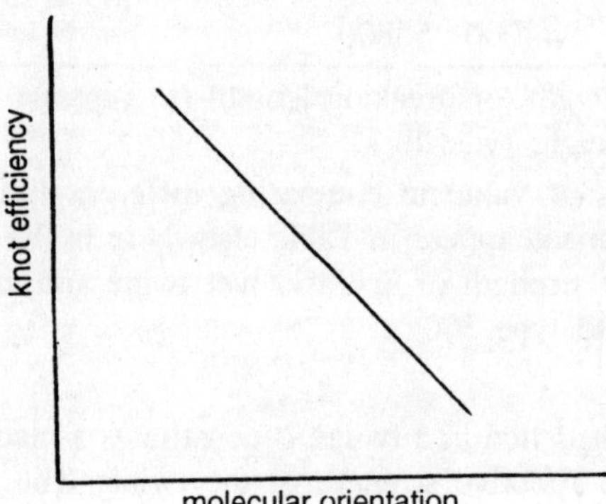

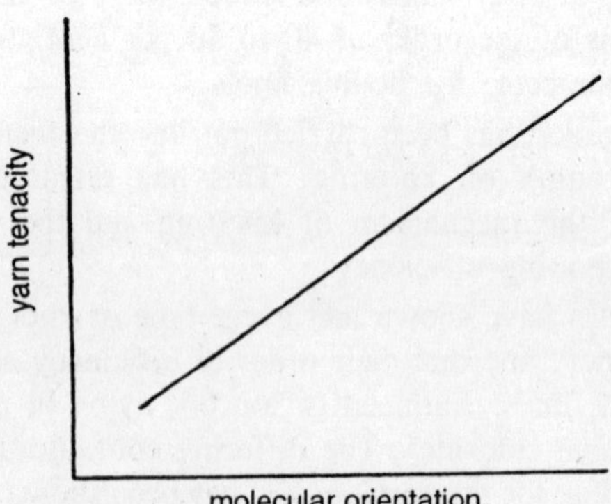

Figure 5.12 : Individual Effect of Molecular Orientation.

the knot strength. The effect of molecular orientation on knotting efficiency was investigated to study the nature of the rupturing forces more closely. Nylon 66 yarn of the same nominal denier and number of filaments was used to tie three types of knots-the overhand, warpers and double weavers. It was shown that with increasing molecular orientation the yarn tenacity increased linearly, knotting efficiency decreased linearly, whilst knot strength assumed a parabolic curve. The shape of the latter could be confirmed by calculation from the two straight line relationships. The degree of orientation of the yarn exhibiting maximum knot strength is of obvious practical importance. The manufacturer of nylon yarn, having the opportunity to tailor-make his fibre, should attach proper importance to this effect.

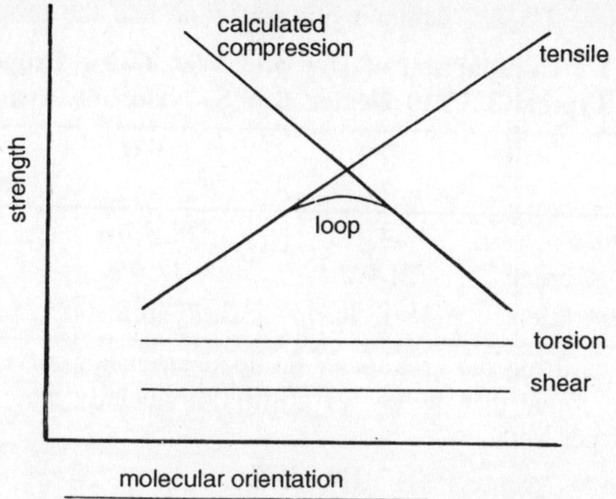

Figure 5.13 : Combined Effect of Molecular Orientation.

The investigation was continued by studying the effect of change of molecular orientation on shear, torsion and compression. No appreciable effect on shear and torsion could be detected, but compressive force, as represented by loop tenacity, described a parabola. From these facts, it was deduced that compressive and tensile forces act in opposition as orientation is changed.

The effect of compressive forces was subsequently shown experimentally by knotting unstretched polythene and examining it visually, and subjecting broken nylon knots to cross-polarised light. Tests on various knots showed that the break always occurred under the looped section.

In symmetrical knots, such as the Blood, two positions of stress concentration corresponding to the two looped sections were observed.

WETTING

When nylon twines are wetted they lose strength but gain in extensibility. The loss in strength (of both type 66 and 6) whether in yarn, twine or net form is of the order of 10 to 15 per cent. The increase in extension at break of twines made from nylon 66 is of the order of 15 to 20 per cent. according to size and construction. The importance of this effect is to counterbalance the loss of tensile strength when assessing the change in energy absorption. It is this latter property which is of critical importance in deciding whether a net will break or not under dynamic conditions. Nylon's present place in fishing nets is largely due to its comparatively high energy absorption under dry and, even more so, wet conditions.

A typical 3/3/210 denier nylon 66 twine had the properties.

Table 5.11 : Comparison of Dry and Wet Twine Properties for a Typical 3/3/210 Denier B.N.S. Nylon 66 Twine

	Dry	*Wet*	*Change on Wetting*
Breaking load	32.4 lbs.	28.6 lbs.	-12%
Extension at break	21-0%	24-9%	19
Energy absorption	3.36 in.lb./in	3.96 in.lb./in.	18%

By measuring the area under the load/extension curves, dry and wet, it was found that the energy absorption actually increased when wet by 18 per cent.

IN USE

Nylon nets, in common with those made from all other fibres, slowly lose strength in use. In certain African fisheries it is now the habit to store gillnets in the water when they are not being fished.

The abrasion resistance, dry and wet, of nylon 66 is extremely good. This can be varied by choice of filament denier. All nylon fishing twines are based on a filament denier of about 6 (as opposed to 1 to 3 for most apparel uses), thus providing a happy compromise between good abrasion resistance and flexibility (which decreases with increased filament denier).

PERLON

Perlon, like nylon, is a synthetic fibre which belongs to the polyamides' group. Like all fibreforming materials, it consists of large elongated molecules. It results from the assembly of a large number of molecules of a homogeneous material, caprolactam.

Perlon:

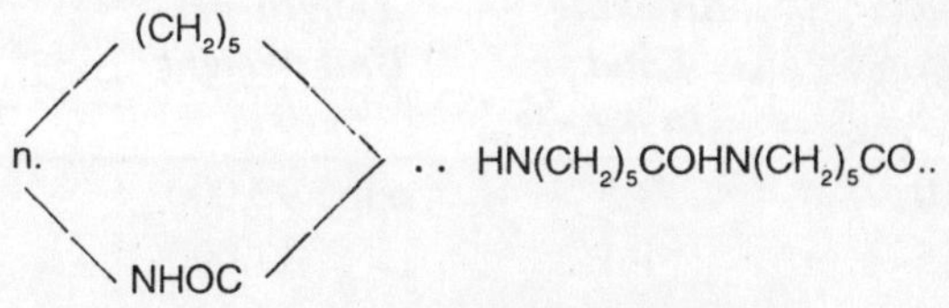

$$n.\left[\begin{array}{c}(CH_2)_5 \\ \\ NHOC\end{array}\right] \rightarrow\ ..\ HN(CH_2)_5COHN(CH_2)_5CO..$$

ε — aminocaprolactam

Both fibres are manufactured in the United States, where the fibre developed by Du Pont is called Nylon 66, as opposed to Nylon 6 for Perlon, a fibre developed by the former IG-Farbenindustrie combine in Germany.

In their technological properties Perlon and nylon are largely identical, except for the melting point which, in the case of nylon, is approximately 30 deg. C. higher.

Both fibres are manufactured by the extrusion process. As a result the fibres are cylindrical and have a smooth featureless surface.

By adding titanium dioxide or other suitable substances during the polymerisation process, a matt fibre can be extruded. This does not affect the physical properties of the fibre except resistance to sunlight which is considerably reduced. Matt fibres should not be chosen for the manufacture of nets.

FORMS OF PERLON

Perlon is available in 3 basic forms for the fishing industry:

Monofil
Continuous Filament and
Staple Fibre.

Monofil

Monofil has the advantage that it can be used for the manufacture of nets without having to undergo further processing (e.g. twisting or plaiting).

Monofils are continuous transparent wires, with a diameter of between 0.1 and 2 mm. (0.004 in. to 0.08 in.). Table I gives average values of breaking load and length per unit weight.

Continuous Filament

Continuous filament has greater strength and less extensibility than staple fibre and is lighter in weight than the equivalent cotton product. It is supplied in fine and coarse qualities.

Table 5.12 : Monoffis

Diameter In mm.	*Breaking Load In kg.*	*Length per Unit Weight m/kg.*	*Nm (Metric Number)*
0.10	0.5	app. 92,000	92.0
0.15	1.1	43,000	43.0
0.17		32,200	33.0
0.20	1.8	25,000	25.0
0.25	2.7	15-16,000	15-16.0
0.30	3.7	10-12,000	10-12.0
0.35	5.1	8,600	8.6
0.40	6.5	6,600	6.6
0.45	8.4	5,300	5.3
0.50	10.0	4,300	4.3
0.55	12.2	3,500	3.5
0.60	14.5	3,000	3.0
0.65	16.5	2,600	2.6
0.70	19.0	2,200	2.2
0.75	22.0	1,900	1.9
0.80	25.0	1,700	1.7
0.85	28.0	1,500	1.5
0.90	31.5	1,300	1.3
0.95	33.5	1,200	1.2
1.00	36.0	1,100	1.1
1.10	42.0	900	0.9
1.20	50.0	760	0.76
1.30	57.0	650	0.65
1.35		600	0.60
1.40	66.0	560	0.56
1.50	76.0	490	0.49
1.60	87.0	430	0.43
1.70	98.0	380	0.38
1.80	110.0	340	0.34
1.90	125.0	300	0.30
2.00	140.0	270	0.27

Table 5.13 : Perlon Continuous Filaments in Fine Deniers.

Denier	*Metric Number Nm.*	*Length per Unit Weight m/kg.*	*Number of Filaments*
210	43.0	42,860	35
630	14.0	14,290	105
750	12.0	12,000	125
840	10.7	10,710	140

Heavier deniers ("cables") is used as the raw material for net twines subject to severe stress, and for the manufacture of ropes.

The choice of 6 and 20 filament denier allows for different degrees of stiffness required, the stiffness of Perlon increasing with the denier.

Consequently Perlon made up to 20 denier filaments should be used where the end product is subjected to a sustained process of kneading and flexing.

Table 5.14 : Perlon Continuous Filaments in Heavy Deniers.

Denier	*Metric Number Nm.*	*Length per Unit/ Weight m/kg.*	*Filament Denier*	*Number of Filaments*
1,050	8.6	8,570	6	175
1,260	7.1	7,140	6	210
2,500	3.6	3,600	20	125
3,000	3.0	3,000	6; 20	500; 150
3,210	2.8	2,800	20	160
4,500	2.0	2,000	20	225
5,000	1.8	1,800	20	250
6,200	1.5	1,500	6	1,035
7,500	1.2	1,200	6; 20 1,	250; 375
9,300	1.0	1,000	6	1,550
10,000	0.9	900	20	500
11,250	0.8	800	6	1,875
12,500	0.7	700	20	625
15,000	0.6	600	6; 20	2,600; 750
30,000	0.3	300	20	1,500

However, where a highly flexible product is required 6 denier filaments are more suitable.

Staple Fibre

Spun staple fibre twines are particularly suitable for inland and coastal waters. The yarns are spun like cotton yarns to attain the highest possible degree of strength or produced from a spinning tow and then spun into yarn by means of a modified schappe spinning process after cutting.

The following types of staple fibre are available in yarn numbers up to Nm. 50:

Cotton spinning process:

2 den., cut length 60 mm.

27 den., cut length 60 mm.

Schappe spinning process:

2, 7 den. average length of staple 100 mm.

The optimum tensile strength of spun staple fibre yarn should be a breaking length of 30 km. in a Nm. 34/1 yarn.

Where net yarns must be stiffened with black varnish, spun staple fibre yarn is superior to continuous filament yarn, since the staple fibre yarn absorbs the stiffening preparation more thoroughly.

PROCESSING INTO NET TWINES, CORDAGE AND ROPES

Perlon Monofils

The appropriate monofil strength for *gillnets* to replace cotton net twines.

For mesh sizes in excess of 30 mm. at least 0.20 mm. diameter should be chosen even for the finest types of nets.

The monofils are available as fishing lines in diameters between 0.1 and 2 mm. The rope making industry is using Perlon monofils as raw material for the manufacture of braided and twisted cordage.

Table 5.15 : Monofils suitable for Gillnets.

Cotton Net twine Nm.	*Perlon Monofils Diameter*	*Length per Unit Weight m/kg.*	*Nm. (Metric Number)*
270/6	15.0 or 0.20 app.	43,000 resp.	43.0resp.25.0
240/6	,, ,,	25,000	,, ,,
200/6	0.20	25,000	25.0

160/6	0.20	25,000	25.0
140/6	0.20	25,000	25.0
120/6	0.25	16,000	16.0
100/6	0.25	16,000	16.0
100/9	0.25	16,000	16.0
85/6	0.30	12,000	12.0
85/9	0.30	12,000	12.0

Table 5.16 : Comparison in length per unit weight of coarser monofils with cotton twine of equal wet strength.

Cotton Net twine		*Perlon Monofils*	
Cotton Metric Number Nm.	*Length per Unit Weight m/kg.*	*Diameter mm.*	*Mean Length per Unit Weight m/kg.*
50/9	5,000	0.35	8,600
50/12	3,600	0.40	6,600
50/15	2,900	0.45	5,300
50/18	2,400	0.50	4,300
20/6	2,900	0.45	5,300
20/9	1,950	0.50	4,300
20/12	1,400	0.60	3,000
20/15	1,100	0.70	2,200
20/18	940	0.70	2,200
20/21	830	0.70	2,200
20/24	680	0.80	1,700

Twines of Perlon Continuous Filament

Twines for equipment exposed to little or moderate strain differ from twines for nets subject to severe stresses.

Net twines are produced from the following materials:

Denier	210	630	750	840
Metric Number Nm.	43.0	14.0	12.0	10.7
Length per Unit Weight m/kg.	42,860	14,290	12,000	10,710

Approximate lengths per unit weight and diameters of net twines made of 210 denier filaments.

Only the finest deniers listed are suitable for bottom-set gillnets; the others, including deniers heavier than those listed can be used for line fishing, baskets, trap nets, bagnets, trawls, seines, purse seines, etc.

Given an equal wet knot strength the weight of nets made of these twines are less than that of the equivalent cotton product.

Table 5.17

Net twine den.	*Length per Unit Weight m./kg.*	*Diameter mm.*	*Net twine den.*	*Length per Unit Weight m/kg.*	*Diameter mm.*
210/2	20,000	0.28	210/18	2,100	0.88
210/3	13,000	0.30	210/21	1,850	0.95
210/6	6,400	0.48	210/24	1,600	1.00
210/9	4,300	0.60	210/27	1,450	1.10
210/12	3,200	0.70	210/30	1,320	1.15
210/15	2,480	0.80			

Continuous filament twines are quite smooth and much less dirt adheres to nets made of such material than to nets made of cotton or spun staple fibre twines. Continuous filament or Perlon monofil is particularly suitable for heavily polluted waters. Braided or twisted twines of 840 denier, 1,059 denier, 1,260 denier, and 3,000 denier are especially suitable for equipment subject to severe strain such as bottom trawls. Perlon continuous filament twisted twines are considerably stronger than manila twines. This makes it possible to manufacture lighter weight nets to reduce the towing drag and allowing the use of less power.

Table 5.18 : Twisted Twines.

Manila-Extra			*Perlon Continuous Filament*		
Length per Unit Weight m/kg.	*Breaking Load (kg.) dry*	*wet*	*Length per Unit Weight m/kg.*	*Breaking Load (kg.) dry*	*wet*
186	120	127	192	260	221
—	—	—	250	208	177
248	95	100	256	195	165
—	—	—	312	167	142

—	—	—	315	173	147
375	65	68	385	130	110
—	—	—	1,248	43	37

Table 5.19 : Perlon Continuous Filament Braided Twines.

Length per Unit Weight m/kg. app.	*Breaking Load (kg.) dry*	*wet*
1,440	40	34
1,270	51	43
1,040	55	47
840	74	62
690	83	71
650	101	81
500	110	94
450	130	110
402	151	128
350	160	136
300	190	162
265	200	170
190	280	238

Cords and Ropes of Continuous Filament

Manila, sisal, hemp, and cotton have up to now been used as raw materials for the cords and ropes used in the fishing industry.

Cords and ropes of natural fibres shrink and swell which renders wet ropes hard and stiff. Natural fibres rot in the water. Coarser qualities dry very slowly and when stored are destroyed by mould. Synthetic fibre ropes should therefore be used with synthetic fibre nets.

Allowance must be made for the extensibility of Perlon which is many times that of the natural fibres. To prevent the ropes from untwisting or developing kinks, it is necessary to adjust the angle and direction of the twist so that the individual yarn constructions reinforce each other within the body of the rope.

Cords and ropes are made of Perlon continuous filament yarn in the heavier deniers.

Table 5.20 : Comparison of Weight and Breaking Load.

Hemp, Manila and Sisal untarred				*Continuous Filament Perlon*			
				Weight % m		*Breaking Load in kg.*	
Din-meter in mm.	*Circutn-ference in inches*	*Weight % m*	*Break-ing Load in kg.*	*Haw-ser laid*	*Cable-laid*	*Hawser-laid*	*Cablein laid and Twisted*
1	5/8	2.6	230	1.5		410	
6	3/4	3.6	380	2.25		580	
8	1	5.8	572	3.7		1,000	
10	1¼	7.4	763	5.9		1,500	
12	1½	11.5	1,143	8.7		2,250	
14	1¾	14.4	1,525	12		3,100	
16	2	18.8	1,970	14.5		3,800	
18	2¼	23.7	2,460	18.5		4,900	
20	2½	29.5	3,000	25.0		5,700	
22	2¾	35.5	3,580	31		6,950	
24	3	42.3	4,210	36		8,400	
26	3¼	49.6	4,870	42		9,750	
28	3½	57.5	5,580	49		11,200	
30	3¾	66.1	6,310	57	52	13,000	11,500
32	4	75.1	7,090	65	59	14,800	12,200
36	4½	95	8,720	82	74.5	18,700	15,700
40	5	117.5	10,500	101	92.5	(23,000)	19,500
44	5½	142	12,500	124	111	(27,800)	23,900
48	6	169	14,700	145	132	(33,000)	29,000
52	6½	198	17,000	170	155	(39,000)	34,600
56	7	230	19,400	197	180	(45,000)	40,800
60	7½	264	21,900	227	208	(52,000)	47,500
64	8	301	24,600	252	236	(59,000)	55,000
72	9	380	30,100	315	299	(75,000)	69,000
80	10	470	36,000	388	367	(92,000)	84,000
90	11¼	570	43,000	493	467	(116,000)	106,00
100	12½	710	50,000	608	577	(144,000)	130,000

The values in brackets merely serve purposes of comparison With these sizes it is advisable to use cable-laid ropes, a construction which has proved particularly efficient.

Perlon Staple Fibre Twines

Two yarn counts: Nm. 50 and Nm. 20 are mainly used for net twines.

Table 5.21 : Cotton and Perlon Staple Fibre Net twines of Equal Wet Strength.

Cotton Net twine		*Perlon Staple Fibre and Net twine*	
Metric Number Nm.	***Length per Unit Weight m/kg.***	***Metric Number Nm.***	***Length per Unit Weight m/kg.***
50/9	5,000	50/6	7,500
50/12	3,600	50/9	5,000
50/15	2,900	50/12	3,600
50/18	2,400	50/15	2,900
20/6	2,900	50/9	5,000
20/9	1,950	50/12	3,600
20/12	1,400	20/6	2,900
20/15	1,100	20/9	1,950
20/18	940	20/12	1,400
20/21	830	20/12	1,400
20/24	680	20/15	1,100
20/27	620	20/18	940
20/30	540	20/18	940
20/32 to 20/36	460	20/24	680

PROPERTIES

Specific Gravity

Perlon has a lower specific gravity than natural fibres:

Perlon	1.14	g/cu. cm.
Cotton	1.54	g/cu. cm.
Hemp, jute	1.48-1.50	g/cu. cm.
Ramie	1-51	g/cu. cm.

Coupled with a high degree of strength, this property enables the industry to make very light nets.

Rot Resistance

Perlon does not. rot as it is not attacked by bacteria or fungi, nor does it need preservative treatment.

Perlon net twines continuously immersed in the brackish water of a North Sea harbour for a period of 3½ years have shown no more than a 10 per cent. loss of strength.

Tensile Strength

The specific tenacity of monofilaments is highest with a small diameter. The breaking length varies between 37 and 47 kilometres. The tenacity of water-saturated Perlon is between 80 per cent. and 90 per cent. of its air-dry tenacity. The knot strength shows an equally close relation to the diameter:

(in % of tenacity)

70-80 % for small diameter	(0.10-0.30 mm.)
60-70% for medium diameter	(0.35-0.70 mm.)
50-60% for large diameter	(over)

The strength of the filaments is determined by the degree to which they have been stretched or "drawn" during manufacture, as shown in an example of Perlon continuous filament of 1,060 denier:

		Ordinary	*Special Pre-stretch*
Breaking Length air-dry	km	55	71
Breaking Length wet	km	47	61
Break air-dry	%	20	17
Break wet	%	22	19
Extension at Knot Strength air-dry	km	44	41
Extension at Knot Strength wet	km	41	36

A high degree of stretching results in a considerable increase in strength coupled with an appreciably reduced extension at break. However, as filament strength is raised, there occurs a reduction in knot strength, particularly when water-saturated.

A comparison in knot strength of filament composition 10,000 denier, single filament denier 20, and of monofils with a hemp string of equal strength:

	Hemp	*Perlon Filament*	*Perlon Monofil*
Strength dry kg/sq. mm.	35	59	52
Strength wet kg/sq. mm.	38	52	42
Knot Strength dry kg/sq. mm.	23	26	26

Knot Strength wet kg/sq. mm.	23	24	20
Relative Knot Strength dry %	66	44	50
Relative Knot Strength wet %	66	41	48

This phenomenon is particularly important when the fibre is subjected to a specially high pre-stretch as for fishing net twines. High strength and a low extension at break are desirable, but can only be achieved at the price of low knot strength.

Experience has shown the knot strength of all fibres to be below their ordinary strength.

Extensibility and Elastic Properties

The extension at break of Perlon monofil varies between 20 and 35 per cent.

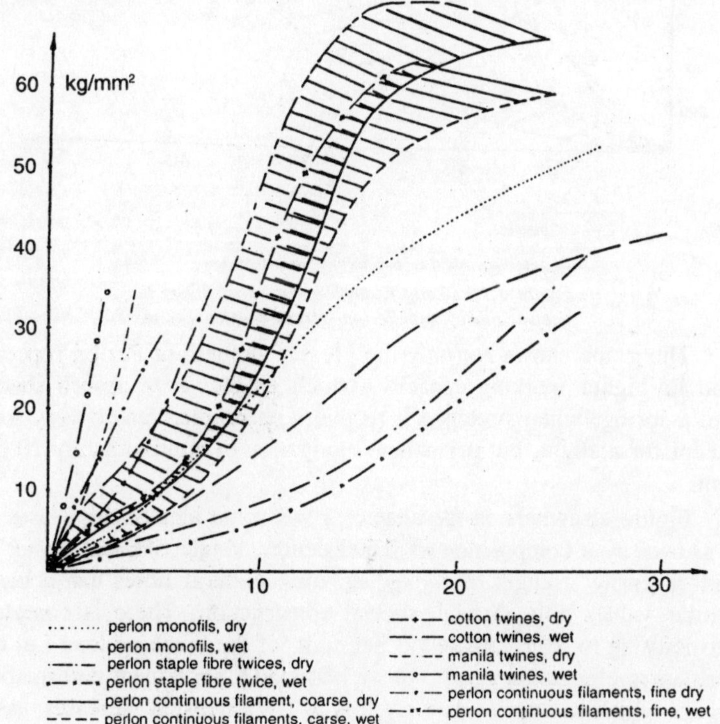

Figure 5.14 : Load/Extension curves of Perlon Products compared with Cotton and Manila twines.

Other characteristic values for the extensibility and elasticity of Perlon monofils are:

Elastic extension	75-90%	of the total extension at
Permanent elongation	10-25%	80% of breaking load

These diagrams show that the load extension curves of Perlon climb at a more obtuse angle than those of natural fibres.

This divergence becomes conspicuous in a comparison of hemp and Perlon ropes with a circumference of 1¼ in.

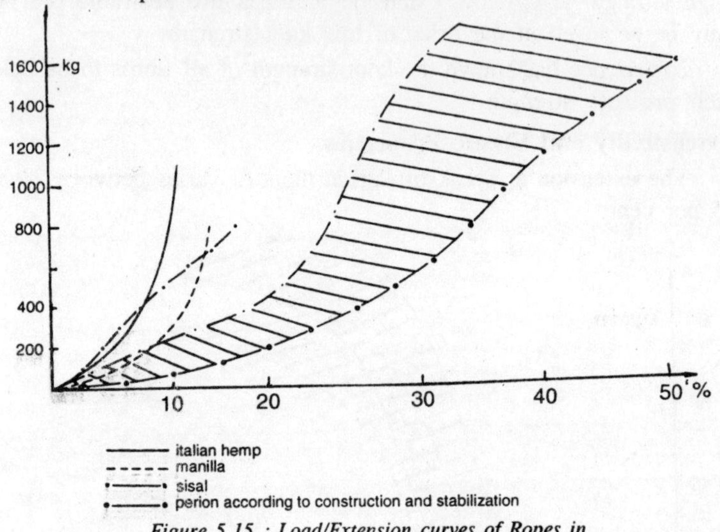

Figure 5.15 : Load/Extension curves of Ropes in Natural Fibres and Perlon, circumference 1¼ in.

The graph shows not only the greater strength of Perlon rope but also its higher working capacity, which enables it to absorb shocks like a spring; when stretched it recovers its original length very soon except for a slight, but permanent elongation of approximately 10 per cent.

Figure elsewhere in the chapter gives some idea of this elasticity as shown by a composition of 7,500 denier, single filament denier 6, with a tensile strength of 64 kg. sq. mm. Natural fibres can achieve similar values only through special construction. There is complete elasticity up to approximately 5 per cent. of the breaking load i.e. the permanent elongation is nil. Above this load a permanent deformation occurs which becomes relatively less as the extension increases. As a result the difference between total extension and permanent elongation increases further.

The relation of elastic extension to total extension gives the elastic ratio, which can be said to hold good for Perlon proportionate to the

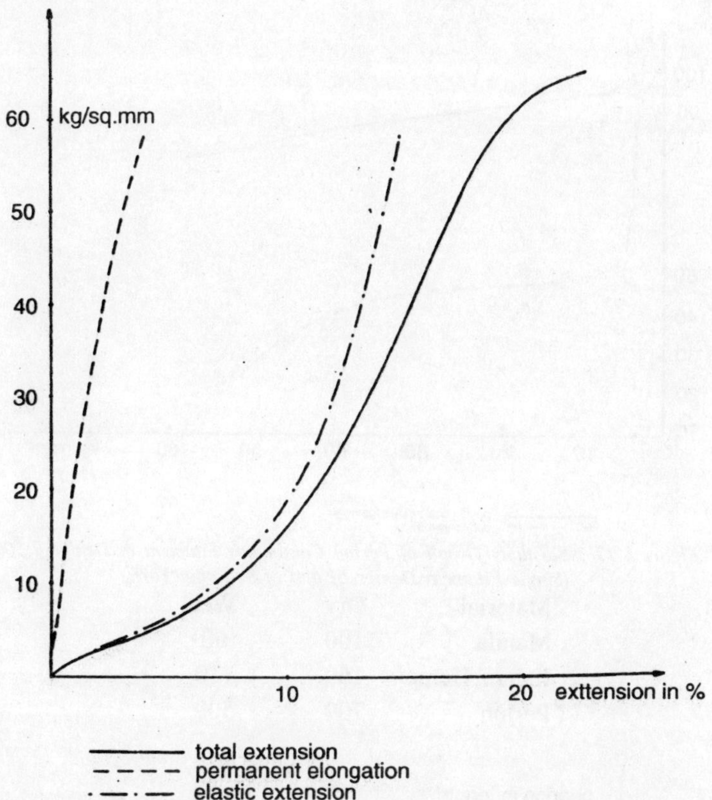

Figure 5.16 : Elasticity Graph of Perlon Continuous Filament in Denier 7,500.

magnitude involved. Figure elsewhere in the chapter compares the elastic ratios of Perlon and cotton yarns.

The load/extension curve has already shown up fundamental differences, which occur once again in a comparison of elasticity. Whilst Perlon shows a high proportion of elasticity within the total extension, with cotton fibres the permanent elongation predominates and 'leads to a much steeper fall of the elasticity graph.

It appears that immediately the load has been applied a high degree of extension ensues, which regains its equilibrium within the next 15 and 30 minutes respectively. The extension is not proportional to the increase in the load, but is relatively greater at low and medium loads than at high loads.

This gives Perlon a springy quality which enables it to absorb kinetic energy as shown below.

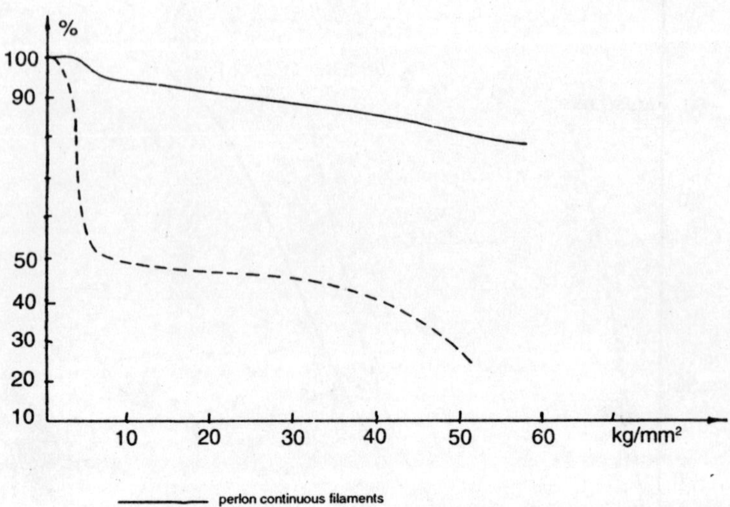

Figure 5.17 : Elasticity Graph of Perlon Continuous Filament in Denier 7,500 (Single Filament Denier 6) and of a Cotton Yarn.

Material	Dry	Wet
Manila	100	60
Italian Hemp	160	110
Perlon	700	500

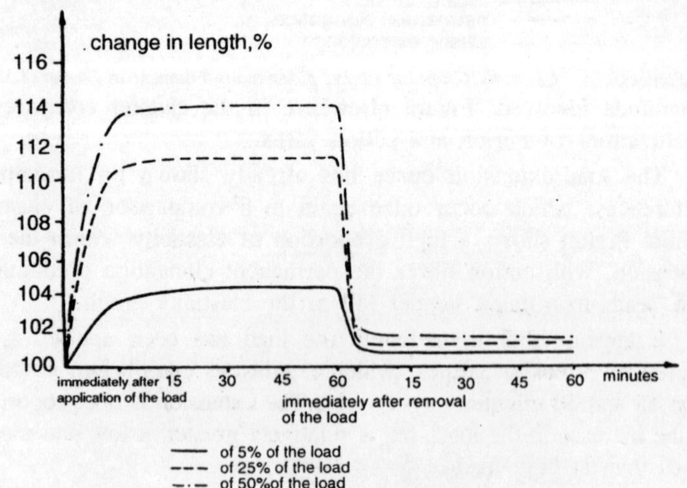

Figure 5.18 : Extensibility and Elasticity of Perlon Continuous Filament (Denier 7,500, Single Filament Denier 6) in Relation to Time and Load.

Flexibility

Net twines made of continuous filament and staple fibre are softer than those made from natural fibres. This applies particularly to their wet state as shown in tables.

Hardness, dry and wet of Net Twines of 2.3 mm. Diameter

Raw Material	*Dry*	*Wet*
Manila	700	410
Hemp	120	110
Cotton	42	140
Perlon Filament	44	21

The higher the figure the harder the net twine.

For some fishing nets, in particular fine gillnets, flexibility is a very desirable quality; for others it is of minor importance. This can be achieved by applying a stiffening preparation which at the same time gives the nets greater resistance to abrasion and sunlight.

Abrasion Resistance

Perlon has a high resistance to abrasion which together with non-rotting quality determine the useful life of nets.

Wet abrasion tests of net twines showed the following results related to their weight per metre:

Net Weight g/m	*Wet Abrasion Rubs Twine*		
	Hemp	*Manila*	*Perlon Filament*
1.5	420	—	660
2.5	630	380	930
3.0	940	400	1,510
4.0	1,210	540	2,120
4.5	1,340	580	2,430

Staple fibre net twines show a lower abrasion resistance than those made of continuous filament but still show considerably greater resistance than those made of cotton fibres. A staple fibre net twine Nm. 20/24 has a wet abrasion resistance of 270 whereas a resistance of 150 was shown by a cotton net twine of the same strength.

A breaking extension test showed a 62 per cent. loss of tensile strength for the manila cord, compared with 19 per cent. for the Perlon braided cord.

Weather Resistance

Both synthetic and natural fibres, are weakened by exposure to sunlight, but monofils display a high degree of resistance to sunlight and weather conditions, superior to that of vegetable fibres, and close to the immunity of poly-acrylonitrile fibres.

Net twines of staple fibre or continuous filament lose more strength than natural fibre twines when exposed to intense sunlight, but because of their high initial strength they remain in the last analysis superior to natural fibres. The bigger the diameter the less noticeable the photodegradation, which is insignificant for thick ropes as the layers below are protected by the degraded outer layer and which is probably no deeper than 1 mm. to which ultraviolet rays can penetrate.

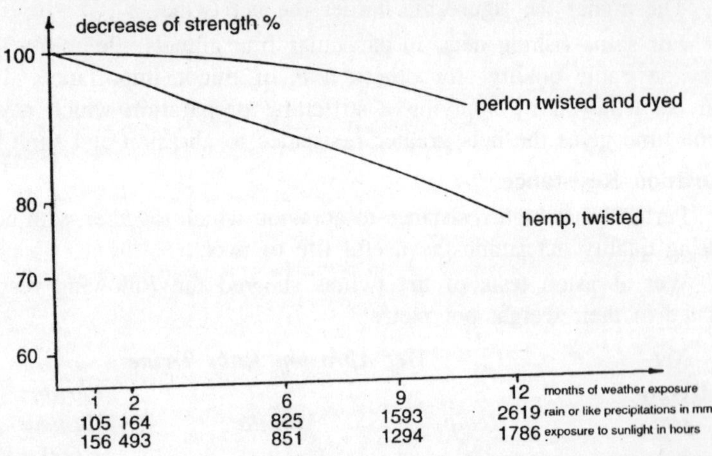

Figure 5.19 : Influence of weather exposure on ropes made of Perlon continuous filament and hemp respectively. Rope circumference: 1½ in. Weather exposure at 6,230 ft. above sea level.

Where Perlon twines are exposed to intense sunlight for long periods, as for example, in the case of fyke nets, it is advisable to dye them. They can easily be dyed with *Perliton* dyes, which dyes must be selected with a view to fastness and with acid dyes-Telon-light and fast dyes respectively, which are remarkable for their fastness in water. Treatment with a mixture of tannic acid and antimony potassium tartrate is particularly durable in seawater. Treatment with a catechu solution has proved an excellent protection against sunlight.

Hygroscopic Behaviour

Perlon has a low moisture regain. An examination of moisture content in air-dry conditions (65 per cent. relative air humidity, 20 deg. C.), and swelling ratio (degree of saturation) shows the following results:

Ram Material	*Moisture Content in Air-Dry-Conditions*	*Swelling Ratio*
Cotton	7.8	45
Bast Fibres	7-13	100-110
Perlon	4.2	12-14

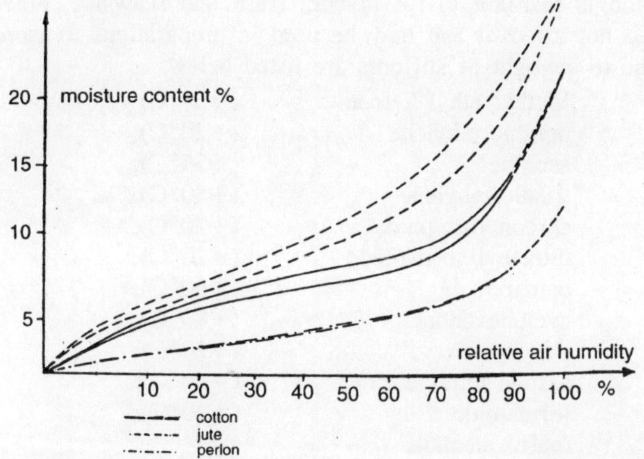

Figure 5.20 : Absorption and Desorption of cotton, jute and perlon.

For Perlon, absorption and desorption are of almost identical magnitude in contrast with the far more marked "swelling hysteresis" of natural fibres and it therefore dries appreciably faster.

The level of moisture regain is known to be closely related to the lateral and longitudinal swelling caused by wetting. Cotton net twines receiving a first wetting of 24 hours, showed lateral swellings of between 5 and 15 per cent.; bast fibre net twines showed an increase of 20 to 40 per cent. in their cross-sectional area whereas Perloti continuous filament net twines contract rather than swell. The following comparison in thickness of a Perlori cable Nm. 0.9, single filament denier 20, with a hemp cord of equal size may serve as an example:

		Hemp	*Perlon*
Diameter dry	mm	1.65	1.42
Diameter wet	mm	2.02	1.38
Variation in Cross-section	%	+22	–3

Net twines made of natural fibres swell considerably and shrink as a result. It amounts to approximately 6 to 9 per cent. for cotton net twines, and to approximately 2 to 8 per cent. for hard bast fibres. Continuous filament twine behaves quite differently. An immersion in water gives rise to an extension proportionate to the cross-sectional shrinkage, in the region of 1 to 3 per cent. The changes in length and cross-section are only slight and consequently assure a constant mesh size.

Chemical Resistance

Perlon is resistant to rot in both fresh and seawater. Solvents which do not affect it and may be used in preparations to increase resistance to sunlight or stiffness are listed below:

Methyl ethyl ketone	(+60°C);
perchlorethylene	(+20°C);
acetone	(+56°C);
trichlorethylene	(+80°C);
carbon disulphide	(+20°C);
dimethyl formamide	(+20°C);
benzipe	(+60°C);
cyclohexanone	(+80°C);
chloroform	(+20°C);
carbon tetrachloride	(+60°C);
formamide	(+20°C);
(base) alcohols	
ethyl oxide	(+35°C);

Perlon solvents include concentrated solutions of formic, hydrochloric and sulphuric acid, as well as phenol (carbolic acid), cresylic acid and resorcin.

Products with a phenol content, such as tars, can in certain cases cause swelling and consequently a loss of strength. Specially treated tars contain only small quantities of phenol derivatives and of phenol itself. In crude tar, however, these substances, "acid oils", may be present in fairly large quantities, although there is a steep variation in the percentage. It seems advisable to draw special attention to this aspect, since tars are also used as stiffeners.

Thermal Properties

Perlon has a high resistance to cold and retains its elasticity even when frozen. There is, in fact, an increase in strength and conversely a loss of extensibility down to a temperature of –40 deg. C. The temperature needed in dyeing processes does not damage the nets, but a certain amount of shrinkage must be expected and therefore ascertained by preliminary tests on small samples.

Visibility

Perlon net twines can be very fine and accordingly inconspicuous in the water.

The translucent monofil is almost invisible under water and this property has increased the catches of monofil setnets and fyke nets.

Processing

Staple fibre net twines can be processed into netting without difficulty, manually or by machinery. They can be tied in non-slip knots in the same way as cotton net twines. With monofil and continuous filament twine, however, slip-proof knots cannot be assured because of the smoothness of the fibre. Tests are about to be completed which aim at giving monofils a rough surface while preserving the inherent strength to increase the knot fastness.

Continuous filament net twines can be roughened by treating them, preferably while they are being twisted, with a preparation insoluble in water, which gives them adhesive properties (bonding).

The resistance to slippage of Perlon continuous filament twine has been improved by blending spun staple fibres.

Another possibility is the heat-setting of the knots under tension, at temperatures of 150 to 180 deg. C. usually applied for *short* periods only.

With nets it is essential to maintain an even tension and twist in the Perlon material. Hawser-laid or cable laid ropes must be heat-set if necessary. No stabilisation is needed, however, if the individual twists reinforce each other. With such a construction heat-setting is only needed where splices have to be made.

Cordage can be set by hot air treatment, saturated steam treatment or boiling. Boiling is preferable for heavier types since neither hot air nor saturated steam can penetrate evenly enough through the rope.

In continuous filament too hard a twist should be avoided as this would reduce strength and increase extension, particularly that proportion which is permanent. Braided twines for fishing nets should be braided with a medium degree of hardness. On no account should a high degree of hardness be employed.

Storage

Perlon nets and ropes should not be left exposed to the sun. They are best kept in dark rooms and can be stowed while still wet. Nets which are freshly treated with preservatives should be handled in the same way as natural fibre nets: they should only be stored after having been used at least once. This applies particularly when stiffening preparations have been used.

Attack by Micro-Organisms

Perlon is immune to attacks by micro-organisms such as bacteria or fungi, nor do molluscs, barnacles and other organisms harm it. However, the larvae of the mayfly which settle particularly on stationary fishing gear in inland waters can cause considerable damage to nets; synthetic fibres are as much affected as cotton nets. Treatment with pesticides (Arkotine, Dieldrin) ensures a high degree of protection.

In running waters which contain organic effluents thick clusters of "sewage fungi" occur which soon cover the nets. Frequent cleaning is needed, or catches are lost. Rhine fishermen are particularly affected by this fouling of their stow nets. Nets made of Perlon filament braided twine are less affected because of their smooth surface and can be cleaned easily and quickly.

USE OF PERLON IN VARIOUS TYPE OF FISHING GEAR

The economics of fishing gear depend, generally speaking, on initial costs, useful life, costs of preparation, maintenance and efficiency.

The initial cost for a given weight of nets is higher for Perlon than for cotton, hemp and especially manila, but the useful life of Perlon is considerably longer because of the high degree of resistance to rot and abrasion. While these nets need little or no preservative treatment nets of cotton, hemp and flax must be treated frequently. The expenditure in money and time involved represents a severe burden.

Maintenance largely consists in preventing unnecessary exposure to sunlight and with monofil even this protective measure becomes unnecessary.

The following paragraphs outline the use of Perlon for some of the more important types of equipment.

Trawls

At the Fishery Industry Fair 1957 in Copenhagen was a still usable bottom trawl made of Perlon braided twine used by a German trawler to catch 128,000 cwt. of herrings having a performance and useful life between 8 to 10 times as great as that of the customary manila trawls. Although Perlon trawls cost two or three times as much as manila trawls, their economic advantages are beyond question.

Their low weight, flexibility and smoothness make them easier to handle and the finer net twines noticeably reduce the tow drag.

Probably the most important advantage is the capacity of Perlon to absorb kinetic energy. The nets also ensure the safe landing of big catches on deck as shown by the big catches of Norway haddock when Perlon codends were used. About half the German cutter fishermen use Perlon for the codends in drag-nets for catching herring. The material in this case was spun staple fibre twine in counts Nm. 20/15 to 20/21.

Parse Seines

This type of equipment plays an important part in the fishing industry of many countries, although not in Germany. Its traditional net material is cotton twine of medium strength. In the Portuguese fishing industry, for example, cotton nets treated with preservatives may have a useful life of between 400 and 500 fishing days.

Large pieces of Perlon spun staple fibre webbing have been employed in Portuguese purse seines for 1,300 fishing days without becoming unusable. These nets were not dyed or prepared in any way. For purse seines Perlon has the following advantages: reduction in the total weight of fishing equipment (essential considering the size of the equipment involved); almost no expenditure on net preservatives; no necessity to keep one set of nets on land for preservation purposes; less labour required to handle the gear; water-saturated nets can be safely stowed and a useful life at least double compared with cotton nets.

Perlon continuous filament rope of 30 mm. in diameter has been used for 1,164 days as a purse line. It should have five times the useful life of a sisal rope to justify its higher price; it has already had nineteen times the life of a sisal rope.

Bottom-Set Gillnets

Set and floating nets used as fine gillnets are highly selective, helping to conserve young stock and supplying fish of high quality. Bottom-set nets are passive rather than active and must be as fine and as soft as possible having the least degree of visibility, requirements met by net twines of Perlon continuous filament yarn in a denier even finer than that quoted in IIIb (100 denier and finer). Monofils are particularly suitable for set gillnets due to their extremely low visibility in water. Now that the problem of providing non-slip knots has been solved and mechanically produced netting can be obtained, set gillnets of monofil assumed considerable importance for fishing in inland waters.

Stationary Fishing Gear

Perlon products have proved their efficiency and economy. when used for fyke and stow nets and traps.

A cotton eel trap such as those used by fishermen in the North German inland lakes, costs about DM 24.00 and lasts 4 years with preservative treatment. A Perlon spun staple fibre eel trap costs DM 17.08, needs no additional expenditure and remains completely efficient for approximately four years.

Monofils proved even more suitable for eel traps; owing to their translucence they catch more eels.

Cotton stow nets have a useful life of about two years if treated between 7 and 10 times with hot tar to preserve and stiffen them. A stow net of Perlon staple fibre twine has been in use for six years now and has been stiffened twice, once with black varnish and once with tar. Total costs for this stow net, which is still in full use, amounts to DM 1,625.00. During the six years, at least 3 cotton stow nets would have had to be purchased, costing DM 3,270.00.

Whaling Ropes and Cordage

In antarctic whaling, ropes of Perlon continuous filament between 100 and 120 metres in length used with the 70 kilogram explosive harpoons have proved their worth as "foregoers".

The strength of the fibres makes it possible to use ropes with a diameter of 33 to 34 mm., as compared to 38 mm. for manila ropes. Perlon ropes do not stiffen in water and hardly ever ice up. The elasticity can absorb the high shock loads involved and greatly reduces the danger of a rope breaking.

Tarpaulins and Protective Covers

Tissues of Perlon continuous filament yarns, PVC= Polyvinyl Chloride coated on both sides are used for tarpaulins, lifeboat covers, etc. They can be folded quickly and require little storage space, They are rot. proof, watertight, tough and can be fireproof.

TERYLENE "POLYESTER FIBRE AND ITS RELATION TO THE FISHING INDUSTRY

Terylene polyester fibre is a new synthetic fibre which has already made a considerable impact on the textile industry and has significantly affected certain fields which have hitherto been dominated by natural fibres. Apart from its merits in the manufacture of wearing apparel and for industrial use, the special properties of Terylene make it suitable for employment in the fishing industry. It has been used not only for netting twines, ropes and lines, but also for lifeboat and hatch covers, sails and tarpaulins.

The fibre is a British discovery made by J. R. Whinfield and J. T. Dickson in the Laboratories of the Calico Printers' Association Ltd.. Later the world patent rights, with the exception of the USA were acquired by ICI Ltd., and the name "Terylene" became a registered trade mark, the property of ICI Ltd. Recently several European companies have been licensed to manufacture the fibre under their own trade names, and it is also being made in Canada by Canadian Industries Ltd.

A new factory at Wilton, Middles brough, Yorkshire, produces over 25 million pounds per annum, and larger outputs are envisaged.

TYPES OF YARN

Both continuous filament yarn and staple fibre are manufactured. The continuous yarn can be divided into two categories. The first is a yarn with a- tenacity of 6 to 7 grams per denier and a corresponding extension at break of 12½ to 7½ per cent. This yarn, which is intended primarily for industrial uses, is being manufactured currently in 125 denier/24 filaments and 250 denier/48 filaments. It enables very fine and strong twines to be produced. The development of a high tenacity extra heavy denier yarn is under way which will enable heavier nets, e.g. trawls, to be produced more economically. All the high tenacity yarns are bright.

The second category comprises a yarn with a tenacity of 4½ to 5½ grams per denier and a corresponding extension at break of 25 to 15 per cent. It is intended mainly for wearing apparel and is produced in deniers ranging from 25 to 150, most of which can be obtained in either bright or delustred forms.

A staple fibre, with a tenacity of 3½ to 4 grams per denier and a corresponding extension at break of 40 to 25 per cent., is currently being manufactured, ranging in staple length from 1½ to 6 in. and from 1½ to 6 denier (that is 1½ in. staple, 1½ denier) for processing on the cotton system, up to 6 in. staple, 6 denier for processing on the flax system. All staple fibre has a heat stabilised crimp and is available with a dull-lustre. Filament yarn is twisted to three-quarters of a turn per inch and supplied to the trade on bobbins.

The fibre is made from polyethylene terephthalate, a condensation product of terephthalic acid and ethylene glycol, both of which are derived through various chemical processes from the products of mineral oil cracking. The polymer is chipped, and the fibre produced by a melt spinning process. Molten polymer is pumped through a spinneret and the spun yarn is subsequently mechanically stretched to develop fibre-like properties. The filament yarn has been exported overseas to be made into twines and nets. In certain countries, where limited doubling facilities are available, twines have been exported by manufacturers in the United Kingdom. Complete nets have also been supplied in certain instances.

The fishing industry is mainly interested in the high tenacity filament yarn on the grounds of high strength and low extensibility, although the staple fibre is used for specialised purposes, such as net mounting ropes.

PHYSICAL PROPERTIES

The most important physical properties which this yarn can offer for use as twines or netting are:

(i) High tensile strength which is unaffected by wetting. In particular, it has a high wet knot strength.

(ii) Low extensibility and high modulus.

(iii) Rot proof and not weakened by mildew.

(iv) Good resistance to sunlight.

(v) Stability on water immersion coupled with low moisture absorption.

(vi) Good resistance to abrasion under wet conditions.

(vii) The smooth nature and transparency of the material.

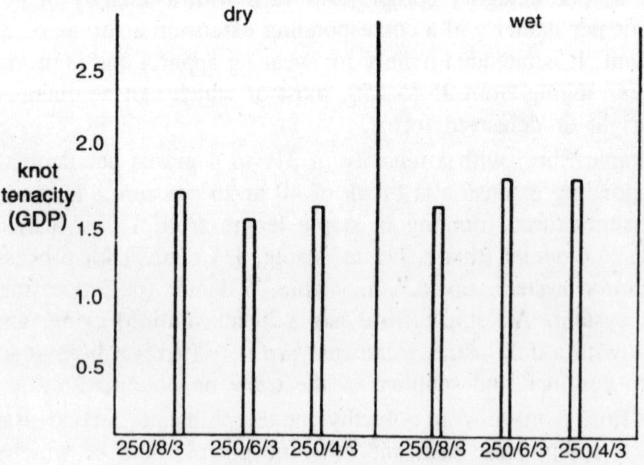

Figure 5.21 : Single knot tenacity of terylene twines.

Other fibres mainly used for netting twines are cotton, flax, manila, sisal and polyvinyl alcohol such as Kuralon, and the polyamides Perlon and Nylon. The general physical properties of a range of Terylene fishnet twines in comparison with twines made from the fibres mentioned.

A study of this table demonstrates the advantages which the fibre has to offer. A loss in strength on knotting is common to all fibres, and synthetic fibres may lose relatively more strength on knotting than natural fibres, such as cotton and linen, but as the initial strength of Terylene is very much higher, its actual knotting strength remains above the level of natural fibres.

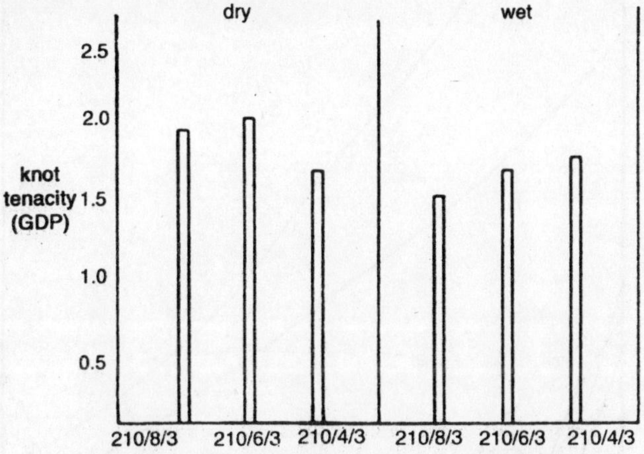

Figure 5.22 : Single Knot Tenacity of Nylon Twines.

In the dry state the twines may have a slightly lower dry knot tenacity than comparable nylon twines. The wet knot tenacities of the two are not significantly different. The wet knot strength of a fishnet twine is obviously of more importance than its dry knot strength. The strength of a twine depends not only on the fibre from which it is made, but also on its construction (twist, number of strands, etc.). The tables also show the low extensibility of Terylene and its resistance to stretch. This facilitates the manufacture of nets with mesh sizes which conform to the required specification and which resist distortion in use.

Terylene twines do not shrink when immersed in water at ambient temperature, but they will shrink at elevated temperatures. If the nets are to be subjected to various heat treatments, e.g. dyeing, prior to use, and the mesh size is critical, an allowance should be made for shrinkage.

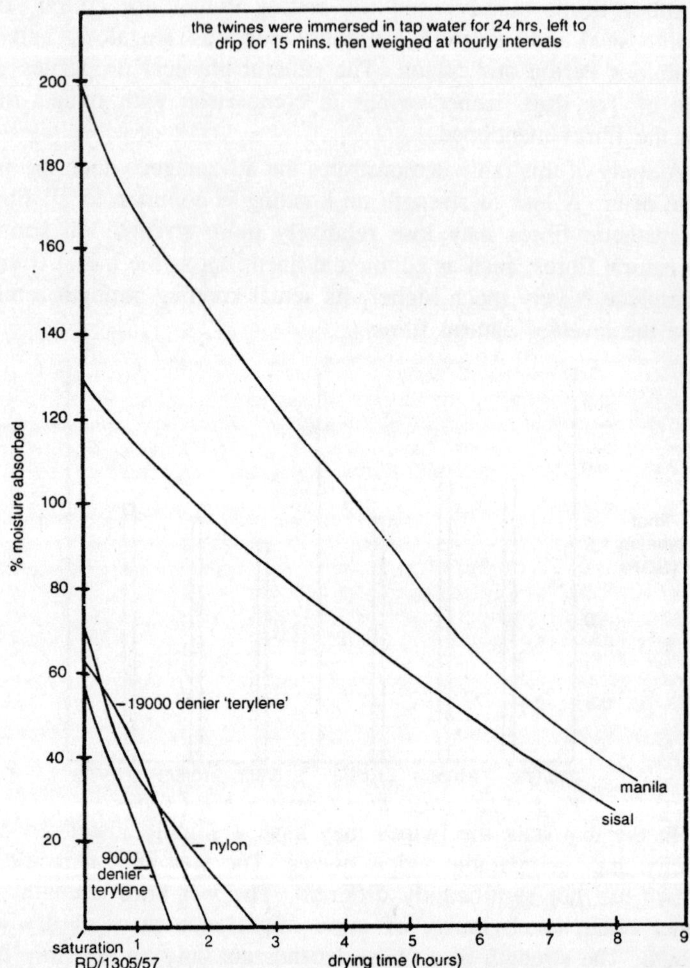

Figure 5.23 : The rate of drying of Manila, Nylon and Terylene Twines.

The abrasion resistance of the twines is superior to those made from cotton and flax. Unlike most synthetic fibres, its abrasion resistance is not appreciably different under wet conditions. The data above were obtained by rubbing both wet and dry twines over a 3 mm. thick hardened carbide steel bar. The abrasion resistance of nylon twines is significantly better than that of Terylene twines, but the difference under wet conditions is not nearly so marked. In practice Terylene fishnets can be expected to show good wearing qualities because of their good wet abrasion resistance. Mention has already been made of the low moisture absorption of the fibre and this is clearly shown in the graph. This means that on immersion in water the twines do not swell, the mesh sizes of the nets remain intact, little change in effective net weight occurs, and the twines dry quickly.

The smoothness, fineness and transparency of the material eliminate air bubbles and contribute to a low order of visibility.

The fibre has very good resistance to chemical attack, in particular to acids and oxidising agents. It is thus resistant to sqa water attack and unaffected by contact with oils, cutch and tar. This has been demonstrated in certain areas where traditional nets were affected by chemical contamination which had no effect on Terylene nets.

In connection with the use of Terylene in twines and nets, the following points are worthy of mention:

KNOT SLIPPAGE

In the manufacture of fishing nets, two types of knots are commonly used, the single sheet bend (single weaver's knot) and double sheet bend (double weaver's knot). The latter type does not slip during use and double knotted twines have a knot strength about 10 per cent. higher than single knotted twines. Both Terylene and nylon single knotted nets are prone to knot slippage, but the tendency is less with Terylene. Tests carried out on samples of Terylene, nylon, cotton and linen single knotted gillnets showed that three out of every four knots slipped in the case of nylon when dry and one in four when wet. One out of every four Terylene knots slipped when dry, but none when wet. No knot slippage occurred with the cotton net and only occasionally with the linen net. However, the resistance of Terylene to single knot slippage is not considered to be good enough and as more single knotted fish netting is produced than any other, knot slippage constitutes a problem. It is said that the somewhat slow rate of production of double knotting machines has been improved and is now about the same as the single knotting machine. . Knot slippage is a result of the smoothness of synthetic filament yarns, and an obvious answer is to increase the coefficient of friction of the twines by the application of a surface coating. The anti-slip agent may be applied to the twine during manufacture of the net, or to the finished net to fix the knots. Some net manufacturers consider both pre- and post-treatments are necessary. The general view seems to be that the nets should be made from bonded twines to enable an undistorted net to be taken off the machine and safely transported to a stretching frame to tighten and fix the knots. In certain cases it is possible to avoid the use of prebonding agents by the application of very high tensions at the back of the loom. These consolidate the knots sufficiently for net handling prior to the stretching treatment.

The bonding agents recommended for Terylene fishnet twine are Colophony resin and Bedesol 76. Colophony resin is applied from a solution in methylated spirits or aqueous ammonia. Bedesol 76 is applied from a solution in 64 deg. over proof methylated spirits. Both bonding agents can be applied either by single-end gumming at a speed of 100 to 150 yards per minute, or by hank dipping for 15 minutes, draining, and, in the case of the ammonia solution of Colophony resin, drying for one hour at 80 deg. C. The percentage solids of resin required on the twines depend on the net making machine used. For a Seriville machine, the required pick-ups of Colophony resin and Bedesol 76 are 1.0 per cent. and 2.5 to 3.5 per cent. respectively. The former is achieved using a 5 per cent. solution and the latter a 6 per cent. solution. For the Zang machine the respective pick-ups are 2 to 3 per cent. and 5 to 7 per cent. In the first case a 7 to 8 per cent. solution is required and in the latter a 12 to 15 per cent. solution. As twines bonded from methylated spirit solution may be slightly sticky and as the solvent is inflammable, bonding from aqueous ammonia should be followed by drying at 80 deg. C. It also has the advantage of keying the resin to the fibre. The differences in optimum pick-up of the bonding agent for the Zang and Seriville machines are a feature of the more critical operating conditions for the Seriville machine.

As already mentioned, the net itself can be treated with bonding agents to prevent knot slippage. Both Colophony resin and Bedesol 76 have been satisfactorily used in concentrations akin to those for bonding twines. The net is opened out as much as possible to ensure that all of it comes into contact with the solutions. It is immersed for a minimum period of 10 minutes, with stirring or agitation to ensure an even application of solids. The net is then removed and dried. A more even application of the solids is obtained if the net is centrifuged to remove surplus solution before drying. Drying at a temperature of 80 deg. C. is preferable to drying in air at room temperature, since the latter process is very much slower, and leaves a net slightly sticky. About 4 per cent. allowance should be made for the shrinkage in drying at 80 deg. C.

During manufacture of the net it is important to ensure that the

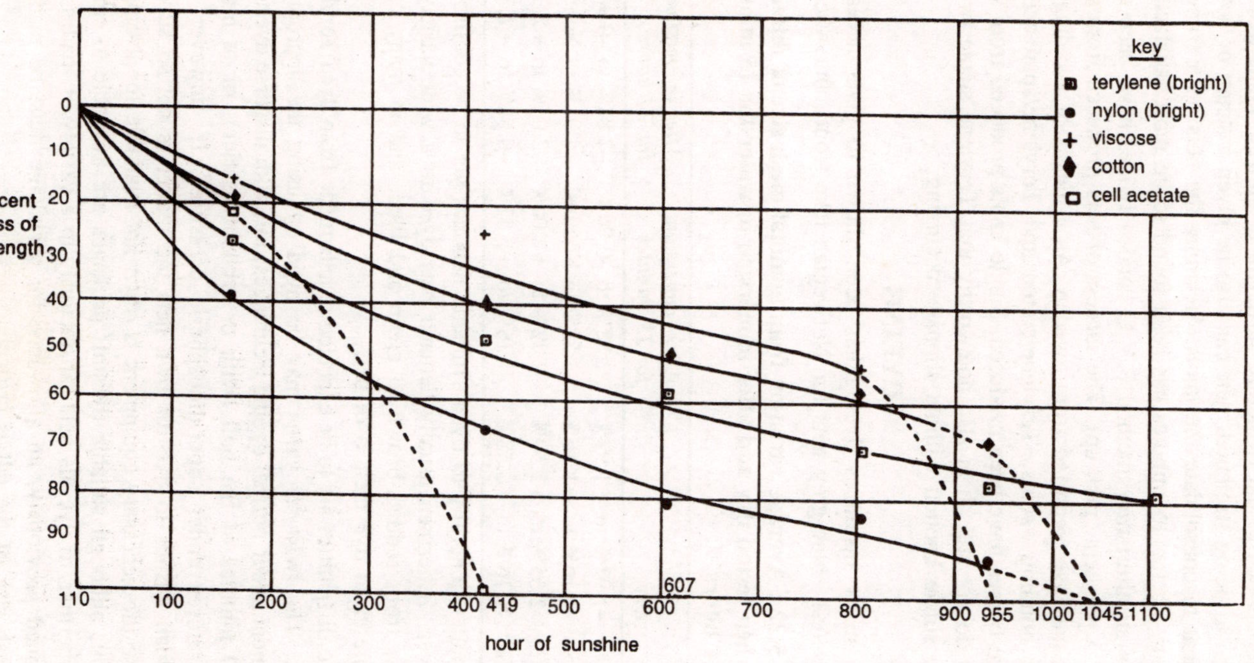

Figure 5.24 : Weather Tendering of Terylene and other Fibres.

tensions applied to the twine are sufficient to pull the knots tight. In addition to being stretched, nets can also be given a steam or hot air treatment to consolidate the knots. Steaming the nets on a frame at 150 deg. C. for 15 minutes has been found to be most satisfactory.

As an alternative method P.V.C bonded twines have been used (about 5 per cent. pick up). The success of nets produced from such twines has been reported from Sweden. As a means of avoiding the use of bonding agents, Terylene/cotton and Terylene/spun acetate mixture twines have been produced. Single knots produced from such twines do not slip. It is also noteworthy that Terylene staple twines can be single knotted without slippage occurring.

DYEING

In many instances dyeing has been shown to give increased catches (e.g. blue-grey nets for Norwegian lake trout), but whether

Table 5.22 : Average variation from nominal mesh size of different nets when tested dry and after immersion in water for 15 minutes and 24 hours.

	Dry	*After immersion for 15 minutes*	*After immersion for 24 hours*
Linen	–1.92% to +1.92%	–2.88% to –0.96%	–2.88% to –0.96%
Cotton	–1.16% to +1.16%	–3.53% to Nominal	–4.05% to Nominal
Nylon	–2.35% to +2.59%	–2.22% to +2.00%	–2.06% to +2.34%
Terylene	–1.16% to +0.58%	–1.16% to +2.32%	–0.58% to +2.32%

this applies generally to dyed synthetic nets has yet to be confirmed. However, fishermen usually demand nets dyed to a wide variety of shades, from reddish-brown to green and blue, and in many cases they prefer to dye their own nets.

No difficulties are to be expected in nets made from dyed Terylene twines. The twist set yarns may be dyed by using the appropriate equipment under normal dyeing techniques, i.e. with disperse dyestuffs for 90 minutes at the boil (with or without carrier), or at higher temperatures under superatmospheric pressures. If, however, the fisherman wishes to dye his own nets the matter is not so simple, because the necessary equipment is often not available for dyeing at the boil, although suitable dyestuff packages are available to enable the fisherman to dye his nets satisfactorily to a variety of shades. As mentioned previously, an allowance for shrinkage must be made if the mesh size of the net is critical.

Tinting of bonded twines at ambient temperature has been successfully carried out and it is possible also to tint the unbonded twine or net at ambient temperature by using disperse dyestuffs dissolved in chlorinated hydro carbons, but some shrinkage of the net may occur. Asphalt based solution can be used to stain and stiffen the net if required.

MESH RETENTION

Some tests have been carried out on the mesh stability of various types of net. The mesh was measured dry at room temperature, before being immersed in water, and also after the nets had been immersed for 15 minutes and 24 hours respectively. The superiority of the fibre in respect of mesh size retention is clearly illustrated. It is interesting to note that after immersion for 24 hours cotton shows the greatest variation with an average shrinkage of 4.05 per cent., followed by linen, nylon and Terylene in that order. It is submitted that a tolerance of 3 per cent. above or below the nominal mesh size is a realistic approach. This excellent mesh size retention of Terylene is of obvious importance with respect to gillnetting.

SUNLIGHT EXPOSURE

The fibre has about the same resistance to daylight and weather as the best of the natural fibres, but because it has a higher strength premium and is rot proof it has a longer useful life. Trials have been carried out with Terylene and nylon twines exposed to daylight and weather in the United Kingdom. Even though the overall time of exposure is comparatively short the superiority of Terylene is clearly demonstrated. Terylene nets and others made from natural and synthetic fibres are now being exposed in several countries, such as Canada, Kenya, India and Scandinavia, but it is too early to draw conclusions from these tests.

DURABILITY

The general toughness of the fibre, its very good mechanical properties, complete resistance to rotting and good resistance to sunlight and weather, ensure a long life for Terylene nets.

EASE OF HANDLING

Nets made of the fibre are easy to handle because thinner twines can be used to give lighter nets. Trawl nets, for example, can be towed more easily. This weight saving factor makes it possible to use larger nets or, alternatively, a small vessel can be used to handle a net. The low moisture uptake means little effective change in weight,

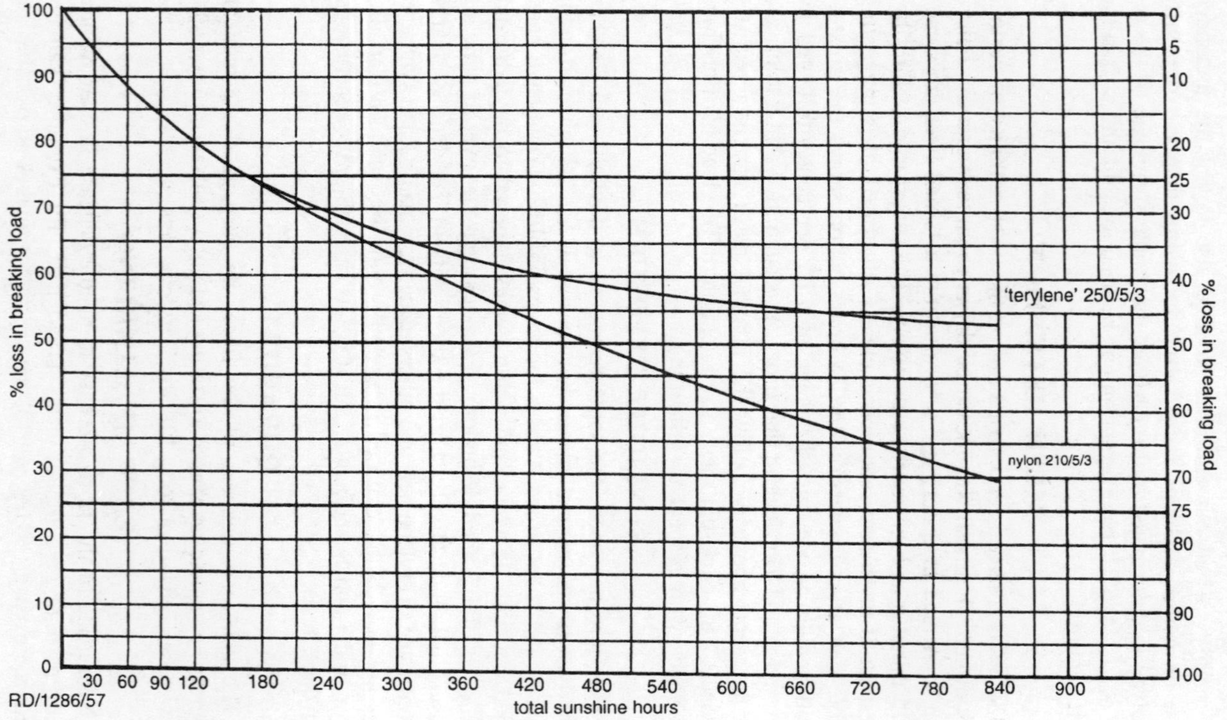

Figure 5.25 : Comparison of 250/5/3 Nylon Twines after Exposure.

and the net is less prone to freeze under icy conditions. The somewhat high specific gravity of the fibre permits a net to sink more rapidly, while the high resistance to stretch is favourable to easy hauling.

COST

The nets are more expensive than natural fibre nets but they provide considerable savings on a price/life basis. In fairness it must be pointed out that the initial cost of the netting is sometimes very considerable and may be more than some fishermen can afford. The risk of accidental damage and loss must also be borne in mind. The difference in price between Terylene and the natural fibres becomes less marked in nets because twines of greater runnage can be used. The price of Terylene compares favourably with that of nylon in terms of pence per pound, but, because of its greater specific gravity, the twines have less runnage than nylon twines of equal thickness. Weight for weight, however, the twines have identical runnage.

CONSTRUCTION

Cords and Twines

The fishnet twines are normally produced from 125 and 250 denier high tenacity filament yarn with plied and cabled constructions. In general, twist is inserted in the singles yarns in twines having plied constructions. The ply twist required to produce a "balanced" twine can be obtained from the formula:

$$\text{t.p.i. strand} = \frac{-\text{t.p.i. singles}}{\sqrt{\text{No. of strands}}}$$

The negative sign indicates opposite twist. No twist is inserted in the singles yarn of twines having cable constructions. The twine twist required to produce a balanced cord is given by the formula:

$$\text{t.p.i. twine} = \frac{-\text{t.p.i. singles}}{\sqrt{\text{No. of twines}}}$$

The amount of twist inserted at each stage in the production of the fishnet twines depends on the hardness of "handle" required and most manufacturers have certain twist factors which enable them to calculate the twist required. In order to produce a twine from Terylene filament yarn which has a balanced construction and a good yarn to twine strength conversion efficiency, twist factors in the range 3.6 to 4.6 are suggested.

Note: Twist factor $= \dfrac{-\text{t.p.i.} \times \sqrt{\text{denier}}}{73}$

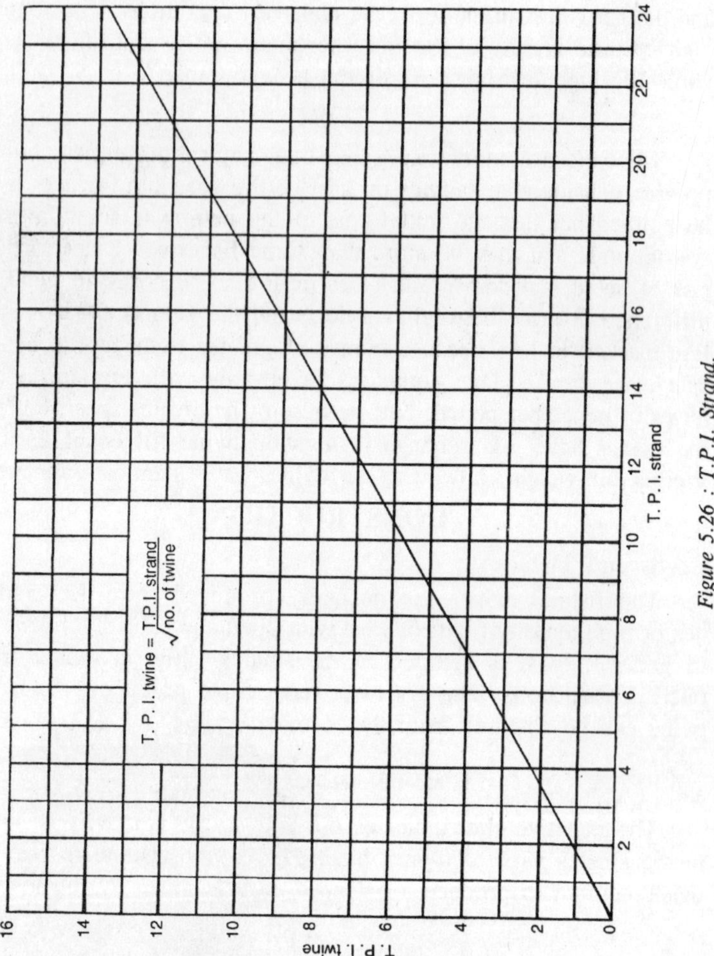

Figure 5.26 : T.P.I. Strand.

The construction for a range of the filament twines.

The information given in this table has been interpreted graphically and from these the twist required for the production of a range of twines having stranded constructions may be obtained. The use of balanced twists gives twines which are completely dead, show no tendency to snarl and have maximum strength.

Some net manufacturers prefer to use twines with unbalanced twists and these must be twist set before net manufacture to consolidate the twist and prevent twine liveliness. The twist setting process is, in essence, one of free shrinkage and as such the properties

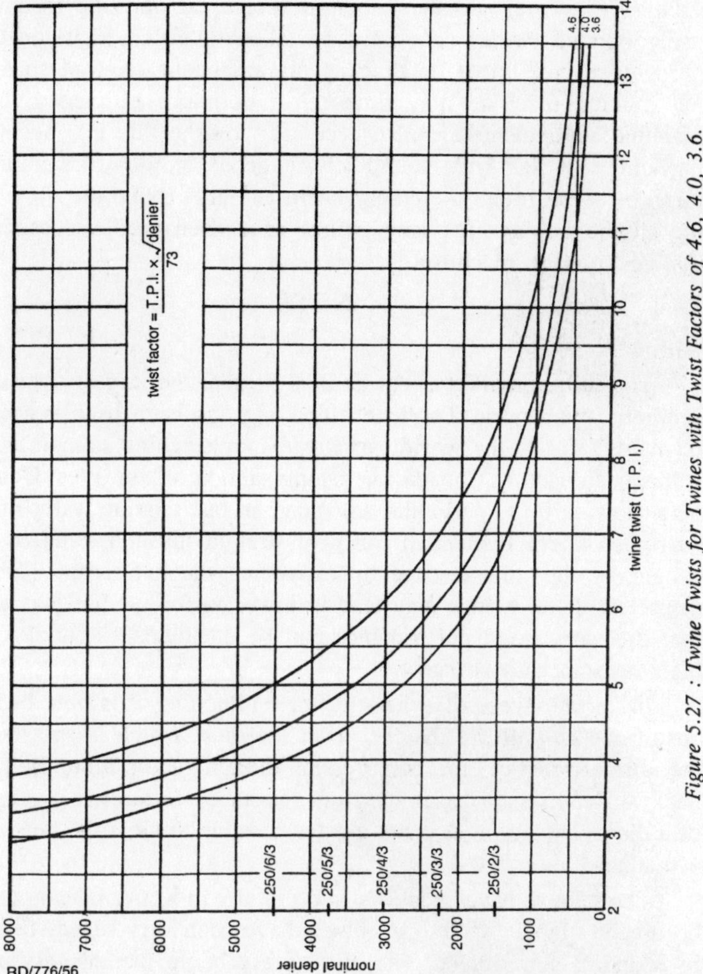

Figure 5.27 : Twine Twists for Twines with Twist Factors of 4.6, 4.0, 3.6.

of the twine alter. The general effect is to increase the denier and extension at break and to decrease tenacity, shrinkage potential and initial modulus of elasticity, the breaking load remaining unchanged. Such changes may detract from the performance of the netting.

Webbing

Terylene filament twines can be handled on traditional net making machines and the nets are said to be easier to make than those of other synthetic fibres, because of the twine's high resistance to stretch. The twines have been processed satisfactorily on single knot machines, such as the Zang and the rather more critical Seriville machine. In using bonded twines care must be taken to ensure that the correct percentage pick-up of bonding agents has been achieved so that the net making machine may function at optimum efficiency. Certain machine adjustments are also necessary to take this into account. In particular, it has been found advantageous to alter the number of turns of twine round the emery beam and also round the drag rod on the shuttle holder. Terylene twines have been used successfully on double knotting machines.

NETS

Gillnets

The gillnet market overseas is at present the largest consumer of synthetic twines and Terylene gillnetting has been used successfully in many parts of the world, principally for catching salmon and cod. It has been used in Canada for fishing salmon off the West Coast, the nets showing up to particular advantage in fast moving water or ocean currents, where lively fish can dive straight through a net or, when caught by the gills, escape by expanding the net mesh. The small diameter of the twines may lead to lower visibility, but it is thought that this very good performance can be attributed essentially to the high resistance to stretch.

In certain types of synthetic fibre gillnetting it is now becoming customary to pull out the fish head foremost which is speedier than the normal practice, i.e. compressing the gills and pulling the fish out backwards. In such cases the high resistance to stretch is somewhat of a disadvantage, as it is not always possible to clear the nets in such a manner.

The gillnets have been used successfully in Kenya lakes for fishing borus and tilapia, while cod gillnets have been very successfully used in Scandinavian waters. The gillnets are being increasingly used in India.

Such gillnets have to date been found more suitable for catching hard rather than soft fish. The twines used have been fine and hard and have cut into soft fish, such as herrings, to such an extent that the fish have become bruised and damaged. When removed from the net by shaking the fish may be decapitated. Chiefly for this reason, Terylene has not yet been extensively used for British home water drift netting. However, such drift netting is being developed and twine constructions have been revised to give thicker and more suitable twines. Further trials are in progress to develop this market.

A mixed high tenacity Terylene filament/spun acetate netting twine is being developed, principally for pilchard nets. It will not bruise or damage the fish and will not slip when single knotted. The principal physical properties of such twines. The use of Terylene core spun cotton yarns is also being considered for similar reasons, but, because of the cotton present, it could not be expected to be so rot resistant as a mixed filament/spun acetate yarn.

Trawl Nets

The filament twine has been used with success for bottom trawls. Trials were carried out some time ago, principally for the codend, because of the shortage of twine, although some complete trawls were also made up. These nets were made by the Great Grimsby Coal, Salt and Tanning Co. Ltd., of Grimsby, and the tests were carried out by distant water trawlers on various fishing grounds. It was found that each trawl net lasted on average about 9 trips and, in one case, 15 trips were made before the net was lost. The normal trawl net is generally good for an average of one trip. It was noted that the Terylene nets had good mesh stability and their resistance to abrasion reduced any chance of the codend bursting as it was hauled in. The nets were much easier to tow through the water and being completely rot-proof, drying was unnecessary. It is understood that cleaner catches were obtained. The ship's crew reported that the nets were more pleasant to handle than the usual nets. The lower moisture uptake was particularly advantageous in the icy operating conditions in Northern waters. Another advantage was that unloading of the codend was sometimes done in fewer operations because of its greater strength.

Risk of loss or accidental damage, of the net is now being greatly reduced by introduction of Decca equipment, which is of special advantage when expensive trawls are used. The longer life and greater security offered by such nets outweigh the occasional losses. Already a number of trawler companies have started their own trials for near, middle and distant water fishing with Terylene trawl nets.

There is perhaps a need for the use of thinner twines to reduce the price of the net and without doubt the introduction of an extra heavy denier yarn will enable nets to be produced more economically.

Appreciable quantities of the fibre are now being used in Sweden for mid-water trawls and trials with such trawls are also being conducted in the United Kingdom by the Ministry of Agriculture, Fisheries and Food.

Purse Seine Nets

Until comparatively recently, little has been done with Terylene purse seine nets, principally because of their cost, which may be as much as £5,000.

Trials have been carried out in Canada with drum seines near Deep Water Bay off the West Coast. The netting, whilst only about a quarter of the weight of corresponding tarred cotton netting and with twines a third as fine, was about 10 per cent. stronger in the wet mesh. This means that smaller vessels can be used or alternatively outsize seines can be carried by vessels which normally work with the smaller seine nets.

The nets wound easily on to the drum, but some difficulty was experienced in playing out, the loose bights of netting tending to get entrapped. The bunt end of the seine was said to be easily held. There was less drag on the net so that it could be closed much more quickly than can normal purse seines. This is a real asset, ensuring a quicker and more efficient fishing operation. The nets were held more easily against the tide, which means, firstly, they can be used in faster waters, thus enabling fishermen to operate in more fishing grounds and, secondly, fishing time can be extended. The nets tended to become entangled, when fish could only be extracted with difficulty, a disadvantage arising from their high order of flexibility. This could be overcome by using coarser twines or applying suitable coating agents.

Despite the disadvantages the preliminary results were most encouraging and, with the suggested modifications, it should be possible to use such purse seines with great profit. It is worthy of note that the somewhat higher specific gravity of the fibre compared with other synthetic fibres was an advantage, since it enabled the net to sink more easily. The lightness in weight of the netting is particularly useful for table seines.

More extensive trials of these purse seines nets are now being made.

Purse/Lampara Nets

Many standard purse/lampara seine nets used in Walvis Bay, South West Africa, have failed prematurely, due it is thought, to chemical contamination, but a preliminary test of Terylene netting for use in these waters has proved most encouraging. Detailed trials are now in progress.

Seine Nets

Trials with Terylene seine nets are to be carried out shortly in the United Kingdom.

OTHER APPLICATION

Other applications include lines and snoods. A fairly substantial market is developing in Norway, where the use of the twines has resulted in increased catches. The high resistance to stretch facilitates pulling in the line and it is easier to tell when the fish are hooked as a more definite response is obtained. Strength stability on wetting and the quick drying properties are also important. The lines are slightly more expensive but this is offset by fishing performance.

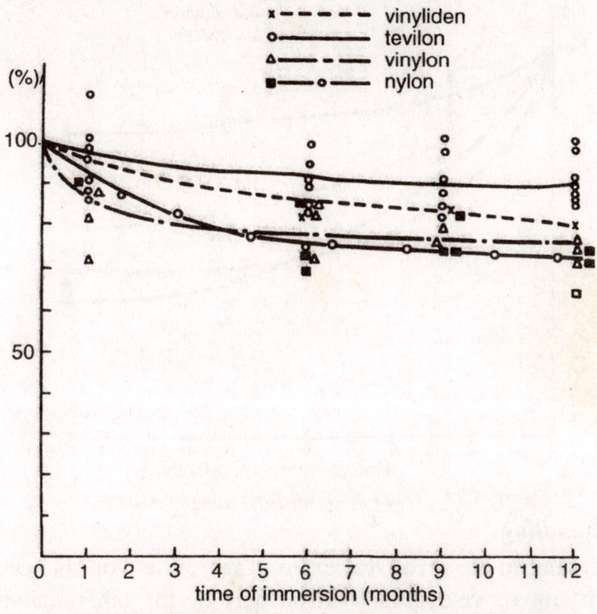

Figure 5.28 : Results obtained from immersion in Sea-water.

The twines are being tried for lobster pots in the United Kingdom, chiefly because of their rot resistance and general toughness.

Fishnet mounting ropes of spun Terylene (flax system) cordage, which are used for supporting the nets, have been tried. The staple yarn is preferred to the continuous filament yarn because its hairy nature minimises slippage of the net and corks. Headlines of the fibre are also being tested.

TEVIRON FISHING NETS

Characteristics of Teviron

Teviron, a polyvinyl chloride synthetic fibre first introduced in October 1956, is produced by Teikoku Rayon. The yarn is available both as filament and as a staple fibre: the former has a silk-like appearance and touch, while the latter resembles wool.

TEVIRON FISHING NETS

Cost

A Teviron net costs less than a net of any other synthetic fibre yarn, and is only 30 per cent. more than a cotton net.

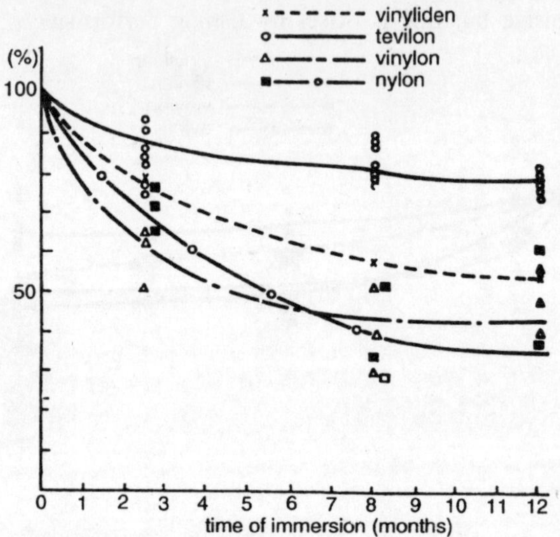

Figure 5.29 : result obtained from outdoor exposure.

Ease of handling

(a) Owing to the great resistance to rot, little work is needed for drying or re-dyeing Teviron nets or for other maintenance services.

(b) As the net does not absorb water, it is very light and can be handled by a smaller crew.

The table below gives the results of an investigation into labour and time factors. The comparison was made between two large fixed nets of approximately similar size, one of Teviron, the other of manila:

	Number of Workers	*Time required for pulling up the net completely from the sea (Minutes)*
Teviron net	44	25
Manila net	60	45

The Teviron net could be pulled up when the current was rapid; the manila net could not be moved.

Suitability for various gear

Gillnets

Teviron twine is flexible, and a minimum of shrinkage assures stability of mesh size. A test of Teviron and nylon drift nets for salmon and trout in northern seas showed that the Teviron net is not inferior to nylon net:

Average number offish caught for each operation

Teviron net	2.27
Nylon net	2.22

Purse seine nets

Teviron needs no drying, and little repair and other maintenance work, allowing more fishing time per day, and longer trips. More fish can be caught because the net sinks rapidly.

Fixed nets

The fibre does not decay even in the warmest season of the year, and has proved a strong and reliable material in rough water.

KREHALON FISHING NETS

General

This vinylidene chloride filament has the greatest specific gravity (1.7) of any synthetic fibre. It is less water absorbent and drains water faster. Nets made with this fibre are pliable and strong and

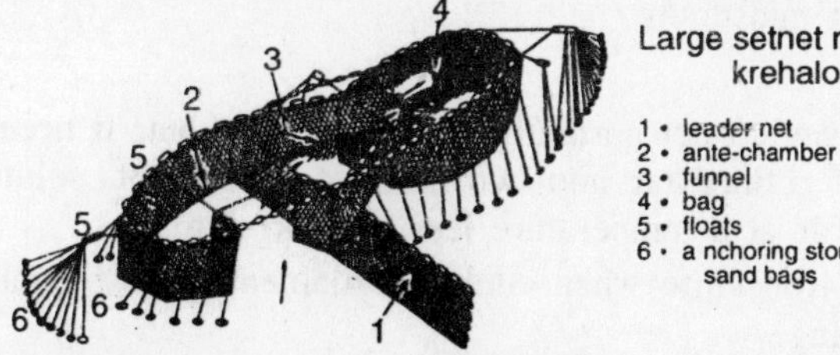

Figure 5.30

retain their shape even in turbulent waters. They sink quickly. The tensile strength of the fibre is 120 to 130 per cent. higher than that of comparable cotton yarn. The flexibility of the fibre gives higher knot strength and greater resistance to impact and friction.

The fibre is particularly suitable for constructing setnets, as fish of any size are unlikely to be injured when trapped because of the pliability of the fibre. Some setnets made of Krehalon have been in use and show no signs of wear. Fishermen say they can withstand the buffeting of the severest typhoons. Stick-held dipnets, surrounding nets, gillnets, longlines and trawlnets made of this fibre show similar durability.

The mesh size of Krehalon nets is usually made slightly smaller than that of nets of natural fibres as it does not shrink in usage. The fibres can be dyed freely in any colour.

Sizes

In multi-filament twines several 180 denier filaments are usually put together to form yarn, twine or line. Mono-filaments are composed of one large filament (i.e. as 1,000 denier). 360 denier is equivalent to cotton yarn of count 20's. Thickness and tensile strength have been taken into consideration. Multifilament nets are more pliable than mono-filament nets.

Nets

The nets are of the same specifications as the conventional cotton yarn or manila twine nets available in knotless or English knot. Smallest are 720 D × 3, mesh size 25 knots per 6 inches.

Ropes

Ropes, lead line, twisted twines, longlines, cord, etc., are manufactured according to specifications.

Cautions regarding use

Long exposure to high temperature can cause a chemical change in the filament. For this reason, a sandy beach or cobblestones are best avoided in midsummer.

Tarring

Krehalon nets do not normally require tarring, but, if necessary, put one part of refined tar into two parts of 5 per cent. solution of neutral soap. Tar at a temperature less than 40 deg. C.

Use a hot iron knife when cutting a filament. This obviates the danger of fraying.

MANILA ROPE

Knowledge of the various intrinsic properties of ropes is indispensable for the right choice and economic use of them. Although experience has resulted in some general knowledge, no definite conclusions based on exact and concrete data are available and very little has been published on this matter. The following article is an effort to contribute some information for fishermen on this subject.

Influence of Basic Manufacturing Procedures

During the process the fibres are twisted successively in opposite directions into yarns, strands, ropes and cable. This results in a decrease in breaking load. A remarkable fact is that even when the same basic fibre material is used, the breaking strength of "Z" yarn is different from that of "S" yarn. It is obvious, therefore, that the actual twisting process itself exercises a very considerable influence on the breaking strength.

Trial I—Influence of the Means of Manufacture

Rope was formerly made exclusively on the rope walk but the process has now been mechanised. Theoretically, it makes no difference how rope is made, although the finger-tips of an experienced ropemaker possess certain qualities which cannot, except with great difficulty, be duplicated in a machine.

To obtain more concrete information on this aspect, a ropemaker was found who agreed to make rope in three different types of machine. He was requested to use different ways from a given yarn, viz. on the rope walk and on two different types of machine. He was requested to do his utmost to ensure that the ropes were as identical as possible. Little difference was apparent in the three ropes at first sight, yet very marked dissimilarities were disclosed by investigation. These differences were not apparent from the slight variations in circumference, which is generally assumed to be an indication of strength when comparing ropes made of the same material.

Table 5.23 : Rope.

Rope closing method	*Averages weight of 1 metre of rope in grammes*	*Total number of turns per strand per metre*	*Average[2] circumference in mm.*	*Averages[1] breaking strength in kg.*	*Breaking length in km.*
Rope walk	90	25.0	38.6	1284	14.27
Machine 1	95	27.3	38.8	929	9.78
Machine 2	95	26.0	37.7	1028	10.82

[1] Average for 16 determinations.

[2] Average for 80 determinations.

This should not be interpreted as an endeavour to expound a theory that rope made on the rope walk is "good" and machine-made rope "bad". Such a sweeping statement could never be made on the basis of a single trial; the only purpose was to prove that the way in which a rope is made can result in significant differences in the breaking strength even when the same fibre gradings are used.

Trial II—Influence of the Lay

Procedure

Three types of rope were made on the rope walk. The conditions, i.e. basic material, equipment, etc., under which the experiments were

Table 5.24 : Soft-laid Rope.

Experiment	*dry*	*%*	*wet*	*%*
Type	3-strand, plain-laid		3-strand, plain-laid	
Total number of yarns	3 × 20=60		3 ×20=60	
Weight per metre in grammes	322.4	100	521.5	161.7
Average number of turns per strand per metre	13.78	100	13.86	100.6
Average circumference in mm.	74.85	100	84.70	113.2
Average breaking strength in kg.	4930	100	5205	105.6
Average breaking length in km. (based on dry weight)	15.30	100	16.14	105.5
Average breaking length in km. (based on wet weight)	9.98	65		
Average duration of test, after application of initial load	6.425 × ½ min.	100	8.125 × ½ min.	126.5
Breaking strength short splice in kg.	4295	87.1	4389	84.3
Breaking strength with overhand knot in kg.	2033	41.2	2333	47.3
Breaking strength with sling around rods 10 cm. in diameter	7344	149	—	—

carried out were identical in all cases. The only intentional deviation was that a number of different lays was selected, so as to produce ropes of hard, medium and soft lay. This was done not only to ascertain possible modifications in the properties of the rope, attributable to the different lays, but also because these modifications are important in some instances in commercial fishing.

The three different ropes were all based on 20 yarns per strand, made from the same fibre mixture.

Table 5.25 : Normal-Laid Rope.

Experiment	*dry*	*%*	*wet*	*%*
Type	3-strand, plain-laid		3-strand, plain-laid	
Total number of yarns	3 × 20=60		3 × 20=60	
Weight per metre in grammes	333.3	100	494.2	148.3
Average number of turns per strand per metre	15.15	100	15.02	99.1
Average circumference in mm.	72.45	100	80.1	110.6
Average breaking strength in kg.	4513	100	4357	96.5
Average breaking length in km. (based on dry weight)	13.54	100	13.06	96.5
Average breaking length in km. (based on wet weight)	—	—	8.82	65.1
Average duration of test, after application of initial load	6.75 × ½ min.	100	8.05 × ½ min.	119.3
Breaking strength short splice in kg.	3652	80.9	3700	94.9
Breaking strength with overhand knot in kg.	1978	43.8	2322	51.5
Breaking strength with sling around rods 10 cm. in diameter	6789	150.4	—	—

Test Methods

During the closing of the rope, 20 bobbins Qrd of the total number used) were marked. The yarn was subsequently examined to establish its various properties. Five lengths were cut off simultaneously from the rope to be tested for breaking strength, etc. The first and third lengths (3.5 metres long) were examined dry, the second and fourth (4 metres long) were examined wet, while the fifth (1 metre long) was untwisted to enable the yarns to be examined dry.

The lengths intended for use in testing the breaking strength at the splice, with overhand knot and as a sling, were then cut off successively.

Breaking strength tests were carried out on a hydraulic breaking-strength machine fitted with clamps. The distance between clamps was 150 cm. and the rate of movement of the straining head approximately 121 cm. per minute.

The ropes examined "wet" were immersed in fresh water for at least 16 hours. Their weight was determined after all non-absorbed water had run off for 30 minutes. The yarn tensile strength test was carried out with 100 cm. between clamps, the rate of movement of the straining head being approximately 25 cm./min.

Results

In the following the results of this experiment are interpreted according to a system developed at the Nederlandsche Visscherij-Proefstation.

Dry Rope

The increase of number of turns per unit length (lay) results in:

(1) increase of weight per unit length (see below)

	relation of weight per unit length	
	yarns (not twisted)	***rope***
Soft laid	1	1.256
Medium laid	1	1.298
Hard laid	1	1.332

(2) decrease in diameter (that means more material per unit diameter)

(3) decrease in breaking strength (that means more material for unit breaking strength)

In the present example the remaining breaking strength of one yarn in the rope in the three different types (1/60 × rope strength) is:

In soft laid rope 82.17 kg. (62.6%)
In medium laid rope 75.22 kg. (57.34%)
In hard laid rope 56.67 kg. (43.20%)

The breaking strength of the yarn, prior to closing the rope, was 131.18 kg. (100 per cent.).

(4) This means that the breaking length, the only accurate tensile strength criterion for rope, cannot be used for the comparison of ropes of different lays.

(5) increase in knot strength.

(6) increase in strength of a loop.

Table 5.26 : Hard-laid Rope.

Experiment	*dry*	*%*	*wet*	*%*
Type	3-strand, plain-laid		3-strand, plain-laid	
Total number of yarns	3 × 20=60		3 × 20=60	
Weight per metre in grammes	340.5	100	473.7	139.1
Average number of turns per strand per metre	16.81	100	16.71	99.4
Average circumference in mm.	65.8	100	69.9	106.2
Average breaking strength in kg.	3484	100	3284	94.3
Average breaking length in km. (based on dry weight)	10.22	100	9.65	94.4
Average breaking length in km. (based on wet weight)	—	—	6.94	67.9
Average duration of test, after application of initial load	5.85 × ½ min.	100	6.70 × ½ min.	114.5
Breaking strength short splice in kg.	3407	97.7	2944	89.6
Breaking strength with overhand knot in kg.	1778	51.0	1878	53.9
Breaking strength with sling around rods 10 cm. in diameter	5878	168.7	—	—

The total extension at break is maximal in normal and less in hard and soft laid ropes.

Wet Rope

Increase in number of turns per unit length (lay) results in:

(1) less water absorption (less increase in weight).

(2) less increase of diameter due to water absorption.

Table 5.27 : Yarn.

	Before closing	*Untwisted from soft-laid rope*	*Untwisted from normal-laid rope*	*Untwisted from hard-laid rope*
Average breaking strength in kg.	131.18	130.76	128.38	121.11
No. of determinations	200	100	100	100
Average weight per 100 metres in grammes	428	434.5	427.4	424.7
Average breaking length in km.	30.65	30.09	30.10	28.45
Percentage of breaking length	100	98.2	98.2	92.8

Table 5.28 : Total Extension.

Tension	*soft rope*		*normal rope*		*hard rope*	
kg.	*dry*	*wet*	*dry*	*wet*	*dry*	*wet*
250	•%	10.8%	5.3%	7.9%	4.0%	6.4%
500	•	13.8	7.4	11	5.6	9.8
1000	10.9	15.8	11.1	13.5	8.4	13.0
2000	13.2	18.0	14.8	16.1	12.8	16.8
3000	14.8	19.6	16.7	17.8	15.0	18
4000	16.1	20.8	17.7	19.2	16.2	0

Contrary to the dry rope, the total extension at break of the wet rope is higher with hard and soft lay than with normal lay. With a tension of 2,000 kg. or less the soft laid rope has the highest extension and the hard laid rope the lowest. Immersion has no remarkable - influence on the number of turns per unit length. There is a certain influence of immersion on the breaking strength but the results available at present are not sufficient to draw reliable conclusions.

Influence of rope manufacturing on the yams

In order to examine the effect of manufacturing, samples of the three types of rope were untwisted and the single yarns tested. It was found that whilst soft and normal lay has only very little influence (loss) on the breaking strength of the single yarn, hard lay results in a certain decrease.

SYNTHETIC YARNS

Synthetic fibre was first used for fishing, as fishing gut. The adaptability of synthetic fibre for fishing gear through cooperation of the Fisheries Agency, Fisheries College, Fisheries Research Institute, fibre makers, fishing net makers, fishermen, etc.

As a result of improvement in quality, advance in net-making techniques and the reduction in cost by mass production in 1953, the demand for synthetic fibre has considerably increased for seine and setnets, and salmon and trout gillnets.

The export of fishing nets has gradually increased , about 1.2 million lbs..

Table 5.29 : Production of Synthetic Fibre Nets.

Unit: Pounds

Fibre	*Type*				
	English Knot	*Reef Knot*	*Knotless Net*	*Moji Net*	*Total*
Nylon	4,489,213	50	82,287	—	4,571,550
Vinylon	1,893,088	3,028,940	810,478	183,394	5,915,900
Polyvinylidene Chloride Fibre	598,034	6,454	1,546,712	—	2,151,200
Polyvinyl Chloride Fibre	70,619	—	104,181	—	174,800
Two Fibres Plied	631,900	—	—	—	631,900
Total	7,682,854	3,035,444	2,543,658	183,394	13,445,350

Owing to the rapid development of different synthetic fishing gears, efforts are being made to determine suitable testing methods for synthetic yarns.

The following methods and standards are proposed.

Method of Testing Twines for Fishing Nets

A—Spun Vinylon. B—Filament Nylon.

C—Filament Vinylidene Chloride and Filament Vinyl Chloride.

In the following text, the materials to which each paragraph refers are shown by the letters A, B and C.

Scope

These standards shall cover the methods of testing twines of the materials shown above.

Definition

Standard Condition in Testing Room (A. B. C.)

Temperature at 20 ± 2 deg. C., and relative humidity at 65 ± 2 per cent. Remark: For determining temperature and humidity, the Assman's Aspiration Psychrometer shall be employed, and the relative humidity obtained from the humidity table by Sprung's formula.

Standard Condition of Test Sample (A. B. C.)

Is when the sample left in a testing room under standard conditions has reached moisture equilibrium.

Moisture Equilibrium. (A. B.)

After pre-drying a test sample at a temperature of 40 to 50 deg. C. It is left in the laboratory under standerd conditions. When the sample has been brought to the constant weight, being steady and uniform in hygroscope state, it shall be considered to be in moisture equilibrium.

Constant Weight (C.)

Weigh the test sample twice successively at intervals of one hour or more at the time of drying. The constant weight shall be considered as that when the difference between the two weighings is under 0.03 per cent.

Absolute Dry Condition (A. B.)

Is reached when the weight of a test sample becomes constant due to being left in a drying oven at 105 to 110 deg. C.

Absolute Dry Weight (C.)

Is the weight under standard conditions of test sample, also called the bone weight.

Remark: Moisture regain under Standard conditions is almost nil.

Constant Weight (A. B.)

Weigh the test sample at intervals of one hour or more for the moisture equilibrium, and at 10 minutes or more for the absolute dry condition. Constant weight is reached when the difference is within 0.05 per cent. of the last respective weights.

Commercial Moisture Regain.

(A.) 5 per cent. of the absolute dry weight (B.) 4.5 per cent. (C.) 0 per cent.

Denier (A. B. C.)

Denier is a unit of fineness, the yarn having a weight of 0.05 gr. per 450 metre length. The denier is equal numerically to the number of grams per 9,000 metres.

Yarn Count (A).

Yarn count to be expressed by the number of hanks (One hank = 840 yards) per pound in weight.

Indication of Yarn Count (A.)

Yarn count to be expressed as follows:

Single Yarn	20 yarn count	20^s
Twine	20 yarn count 2 ply	$20/2^s$
	3 strands of 20 yarn count 2 ply	$20/2/3^s$

Indication of Denier (B.)

Denier to be expressed as follows:

Twine 3 strands of 15 filament 210 denier 5 ply 210D/15f x 5 x 3

Indication of Denier (C.)

Denier to be expressed as follows:

Twine 3 strands of I filament 1000 denier 8 ply

1000D/If × 8 × 3

3 strands of 10 filament 1500 denier 6 ply

1500D/10f × 6 × 3

Indication for Direction of Twist (A. B. C.)

The direction of twist to be expressed by S and Z.

Figure 5.31

The direction of upper twist, middle twist and lower twist to be expressed as follows:

Upper Twist Direction × Middle Twist Direction × Lower Twist Direction.

Indication for Number of Twist (A. B. C.)

The number of twist will be indicated by the numerical value of turns per metre (t./m.) or turns per inch (t./in.). In the case of twine the number of upper twist, middle twist and lower twist to be indicated as follows:

Lower Twist Number × Middle Twist Number × Upper Twist Number.

Example: Upper twist Z 120 turns per metre, middle twist S 5 turns per inch and lower twist Z 10 turns per inch-Z 10 t./in. × S 5 t./in. × Z 120 t./m.

Standard Initial Tension (A.)

Standard initial tension is the first tension in which the yarn is suspended without any extension. In cases where tension affects mean "thickness", "ply number", "apparent yarn count", "twist", "tensile strength and extensibility", "knot strength", "elastic recovery" and "shrinkage in water", the initial tension is given to a yarn of 20s with a load of 5 grams. An initial tension other than the standard one shall be indicated.

Standard Initial Tension (B. C.)

In cases where tension affects mean "thickness", "ply number and filament denier", "denier", "twist", "tensile strength and extensibility", "knot strength", "elastic recovery" and "shrinkage in boiling water" the initial tension used is 1/30 g. of the nominal denier (filament denier x ply number). An initial tension other than the standard one shall be noted.

Sampling and Preparation (A. B. C.)

The test sample is taken by cutting off 5 m. from the end of the yarn. Care must be exercised to prevent change in twist, and no tension given. In case of testing for "knot strength" and "elastic recovery", the test sample will be left in the testing room under ordinary conditions until it reaches the constant weight. When the testing room cannot be kept in standard condition, the test sample will be put in a closed vessel (of 36 per cent. sulphuric acid) and the temperature be kept at the constant degree (20 deg. C.).

Test Items (A. B. C.)

(1) Corrected weight

(2) Moisture regain

(3) Standard weight

(4) Thickness

(5) Ply number

(6) Yarn count or denier

(7) Twist

1. Upper twist number
2. Middle twist number
3. Lower twist number
4. Twist shrinkage

(8) Twist setting

(9) Tensile strength and extensibility

1. Dry tensile strength and extensibility
2. Wet tensile strength and extensibility

(10) Knot strength

1. Wet strength of reef knot
2. Wet strength of English knot

(11) Elastic recovery

(12) (a) Shrinkage in cold water; (b) in boiling water

(13) Moisture absorption

(14) Sinking speed

(15) Weathering resistance

Methods of Testing

The test of "standard weight", "tensile strength and extensibility", "knot strength" and "elastic recovery", will be carried out in a testing room under standard conditions. When the testing room cannot be kept at the standard temperature, the temperature at the time of the test will be noted.

Corrected Weight

Find out the weight of gross and tare of two samples and get the corrected weight from the following formula and indicate the average number.

$$\text{Corrected weight} = W = \frac{100 + R_c}{100 + R}$$

where: W = weight of the test sample (gross weight= tare weight)

R = moisture regain measured (per cent.)

R_c = commercial moisture regain (A = 5 per cent., B = 4.5 per cent., C = 0 per cent.)

Moisture Regain

(*A*. B.) or *Absorbed Moisture (C}* Weigh two test samples both before and after absolute dry condition. The average moisture regain is obtained from the following formula to (one place of decimal):

$$\text{Moisture Regain (per cent.)} = \frac{W - Wd}{Wd} \times 100$$

where W = weight before drying the test sample.

Wd = weight of absolute dry test sample.

Standard Weight (A. B. C.)

Suspend a test sample 2 m. or more in length in a perpendicular position. Then find out the weight of a standard length and indicate the average weight.

Thickness (A. B. C.)

Take five test samples and wind them 20 times closely, parallel to each other, and with a standard initial tension, around a cylinder about 5 cm. in diameter. Measure the breadth and divide by 20. The average number is indicated in millimetres (to one place of decimal).

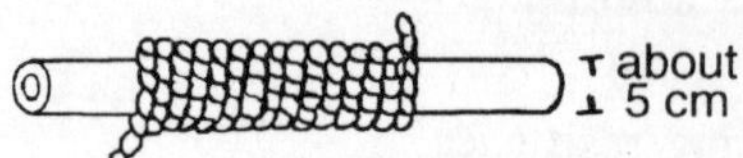

Figure 5.32

Ply Number (A.) Ply Number and Filament Denier (B. C.)

After untwisting the twine the ply number and filament denier are measured.

Yarn Count (A.) or Denier (A. B. C.)

Give the testing sample the standard initial tension and cut it into ten lengths of 30 to 90 cm. After weighting the yarn count or denier is obtained from the following formula:

$$\text{Yam count (s)} = 0.5906 \times \frac{L}{W}$$

$$\text{Denier (D)} = 9000 \times \frac{W}{L} = \frac{5315}{s}$$

where: W = weight of test sample (g.)

L = total length of the test sample (m.)

Remark: The corrected denier shall be calculated by the following formula:

$$\text{Corrected denier (B)} = D \times \frac{100+4.5}{100+R}; \text{(C)} = D \times \frac{100}{100 \times R}$$

where: D = denier measured

R = moisture regain measured (per cent.)

Twist (A. B. C.)

Apply a standard initial tension to the yarn, with 25 cm. (10 inches) between the clamps, with a yarn twist tester, and test for twist as follows, taking an average of ten or more tests

Twist Direction (A. B. C.)

After untwisting the test sample, the twist directions are examined from upper twist, middle twist, and lower twist.

Upper Twist Number (A. B. C.)

Untwisting the upper twist thoroughly, the untwisted number is converted into the number corresponding to one meter or one inch, and this figure is indicated as the upper twist number.

Middle Twist Number (A. B. C.)

All but one strand of the thoroughly untwisted strands of the upper twist are cut out, and then untwisted. This untwisted number is converted into the number corresponding to one metre or one inch, and this figure is indicated as the middle twist number.

Lower Twist Number (A. B. C.)

All but one yarn of the thoroughly untwisted strand of middle twist are cut out, and then untwisted. This untwisted number is converted into the number corresponding to one metre or one inch, and this figure is indicated as the lower twist number.

Twisting Shrinkage (A. B. C.)

After untwisting the test sample, measure the length of the yarn. The shrinkage percentage is measured from the following formula:

$$\text{Twisting Shrinkage (per cent.)} = \frac{L^1 - L}{L} \times 100$$

where: L = Length of the test sample

L^1 = Length after untwisting

Twist Setting (A. B. C.)

Take ten pieces of any test specimen each one metre long, pick up both ends and put them together and count their twisting number.

Tensile Strength and Extensibility

Dry Tensile Strength and Extensibility (A. B. C.)

Employing a suitable "Tensile Strength Tester", and exercising care not to untwist the test specimen, grip one end in the upper clamp, and after applying a standard initial tension, grip the other end in the lower clamp, the clamps being 25 cm. apart, and tension speed being 30 cm./ min. Then measure the tensile strength and extensibility (kg. and per cent.) at the time of breaking. Take the mean of ten or more tests. (Carry to three figures.)

Wet Tensile Strength and Extensibility (A. B. C.)

The test specimen is immersed in water at room temperature, and after it has thoroughly absorbed water, the wet tensile strength and extensibility is measured in a similar manner.

Knot Strength

Reef Knot-Wet (A. B. C.)

The standard initial tension is applied to the test specimen, and a reef knot is made as shown in figure 3. Then it is immersed. After the test specimen has thoroughly absorbed water, it is gripped in the clamps, keeping the knot in the middle, and the wet strength (kg.) is measured.

The average of ten tests is taken.

English Knot-Wet (A. B. C.)

A standard initial tension is applied to the test specimen and an English knot breaking strength is measured. The average of ten tests is taken.

Elastic Recovery (A. B. C.)

Employing a suitable "Tensile Strength Tester", the test specimen is extended to 12.5 mm. (5 per cent. of the original length). Then the load is removed for two minutes, and again the standard initial tension is applied and the remaining elongation is measured. The elastic recovery is measured from the following formula:

$$\text{Elastic recovery (per cent.)} = \frac{12.5 - L}{12.5} \times 100$$

Where: L = remaining elongation (mm.)

The average of ten tests or more is taken.

Shrinkage in Water (A.)

A standard initial tension is applied to the test specimen, and a section one meter long is marked. Then a loop is made by tying both ends together outside the marks. Immerse and after the specimen has thoroughly absorbed water, dry it in the air. Apply an initial tension again and then measure the length of the marked section. The shrinkage in water is obtained from the following formula (to one decimal):

$$\text{Shrinkage (per cent.)} = \frac{1000 - L}{1000} \times 100$$

where: L = length (mm.) of the air dry test specimen after treatment.

Shrinkage in Boiling Water (B.)

Take the test sample of 1 m. or more in length, fix both ends of the yarn together to overlap two-fold, and then measure a length of 50 cm., marking the two end points. Apply the weight to give the standard initial tension. Remove the weight, immerse the test specimen in boiling water for 30 minutes, and then take out of water to allow to dry in the air. Applying the same weight again, measure the length of the air dried sample. Calculate the shrinkage according to the following formula and take mean value of 5 tests or more (to one place of decimal).

$$\text{Shrinkage in boiling water (per cent.)} = \frac{500 - L}{500} \times 100$$

where: L = length of air-dried test specimen after immersion (mm.)

Moisture Absorption (A. B. C.)

Take about 2 m. length (if this weighs less than 2 g., take about 2 g.) of the test specimen, and after measuring the weight in air dry condition, immerse it in water at room temperature. Then take the specimen out of the water and allow the water to drip for two minutes; take the weight, and calculate the moisture absorption from the following formula:

$$\text{Water absorption (per cent.)} = \frac{W^1 - W}{W} \times 100$$

where: W = air dry weight of test specimen

W^1 = water absorption weight of test specimen

Sinking Speed (A. B. C.)

Take a piece of twine of 2 cm. in length with a knot in the

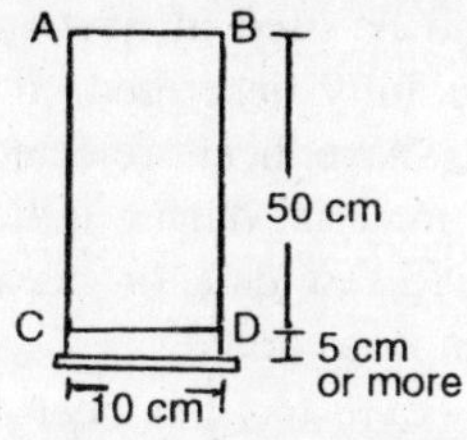

Figure 5.33

middle, let it sink from the surface of water at 20 deg. 5 deg. C. contained in a glass beaker. Measure the speed per second from AB to CD, a distance of 50 cm. The test specimen should be previously deaerated and immersed in clear water. Take the average of three or more tests.

Remark: Tests in which the specimen sank in a diagonal way or sank close to the wall of the vessel should be rejected.

Weathering Resistance (A. B. C.)

Take two test specimens at random, fix them to a textile testing board, and measure their strength (ten times × 2) by exposing them under the condition mentioned below for 20 hours with the weathering resistance tester made in the form of the weather-O-meter. This strength is compared with that of the control sample and the average value is taken in per cent.

Arc 130 to 145 V. 50 to 60 c/s 15 to 17A to 2 sets

Carbon for Arc " 70 (Solid) and " 20 (Core), or other corresponding types.

Temperature in the tester	40 to 50° C.
Revolving Speed	once per min.
Exposing time	102 min.
Spraying time	18 min.
Pressure of spraying water	25 to 30 lbs./in.^{2}
Water requirement for spraying	20 to 30 gal./hr.

PROPERTIES OF NETTING AND TWINES

During the past decade, many new materials, particularly synthetic fibres, have been introduced into commercial fishing, usually being presented as substitutes for other materials. If these new materials are to be substituted rationally, certain physical properties should be determined quantitatively a *priori,* although the data describing the new and the conventional materials need only be relative. But where these materials are to be used in new applications, then the physical properties should be fully described. It is the author's opinion that many new materials have been rejected as unsatisfactory, through improper use of the material during initial trials, a result of complete neglect of quantitative test data or because only irrelevant physical properties have been considered.

The misuse of test results has been illustrated in the selection of twine sizes for salmon gillnets. Because fish must be caught from water, wet strength is more important than dry strength even though, for convenience, many people measure only the latter. At one time all salmon gillnets on the British Columbia coast were made of premium-grade linen which increases about 50 per cent. in strength when wet. In contrast, nylon loses about 15 per cent. strength in water. Therefore, if a nylon twine is chosen to give the same dry strength as the linen it is to replace, the wet strength of the nylon net will be little more than half that of the linen net.

The manufacturers of nylon gillnets were aware of this and, for the first experimental nets, selected twine sizes of sufficient wet strength to carry normal fishing loads. The results were satisfactory. Nevertheless some net men and fishermen still select their nylon gillnets on the basis of hand tests applied to dry netting, with the result that nets are chosen too light for their loads. They 'are torn more easily, and their owners erroneously claim that the quality of nylon is becoming poorer. Because wetting affects the strength of different materials in different ways, wet strength tests are much more significant to fishing gear design than are dry strength tests.

Another example of the improper use of test results is in the choice of the physical property which the test evaluates. Because twine must be knotted to form netting the strength of the knot or the mesh is more important than the strength of the straight twine. Soon after the introduction of nylon 66 multifilament gillnets into the British Columbia salmon fishery, nylon 6 multifilament gillnets began to appear. On one occasion at least, the tensile strength of twine from the latter nets was found to be about 40 per cent. weaker than nylon 66 twine of the same weight, and nylon 6 was rejected as being unsatisfactory for salmon gillnets. However, because these two nylons react differently to knotting, the mesh of a nylon 6 is only about 20 per cent. weaker than that of a nylon 66 net of the same weight. Considering their lower price nylon 6 nets do have a place in the British Columbia salmon fishery in competition with the nylon 66 nets. Because knotting affects the strength of different materials in different ways, knot strength or mesh strength tests are much more significant to fishing gear design than are tensile tests on the straight twine.

Many new materials have been made available, unaccompanied by specific test data describing their physical properties. Inquiries addressed to the suppliers often elicit no further information, or bring data which have little significance to fishing gear applications. It has therefore been necessary for the Fisheries Research Board of Canada to perform its own tests prior to recommending how these materials may be used to greatest advantage. Obtaining our own test data has the threefold advantage that: (1) pertinent properties may be measured; (2) the test data are reliable to the best of our ability; and (3) by using consistent test procedures, data on all materials, both conventional and new, may be compared.

This paper presents the results of our many tests on seven different fibres in eighteen different forms suitable for use in commercial gear for the British Columbia fishery. Because different materials were designed for different applications, not all properties of all materials were measured, and the accompanying table is not complete. However, except for the mesh strength of hemp and medium-laid cotton, test procedures were consistent throughout and the test data presented are comparable between different materials. All test results have been reduced to "parametric" form, that is, to properties which are reasonably constant over the complete twine-size range of a given material and style. Sometimes corresponding materials from different manufacturers have properties in which measured differences are statistically significant, but these have been grouped into the overall averages presented.

MATERIALS TESTED

The cotton netting and twine, the Manryo twine, and the twisted, spun, nylon 66 twine, were of "cable" construction: i.e., the twine contains three or four plies and each ply contains several yarns. The yarn style is identified by the British system of hanks (1 hank= 840 yards = 768 metres) per pound (453.6 grams). The twine size is identified by two numbers, the first being the number of yarns per ply and the second the number of plies in the twine. The total number of yarns in the twine is the product of these two numbers. In the mediumlaid twines the helix angle of the plies was about 54 deg., in the soft-laid twines about 28 deg., and in the extrasoft-laid twines about 22 deg.

The linen, ramie, and hemp twines were of plied construction, i.e., the yarns were doubled directly into the twine. The linen yarn is identified according to an arbitrary commercial numbering system, but the approximate number of leas (1 lea = 300 yards = 274.3 metres) per pound (453.6 grams) is quoted in parenthesis. The ramie yarn is identified according to this same leas-perpound system, and the hemp yarn is identified by the number of pounds (1 pound = 453.6 grams) per spindle (14,400 yards = 13,167 metres). In all cases the twine size is identified by the number of yarns which have been twisted into the twine.

The monofilament twine is identified by its nominal diameter.

The continuous multifilament yarns (nylon and Terylene) are identified both as to weight or total denier per yarn and as to the number of filaments per yarn. The twisted twines were of "cable" construction and are described by two numbers in the same way as are the cotton twines.

The braided twines are described by the total number of yarns in the braid.

PROPERTIES AND TEST PROCEDURES

The nett weight (gravitational force less buoyant force) of the netting in the water is an important factor in determining how it will lie or move in the water and how much lead and what float capacity is required for a given piece of equipment. For example, the low density of nylon contributes toward the fishing efficiency of nylon gillnets but creates handling problems in nylon purse seines. The nett weight was measured by weighing an air-dry sample of netting in the air, by soaking the sample overnight in distilled water at 70 deg. F. (21.1 deg. C.) and then reweighing while still immersed in the water. The weight in water was then expressed as a per cent. of the air-dry weight. This weight per cent. is numerically equal to the weight in water in pounds (or kilograms) per hundred pounds (or kilograms) air-dry in air. The specific gravity was calculated from the weight per cent. in water. The nett weight (per cent.) of the netting in salt water or in fresh water and at any temperature may be estimated by the relation:

$$\left.\begin{array}{r}\text{Weight (\%)}\\ \text{in sea water}\end{array}\right\} = \left[1 - \frac{\text{specific gravity of sea water}}{\text{specific gravity of material}}\right] \times 100$$

The change in length due to wetting must be known where the mesh length is specified only for dry netting. In many types of fishing gear, such as gillnets and trawls, the mesh length of the web in the water is an important factor in determining the fishing efficiency of the gear or in releasing under-size fish for conservation purposes.

Different materials react differently to wetting, e.g., most natural fibres shrink and most synthetic fibres extend, so that a quantitative measure of this reaction is needed before dry mesh length can be properly specified for a given application. The change in length (per cent.) due to wetting was determined by measuring the length of meshes in the sample when air-dry, by soaking the sample overnight in cold tap water, by measuring the length of meshes in the same sample while wet, and by expressing the change in mesh length which occurred during wetting as a per cent. of the average mesh length of the dry sample. Shrinkages are prefixed by the negative (–) sign and extensions are prefixed by the positive (+).

The linear density is a convenient measure of the size of the twine. Often twine must be purchased by weight but is required for use by length. A knowledge of linear density therefore permits estimation of the quantity (weight) of twine required to construct a given piece of equipment. In keeping with common Canadian practice, the indirect system (length per unit weight) is used. Ten one-yard (1 yard = 0.9144 metres) pieces of air-dry twine were cut under four ounces (113.4 grams) tension and were weighed collectively to the nearest milligram. The linear density of the twine (1 yard per pound = 2.016 metres per kilogram) was calculated by dividing the weight of the ten-yard sample in grams into 4536. The effective linear density per yarn was then calculated by multiplying the linear density of the twine by the total number of yarns in the twine. Further division by 1,000 reduced the number of integers required for the table and changed the length unit from yards to kiloyards (1 kiloyard = 1,000 yards= 914.4 metres).

The diameter of the twine is a major factor in determining how strongly moving water pulls the netting, how easily swimming fish can move the netting, and what power is required or speed attained while towing a given piece of equipment. The diameter was measured by placing four parallel strands of air-dry twine under the 1 -25 inch (3.15 centimetres) diameter circular foot of a dial gauge. The total force exerted diametrically on the twine by the foot during measurement was 6 ounces (170-1 grams). The separation of the foot from the supporting anvil caused by the interposed twine was indicated by the dial gauge in mils (1 mil = 0.001 inches = 0.0254 millimetres) and was reported as the twine diameter. This method of measurement has certain defects when used to determine the diameter of twine for fishing gear. Twines made from staple fibres have loops and fibre ends protruding from the twine which increase the resistance to movement of the twine through water but which are easily pressed to the twine during diameter-measurement; and soft-laid twines are pressed out of round by the measuring device more than are harder laid twines. However, when interpreted in the light of these effects, twine diameters measured in this way can be useful.

The effective cross-section area per yarn was calculated as a dimension which is a function of the diameter and yet is reasonably constant over all twine sizes for each yarn style and twist hardness. The effective cross-section area of the twine in circular mils (1 circular mil = 0.0005067 square millimetres) is the square of the twine diameter in mils (1 mil = 0.001 inches = 0.0254 millimetres). Further division by the total number of yarns in the twine gives the effective cross-section area of each yarn. The diameter of any size twine of the same style may then be estimated by multiplying the effective yarn area by the number of yarns in the twine and by deriving the square root of this product.

Total extension and breaking load of twines were measured simultaneously on a Scott Model J2 recording tester of 100 pounds and 400 pounds maximum capacities, using type X-1 clamps. The tester was set so that 10 inches (25.4 centimetres) of untensioned twine lay between the nips of the clamps prior to test. For knot strength tests, the twine was cut and the appropriate knot, single weavers' knot or double weavers' knot, was tied in the twine by hand.

For testing the mesh strength of netting for seines and trawls (medium-laid cotton and hemp), small panels of netting were cut from the sample, all the meshes along two opposite sides of the panel were threaded respectively onto 4 inch (6.35 millimetres) diameter rods on the tester, and the sample was stressed to rupture through separation of the rods by the tester. The mesh strength was then calculated by dividing the panel strength by the number of meshes in the width of the panel. The netting was tested in both directions relative to the selvedge and the overall average is reflected in the data quoted here. Netting in seines and trawls is stressed in use over several meshes at a time rather than on single meshes; therefore, by testing panels as above, irregularities in the physical dimensions of the netting result in lowered mesh strength as is experienced in use. Netting in gillnets is more commonly stressed on individual meshes so that these materials (linen, ramie, nylon, and Terylene) were tested one mesh at a time. The mesh was placed around two one-inch-diameter (2.54 centimetres) drums so that no knots touched either drum, and the drums were separated by the tester until the mesh broke. For dry strength, the materials were tested in their natural, air-dry state, and for wet strength, the materials were soaked overnight in cold tap water.

Specific strength is dimensionally identical with tenacity, the only difference being in the units employed. Arithmetically it is the test strength per twine (i.e., twine strength, knot strength, or half the mesh strength) in pounds (1 pound = 453.6 grams) multiplied by the linear density of the air-dry twine in kiloyards per pcund (1 kiloyard=1,000 yards=914.4 metres). This specific strength'may be converted into tenacity in grams per denier by further multiplying by 0.1016. For specific strength units, the weight or gravitational force was cancelled against the strength or breaking force, leaving length as the unit for specific strength. This permits an easier understanding of the physical significance of specific strength or tenacity than is possible with the metric units, grams per denier. Specific strength (and tenacity) is the length of twine the weight of which equals the breaking strength of that same twine. For the same strength-quality, an increase in weight per unit length causes an identical increase in strength, leaving the specific strength (and tenacity), which could well be called breaking length, constant. For a given material of a given quality the specific strength is reasonably constant over all twine sizes; and the breaking strength of a particular twine may be estimated from the specific strength by dividing this latter by the linear density of the twine. Specific strength figures may be used directly to compare the strength of different materials on an equal weight basis, e.g., high tenacity nylon 66 when wet is seen to be slightly weaker than wet linen, a relation often obscured by the fact that nylon 66 and linen twines are not made in identical weights. Specific strength may also be used as a measure of strength-quality, the specific strength of different specimens being directly comparable even though the yarns may be spun to slightly different weights. Finally, the tensile strength (breaking force per unit area normal to the direction of the force) of the fibre itself is related to the specific strength through the fibre density so that the usable tensile strength of the fibre in pounds per square inch (1 pound per square inch = 70.307 grams per square centimetre) may be estimated from the relation:

Tensile Strength = 1,300 × Specific Gravity × Specific Strength.
Comparison of this usable tensile strength with data available elsewhere for the virgin fibre gives an idea of the manufacturing and geometrical efficiencies realised in these fishing gear materials.

The load-extension characteristics on initial loading of each twine, both dry and wet, were drawn autographically by the Scott tensile tester. The average of ten replicate test plots was redrawn showing the per cent. extension (increase in length under load expressed as a per cent. of the unstressed length) as a function of specific load. Specific load, analogous to specific strength, is the actual load in pounds (a pound = 453.6 grams) multiplied by the linear density of the air-dry twine in kiloyards per pound (1 kiloyard = 1,000 yards = 914.4 metres). All these average curves for all twine sizes of a given material and style were drawn against the same reference axes and an average of the curves was then drawn. The stretch data given table 1 were read from this last average plot against the corresponding specific strengths of each material in each state. These data do not include elongations consequent upon tightening of knots, which are functions of the initial knot tightness; but they do include irreversible elongations in the twine itself. Undoubtedly, the load-extension characteristics under cyclic loading are more significant to fishing gear design than are initial loading characteristics, but the former cannot be obtained with our Scott tensile tester.

Data describing the tensile strength of the dry, straight twine often have little application to fishing gear design; in fact, they can lead to serious error. Even so, such data are frequently all that are available. Th- change (increase or decrease) in tensile strength which results from wetting the materials is reported as a per cent. of the dry strength to assist in estimating the strength in water where only dry strength data are available. Similarly, the per cent. of the twine strength which is; retained in knotted structures is reported to assist in estimating the knot or mesh strengths where only twine strength data are available. These per cent. changes in strength to wetting and per cent. strength efficiencies of knots and meshes were calculated from the specific strength data.

The tensile strength of the fibrous materials used in fishing gear can be measured relatively easily, so this property is often the only one evaluated, and is sometimes blindly adopted as the only measure of quality and performance. Where the material is to be subjected only to dead loads, tensile strength is an adequate measure of quality. However, fighting fish are not dead loads, and it is more important that the material be able to absorb the energy inflicted on it by the fish than merely carry the weight of the fish. A dramatic illustration of this is that nylon gillnets for salmon need not be so strong as linen gillnets to hold the same fish, because the greater elastic extension of the nylon imparts greater ability to absorb the energy expended by the salmon. The "toughness" used here is a rough measure of the energy required per unit length of twine to break that twine, whether it be straight, knotted, or in a mesh. Arithmetically it is one-half the product of the breaking load and the fractional extension from unstressed length to rupture. This estimate is in error to the extent that the load-extension curve is assumed to be linear; but, whereas most fishing gear materials deviate from linearity toward the extension axis, the error is in the same direction in most cases and the figures are reasonably comparable. A further loss of correlation between this calculated "toughness" and performance toughness originates from use of the load-extension relation under initial load-permanent elongation is included. However, despite approximations in its estimation, "toughness" is more important in the design of some types of fishing gear than is tensile strength. "Toughness" is reduced to the parametric form, viz., toughness index, by multiplying it by the linear density of the air-dry twine. This toughness index is thus the energy required per unit weight of twine to break that twine, and is reasonably constant for all twine sizes of the same material in the same style.

DEFINITIONS AND DIMENSIONS

A. Parametric properties, quantitatively reported in the table, which are reasonably constant for all twine sizes of the same material and style.

1. Weight (per cent.) in water is the nett weight (gravitational force less buoyancy) of the material when immersed in distilled water at 70 deg. F. (21.1 deg. C.), expressed as a per cent. of the air-dry weight.

$$\text{Weight (\%) in water} = \frac{\text{Nett weight in water}}{\text{Air-dry weight in air}} \times 100$$

2. Specific gravity is the density of the material relative to the density of pure water at 4 deg. C. 39.2 deg. F.).

$$\text{Specific gravity} = \frac{0.998}{1 - 0.01\ (\text{weight (\%) in water})}$$

3. Change in length (per cent.) to wetting is the increase (+) or the decrease (–) in length caused by soaking the material in water, expressed as a per cent. of the air-dry length.

$$\begin{array}{l}\text{Change in length} \\ \text{(\%) to wetting}\end{array} = \frac{\text{Wet length} - \text{Dry length}}{\text{Dry length}} \times 100$$

4. Linear density is the length per unit weight (indirect system).

$$\begin{array}{l}\text{Linear density}\\ \text{per yarn}\\ \text{(kyd./lb.)}\end{array} = \frac{1}{100} \times \begin{array}{c}\text{Linear density of}\\ \text{twine (yd./lb.)}\end{array} \times \begin{array}{c}\text{Total number}\\ \text{of yarns}\\ \text{in the twine}\end{array}$$

1 yd./lb. = 2.016 m./kg.

1 kyd./lb. = 1000 yd./lb.

= 2.016 m./g.

5. Effective cross-section area is the area of the circle whose diameter equals the measured diameter of the twine.

$$\begin{array}{l}\text{Effective cross-section area}\\ \text{per yarn (circular mils)}\end{array} = \frac{(\text{Twine diameter (mils)})^2}{\text{Number of yarns in twine}}$$

1 circular mil = area of circle 1 mil (0.001 in.-0.0254 mm.) in diameter.

= 0.0005067 sq. mm.

6. Total extension (per cent.) at break is the increase in the length of the twine caused by the load required to break the twine (straight, knotted, or mesh) expressed as a per cent. of the unstressed length.

$$\text{Total extension at break} = \frac{\text{Twine length at break} - \text{Unstressed length}}{\text{Unstressed length}} \times 100$$

7. Specific strength (kyd.) is the effective tenacity.

$$\begin{array}{c}\text{Specific strength}\\ \text{(kyd.)}\end{array} = \frac{1}{1000} \times \begin{array}{c}\text{Linear density}\\ \text{of air dry}\\ \text{twine (yd./lb.)}\end{array} \times \begin{array}{c}\text{Measured}\\ \text{strength}\\ \text{(lb.)}\end{array}$$

1 kyd. = 1000 yd.

= 0.1016 g./den.

8. Strength change (per cent.) to wetting in the increase (+) or decrease (–) in tensile strength caused by soaking the material in water, expressed as a per cent. of the dry strength.

$$\begin{array}{l}\text{Change in strength}\\ \text{to wetting}\end{array} = \frac{\text{Wet strength} - \text{Dry strength}}{\text{Dry strength}} \times 100$$

9. Strength efficiency (per cent.) is the per cent. of the straight twine strength which is retained in knotted structures.

$$\begin{array}{c}\text{Strength efficiency}\\ \text{of knotted twine}\end{array} = \frac{\text{Knot strength}}{\text{Straight twine strength}} \times 100$$

$$\begin{array}{c}\text{Strength efficiency}\\ \text{of mesh}\end{array} = \frac{\text{Mesh strength}}{2 \times \text{Straight twine strength}} \times 100$$

10. Toughness index (kyd. lb./lb.) is an approximate estimate of the energy per unit weight of twine required to break the twine.

$$\text{Toughness index} = \frac{1}{2} \times \frac{\text{Total extension (\%) at break}}{100} \times \begin{matrix}\text{Specific}\\\text{strength}\\\text{(kyd.)}\end{matrix}$$

1 kyd. - lb./lb. = 3000 ft. - lb./lb.

= 91440 g. - cm./g.

B. Specific properties, variable with twine size, which may be estimated from the values of the parametric properties given in the table.

1. Twine weight (yd./lb.) $= 1000 \times \dfrac{\text{Linear density per yarn}}{\text{Number of yarns per twine}}$

1 yd./lb. = 2.016 m./kg.

2. Twine weight (mils). $= \left(\begin{matrix}\text{Effective}\\\text{cross-section}\\\text{area per yarn}\end{matrix} \times \begin{matrix}\text{Number}\\\text{of yarns}\\\text{per twine}\end{matrix}\right)^{\frac{1}{2}}$

1 mil. = 0.001 in.

= 0.0254 mm.

3. Twine strength (lb.) $= \dfrac{\text{Specific strength of twine} \times \text{no. of yarns}}{\text{Linear density per yarn}}$

1 lb. = 453.6 g.

4. Mesh strength (lb.) $= \dfrac{2 \times \text{Specific strength of mesh} \times \text{no. of yarns}}{\text{Linear density per yarn}}$

1 lb. = 453.6 g.

5. Toughness (ft.-lb./ft.) $= \dfrac{\text{Toughness index} \times \text{Number of yarns}}{\text{Linear density per yarn}}$

1 ft.-lb./ft. = 453.6 g.-cm./g.

NET TWINES AND NETS

The use of synthetic yarns during the last few years as material for nets has caused a growing interest in testing methods for twines and nets. This is expressed by more exact measuring and by development of new measuring methods to deal with the special problems which have arisen.

The "Bureau International pour la Standardisation de la Rayonne et des Fibres Synthetiques" (BISFA) with a membership of all the main North and West European producers of man-made fibres, and connected with the International Organisation of Standardisation (ISO), is engaged in standardising the different testing methods for textile and tyre yarns, and for this purpose publishes the "BISFA-rules". In the USA similar work is done by the American Society for Testing Materials (ASTM).

This contribution discusses various testing methods for net yarns and twines, and nets. Some of these methods are already known in the textile industry, some, of specific interest to the fishing industry, relate to testing methods developed in our own laboratories. The BISFA rules are quoted with regard to the first group, but proposals are made for the second primarily from the point of view of the yarn producer.

The term "yarn" is understood to mean a continuous strand of fibres and/or filaments, from which twist, if any, can be removed in one operation; strand is the product obtained by twisting together two or more twisted yarns, twine is the product obtained by twisting two or more strands together.

METHODS OF TESTING YARNS AND TWINES

Yarn Number

Several units are used for the numbering of net yarns and twines. The ISO Conference, May 1956, recommended the use of the combination of grammes per kilometre of yarn, called "tex", as a unit for yarn numbering and we strongly advocated the use of the same unit for net yarns and twines. The numbers in tex, denier and m/kg. for some yarns.

Table 5.30

m/kg.	100	500	1,000	5,000	10,000	50,000	1,000,000
Td (g./9000 M.)	90,000	18,000	9,000	18,000	900	180	90
Tex (g./1000 m.)	10,000	2,000	1,000	200	100	20	10

Twist

Since the determination of the twist of net yarns and twines is the same as that of textile yarns only a few points will be mentioned..

The direction of twist is normally indicated by the letters Z or S. In order to maintain the metric system we recommend that the twist shall be expressed as the number of turns per metre of twisted yarn or twine (t/m.).

Below is illustrated the indication of the twist construction of a twine:

100 tex Z 400 × 2 S 300 × 3 Z 100 means a twine composed of 6 yarns: every yarn is first twisted to 400 t/m. in Z-direction, then two of these twisted yarns are twisted together to 300 t/m. in S-direction, and finally 3 of these strands are twisted in Z-direction to 100 t/m.

Strength and Extensibility

The stress-strain curve, and the strength and total extension at break, are determined by the load-extension tester or dynamometer. The many types of dynamometer can be classified according to the time conditions under which the determination of strength and extension are performed. The time factor plays an important part in this determination for the materials used in fishing nets. The velocity and the way of increasing the strain have an influence on the shape of the stress-strain curve and the values of strength and extension at break. Efforts have been made to realise one of the following principles in making the dynamometers:

1. the length of the test object increases in proportion to the time; thus the rate of extension is constant;
2. the force on the test object increases in proportion to time; constant rating of loading.

The electronic recording dynamometers which are characterised by their stability, accuracy and versatility, are of recent design.

One clamp is moved at constant speed while the force is measured by the other clamp, connected with an electric element for measuring the force.

These meters have some specific properties, viz., the possibility of recording both a decrease and an increase in stress, and the many and wide measuring ranges that can be covered by a single instrument. These make them particularly suitable for testing twines and nets.

In practice, the wet strength and total extension at the breaking point of the twines are of primary importance. As regards modern, synthetic materials, such as e.g. nylon 6, nylon 66 and polyesters, the wet stress-strain properties do not differ very much from those measured in dry condition, so it suffices to determine the latter only for routine measurement.

The BISFA rules describe the measuring conditions as "the length of the test specimen between the jaws shall be 50 cm., and the mean time to break shall be 20 ± 2 seconds". From a scientific point of view this may not be the best possibility, but this prescription of the time was the only solution, owing to the many different working principles of the dynamometers which are in use in the various laboratories.

In consequence of the growing use of the electronic dynamometer, however, more and more laboratories have begun to apply a constant rate of extension, generally 1 per cent. of the original length between the clamps per second.

To avoid slipping or breaking of the test specimen in the clamps, it is advisable to use clamps of such a shape that a very sharp kink is avoided. Usually only the strength and total extension at break are given of the stress-strain properties.

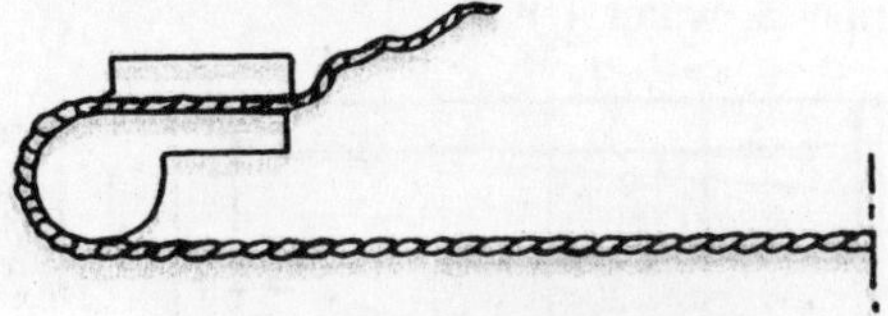

Figure 5.34 : Callaway clamp.

In practice, the shape of the stress-strain curve, or the modulus, can, however, be of great importance to the behaviour of the twine. It is therefore advisable to determine some points on this curve, for example, the extension at 25 per cent. of the breaking load.

Shrinkage

The amount of shrinkage, either in hot or in cold water, can be important as it may effect a change in size of the mesh when the nets receive a final treatment in hot water or when they are dyed.

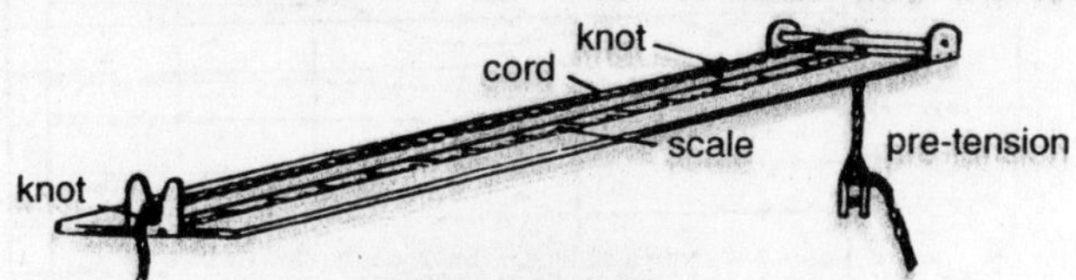

Figure 5.35 : Apparatus for measuring the shrinkage.

The method we have used is as follows: 5 knots were laid in a twine at distances of about 1 m., these distances being measured to an accuracy of 1 mm. (l_0) on an apparatus. The pre-tensions were taken according to the BISFA rules. The sample was then placed in cold (20 deg. C.) or in boiling water, after which the distance between the knots was measured again (l_1), wet or after conditioning in a standard atmosphere according to the BISFA rules : temperature 20 deg. ± 2 deg. C., relative humidity 65 per cent. and to the ASTM standards

70 deg. ± 2 F., 65 per cent. The value $\frac{l_0 - l_1}{l_0} \times 100$ then gives the shrinkage in per cent.

In this way we can determine 4 values of the test sample:.

(a) cold water shrinkage, measured in wet condition.

(b) cold water shrinkage, measured after conditioning.

(c) hot water shrinkage, measured in wet condition.

(d) hot water shrinkage, measured after conditioning.

For some twine samples of synthetic material as a function of the immersion time in water (t).

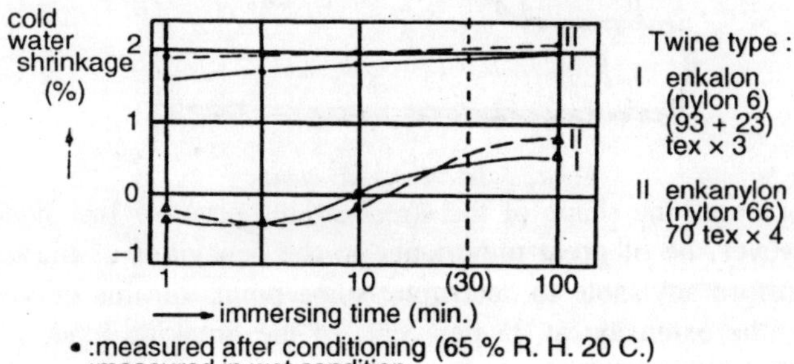

Figure 5.36 : Influence of the immersing time on the cold water shrinkage.

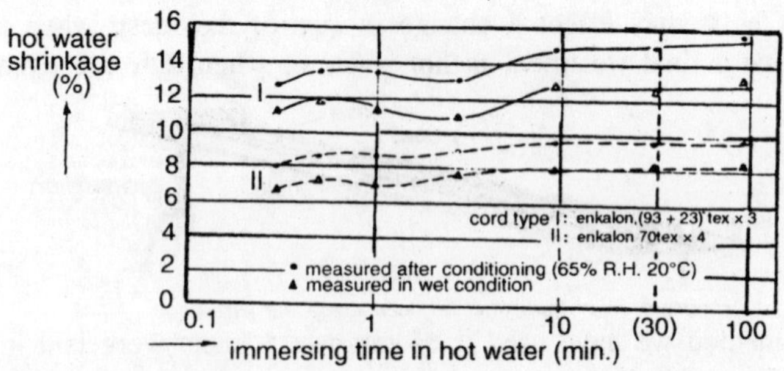

Figure 5.37 : Influence of the immersing time on the hot water shrinkage.

It can be seen that especially in the first minute of t. the shrinkage strongly increases, and that for t.=30 min. a nearly constant value is reached for all the samples. An immersing time of 30 min. should be accepted as the standard immersing time for routine measurements.

Stiffness

For measuring the stiffness of twines we used the following method:

Around a rod, 4 cm. in diameter, we wound 20 turns of the twine close to each other, held together by a narrow strip of adhesive tape at the turns. The coil thus formed is taken off the rod.

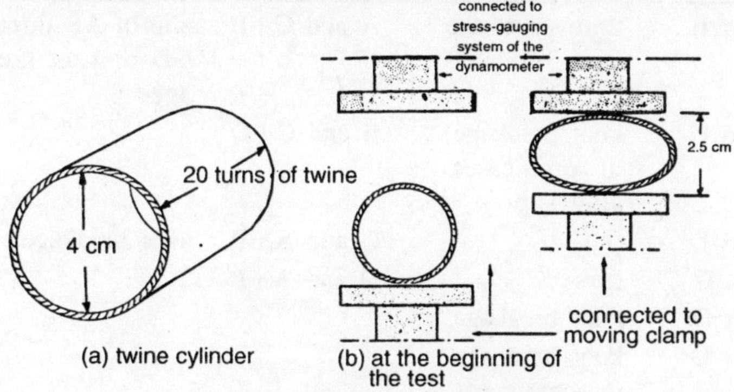

Figure 5.38 : Stiffness test of twines.

Two flat plates of about 10 cm. diameter are attached to the clamps of the electronic dynamometer, the distance between the two plates in the highest position of the moving clamp, being 2.5 cm. The moving clamp is not lowered a few centimetres and the cord cylinder is placed on the lower flat plate. The lower clamp is then moved upwards until the extreme position is reached, when the cylinder becomes an ellipse with a short axis of 2.5 cm. The force needed for this is taken as a measure of the stiffness.

METHODS OF TESTING KNOTS

General

The different methods of test on knots, dealt with further on, refer to the single-knot.

In the following the knot-ends will be as indicated according to the letters given in this figure.

There are various methods of loading a knot according to the twine ends which are clamped on the dynamometer. These different

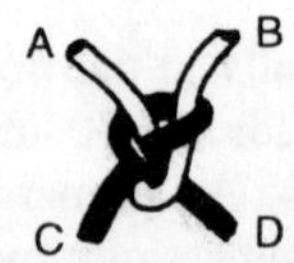

Figure 5.39 : Single knot.

possibilities of clamping, with the results obtained on loading the knots with an increasing force.

Table 5.31 :

Clamped Twine-ends	*Result*	*Clamped Twine-ends*	*Result*
A and B	tip-over of the knot	A and C+D	slip of AB through CD or knot break age
A and C	knot breakage, in some cases after slip	B and C+D	„
A and D	„	C and A+B	knot breakage
B and C	,,	D and A+B	„
B and D	knot breakage		
C and D	knot breakage in some cases after tip-over of the knot.		

Knot Strength

The knot strength, and especially the wet knot strength determines the strength of the nets. The twine ends B and D are clamped in on the dynamometer to determine the knot strength.

The stress at which the knot breaks-expressed in kg. or in per cent. of the breaking load of the twine-is called the (wet or dry) knot strength. We investigated the influence of:

(a) the load with which the knots were tightened initially (in the range from 30 per cent. to 60 per cent. of the twine strength).

(b) the velocity of the moving clamp (range: 1 per cent. to 4 per cent. rate of elongation).

(c) the waiting time, i.e. the time between the tightening of the knots and the beginning of the test (range: 6 seconds to 24 hours).

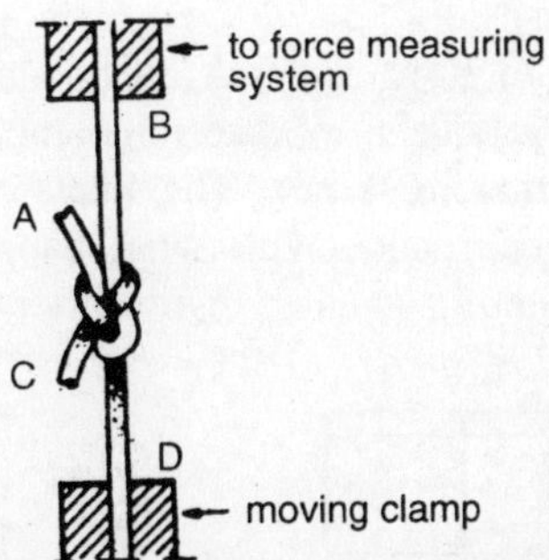

Figure 5.40 : Determination of the knot strength.

The factors. were not found to influence significantly the knot strength.

To find agreemept between the values for the knot strength, an exact description of the circumstances under which the test is performed is therefore unnecessary.

Tip-Over Resistance

Generally, there are two causes of changes in the size of the meshes, viz. the tip-over, and the slip of the knots. In the tip-over the structure of the knot changes, after which the twine AB can easily shift through the knot.

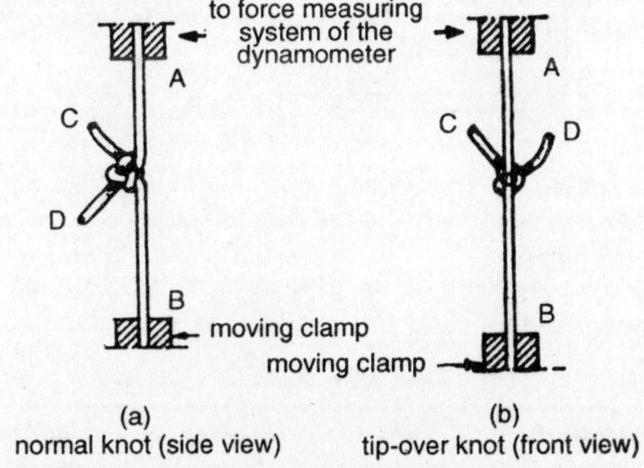

Figure 5.41 : Determination of the tip-over resistance.

The testing method is performed as follows : the twine ends A and B are clamped on the dynamometer and the force at which the tip-over of the knot takes place is determined. This force, in kg. or in per cent. of the knot strength, is called the tip-over resistance.

When the tip-over begins, the stress-strain curve mostly shows that for a short time the force is constant or decreases somewhat.

In some cases, e.g. in research on bonded twines, it can be of importance to the producer to determine the tip-over resistance of the knots without making a net. The knots must be tied manually and tightened in accordance with what happens on the stretching apparatus. Subsequently, whether or not after fixation, the knots are measured.

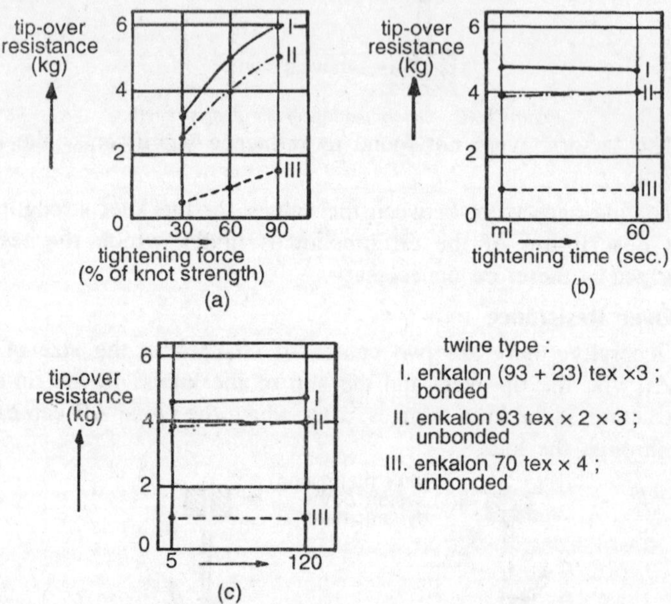

Figure 5.42 : Influence of (a) tightening force, range 30-90 per cent. of the knot strength; (b) tightening time, range 1-60 seconds (c) waiting time, range 5-120 min. on the tip-over resistance.

The tip-over resistance is given as a function of the three abovementioned factors, and the analysis of variance.

Table 5.32

Source of variation	*Sum of squares*	*Degrees of freedom*	*Mean square*	*Variance ratio*
	Main Effect			
A (tightening force)	1334	2	667.00	154
B (tightening time)	12	1	12.00	2.76
C (waiting time)	9	1	9.00	2.07
Interaction				
AB	2	2	1.00	0.23

AC	21	2	10.50	2.42
BC	1	1	1.00	0.23
ABC	3	2	1.50	0.34
Residual	469	108	4.34	—
Total	1851	119		

It appears that in the applied ranges of the factors only the tightening force has a significant influence.

We have chosen for the tightening force a fairly arbitrary value, viz. 20 per cent. of the knot strength. For the rest the standard conditions were:

rate of extension : 1 per cent. per second
tightening time : 10 sec.
waiting time : 5 min.

As a chemical preparation is applied to synthetic materials to prevent the knots slipping, it is advisable to determine both the wet and the dry tip-over resistance.

Knot-Slip Resistance

The slippage of the knots, while retaining the knot structure also causes a change in mesh size.

To determine the resistance to this slippage, the knots are clamped.

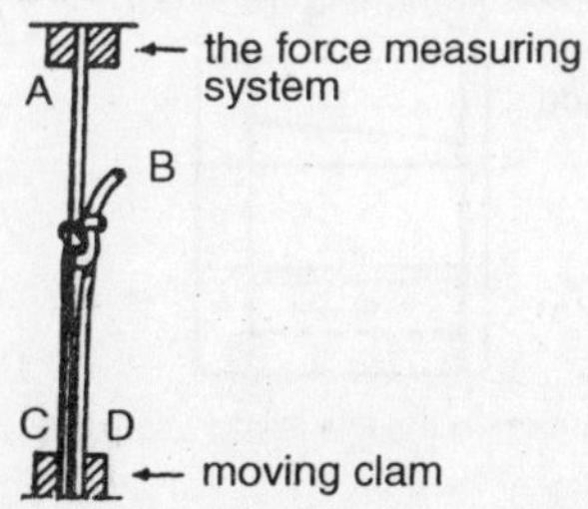

Figure 5.43 : Determination of the knot-slip resistance.

The force, in kg. or in per cent. of the knot-strength, at which the slip begins is a measure of knot-slip resistance. The slippage can occur either suddenly or gradually, depending on the specimen type. The stress-strain curve which can be obtained, where S in the knot-slip resistance.

Curves a, b and c show a sort of slip-stick effect, curve d shows the slippage which takes place gradually. Sometimes only a slight slippage occurs.

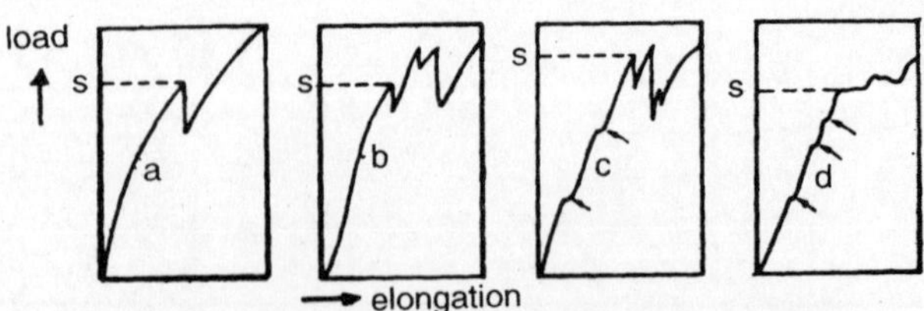

Figure 5.44 : Stress-strain curves which can be obtained with the determination of the knot-slip resistance.

In this connection we decided to take only that force as knot-slip resistance, in which the slippage occurs for longer than I second under the measuring conditions mentioned below.

We investigated the influence of various factors on the knot-slip resistance. First the influence of the angle (a), under which the twine ends c and d can be clamped. We determined the knot-slip resistance at a=0 deg., 60 deg. and 120 deg. In all cases it was lowest at a=0 deg. Further, just as for the tip-over resistance, the influence of tightening force, tightening time and waiting time were investigated.

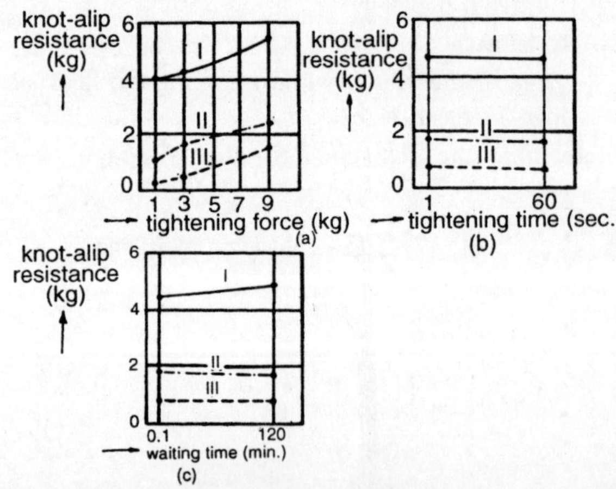

Figure 5.45 : Influence of (a) tightening force, range 1-9 kg. (b) tightening time, range 1-60 seconds; (c) waiting time, range 0, 1-120 min. on the knotslip resistance.

In all cases the influence of the tightening force is significant, and that of the tightening time is not. In the range investigated, the influence of waiting time is only significant for some samples, so the influence of this factor was investigated on a somewhat larger scale.

From which it appears that for a waiting time exceeding 5 min. there is no further influence on the knot-slip resistance.

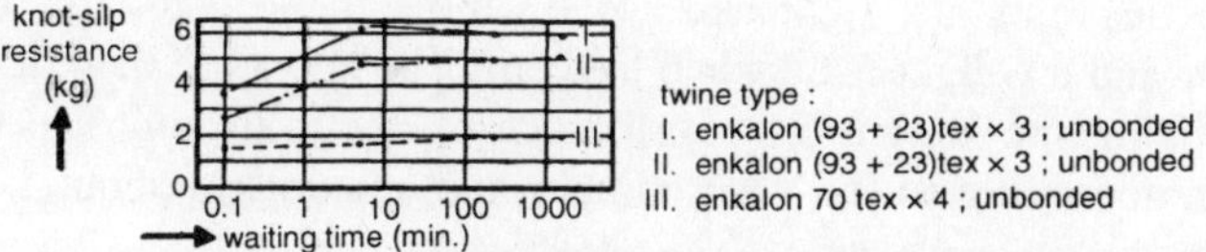

Figure 5.46 : Influence of the waiting time on the knot-slip resistance range 0, 1-1440.

Table 5.33

Source of variation	***Sum of squares***	***Degrees of freedom***	***Mean square***	***Variance***
	Main Effect			
A (tightening force)	37773	2	18886	176
B (tightening time)	2	1	2.00	0.02
C (waiting time)	4	1	4.00	0.02
	Interaction			
AB	22	2	11.00	0.10
AC	401	2	200.50	1.87
BC	154	1	154.00	1.44
ABC	384	2	192.00	1.79
Residual	11571	108	107.14	—
Total	50311	119		

On the basis of the above-mentioned results, we have chosen the following standard conditions for routine measurements:

angle between the clamped specimen ends C and D: 0 deg.;

rate of elongation: 1 per cent. per second, and, if the knots are tied manually:

tightening force: 20 per cent. of the knot-strength

tightening time: 10 seconds

waiting time: 5 minutes

For some twine types the knot slip resistance appears to be higher and for other types lower than the tip-over resistance. Hence, for a complete evaluation of a twine, both measurements must be carried out.

Open-Up Resistance

In the use of synthetic materials, one difficulty which may arise is that the knots loosen just after they have been made on the netting machine. To evaluate a net twine in this respect, the following testing method has been developed: in two twines (length about 20 cm.) a knot is tied and just tightened. Two of the twine ends are brought together and a weight is attached to them. The knots are then tightened by dropping the weight over a distance of about 20 cm. The weight is taken according to the yarn number and amounts to about 1 g. per 5 tex.

Five knots are tested with an apparatus as shown in figure. The knots are placed in a wooden box, which is rotated with a velocity of 60 r.p.m.

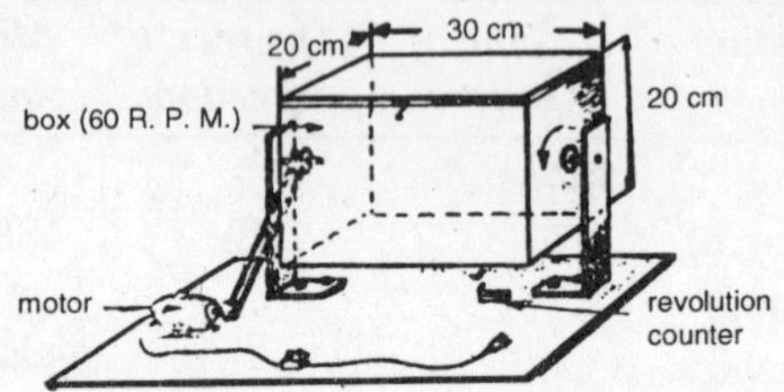

Figure 5.47 : Apparatus for determining the open-up resistance.

This process is regularly interrupted to check on how many, if any, of. the knots have loosened.

From the values found, i.e. the number of revolutions after which each of the knots had loosened, we deduced that the frequency distributions of these values approximately correspond.

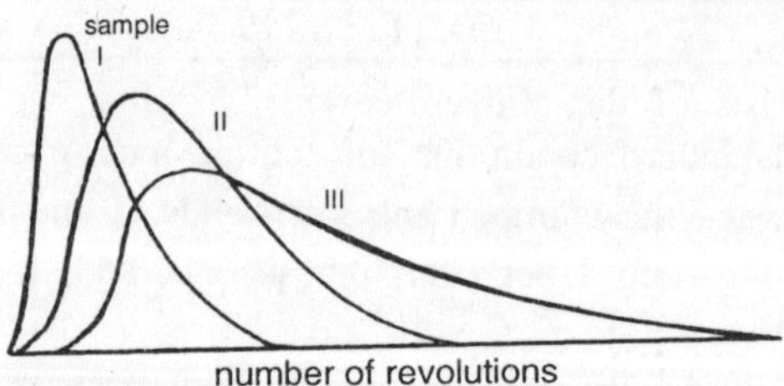

Figure 5.48 : Frequency distribution of the number of revolutions after which the knots have loosened.

In such a frequency distribution, the most characteristic value when taking a random sample is the middle or so-called median value. In this case, for example, if the number of revolutions at which the five knots loosen, is arranged in order of size: X_1, X_2, X_3, X_4 and X_5, then X_3 is taken as the determining value, called the open-up resistance. Generally, the measurements are repeated several times to find an average value of X_3, (X_3), and a measure of the spread (standard deviation sX_3).

The X_3 values are given for some samples of the same twine,

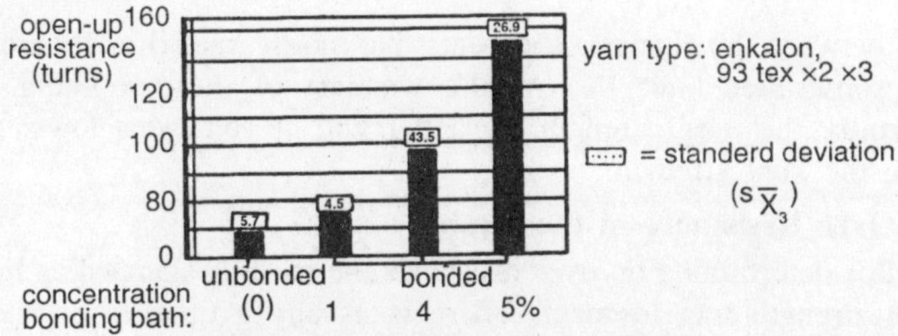

Figure 5.49 : Some values obtained with measurements of the open-up resistance.

each dipped into a bath of different concentration and containing a bonding preparation. Tests were made of 6 × 5 knots of each sample.

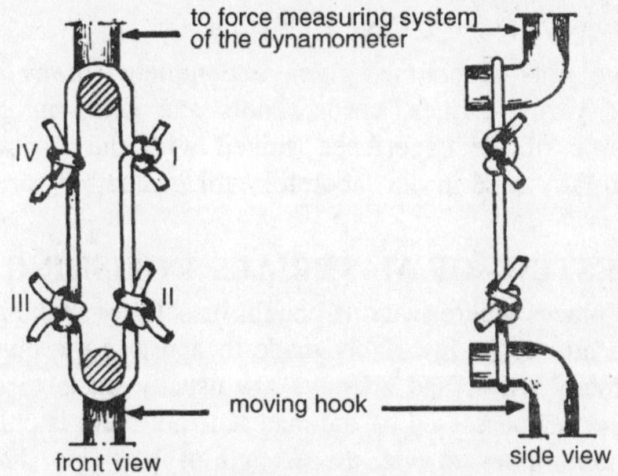

Figure 5.50 : Determination of the mesh strength.

It appears that the difference between the X_3 values are relatively large as compared with the spread S_{X3}.

METHODS OF TESTING MESHES

General

In principle, some of the methods mentioned above for testing knots can be applied also to testing the meshes for mesh strength and tip-over resistance of the mesh. As the remarks, made in 3.2, and 3.3 also apply partly to these two qualities, only a few points of the testing methods will he dealt with.

Mesh Strength

The meshes are clamped. The load at which the mesh breaks is called the mesh strength, and is given in kg. or in per cent. of twice the twine strength (because, apart from the four knots, there are two ends between the clamps). It appears that mostly knot II or III breaks, and sometimes knot IV. As the weakest of these 3 knots will determine the mesh strength, the latter will be somewhat lower than twice the knot strength.

Tip-Over Resistance in the Mesh

For determining tip-over resistance the mesh is clamped as in the mesh strength measurement. There is usually a tip-over of knot 1 before the mesh breaks. The load at which this happens is called the tip-over resistance of the mesh and is given in kg. or in per cent. of the mesh strength. The value is about twice as high as the corresponding tip-over resistance of the knot.

CONCLUSION

We have not, of course, given a complete survey of testing methods for yarns, twines, cords, knots and nets, but have only recorded some of the experience gained with the various testing methods normally used in our laboratory for evaluating yarns and net twines.

TESTING OF MATERIALS IN FISHING

In most places where twine is bought or sold for making webbing and nets, an attempt is invariably made to appraise the quality. This is done in many ways, and attempts are usually made to define the various properties possessed by the material, but there is one property that is invariably assessed, viz. the strength of the twine. The average fisherman tests the strength by trying to break the twine with his hands; the more well-to-do arrange for it to be examined at a research laboratory under ideal test conditions (20 deg. C. ± 2 deg. and 65 per cent. relative humidity).

The greatest value is attached to the breaking strength and it is taken for granted that strong twine makes strong webbing. When various twines are to be compared qualitatively, this same criterion is adopted, so that the prospective purchaser inevitably assumes that the strongest twine is most advantageous.

A fundamental misconception underlies this appraisal method which frequently leads to faulty conclusions, with all the attendant disadvantages to the fisherman, because the strongest twine is not always synonymous to the strongest net.

Twine intended to be worked up into webbing is usually tested dry to determine its tensile strength. The way in which this is actually done is of secondary importance, as the process is used simply to obtain data on the dry breaking strength or dry breaking length of the twine. But the only correct basis for examining a twine's properties is to test it under wet conditions, i.e. wet and knotted, since it will be used in wet condition, and the properties when wet are decisive when it comes to comparing various twines for webbing.

It would be of great help if both national and international standards of testing could be established and the results passed on to the fishermen; this would be-of practical value to them when buying materials.

As a supplement, a list of tests are given here which have been carried out at the Netherlands Fishery Experimental Laboratory as a routine investigation of the qualities of the rhaterials bought and for comparing properties of materials. It is not considered that these are the only suitable tests, but they are based on experience and have met requirements so far.

The figures quoted for the breaking strength are the mean averages of 10 test samples, as customary in routine investigations. The usual sample lengths and rates of traverse were adopted. The material for the "wet" tests was previously immersed in water for at least 12' hours.

EXAMINATION OF HEMP TWINE

The tests were carried out in accordance with the normal textile method. The only possible conclusion under these conditions was that Quotation I was the best, as the breaking length of this material was considerably superior to that of Quotation II.

Table 5.34 : Hemp Net Twine.

Quota-tion No.	*Analy-sis No.*	*Make lays (1)*	*Theor-etical No. of per kg.*	*Theor-etical weight per m.*	*Actual no. of metres per kg.*	*Actual weight in gr. per m.*	*Dry breaking strength of twine*	*Break-ing length in km. (2)*
I	24	100	340	2.941	309	3.240	65.2	20.6
I	23	130	442	2.262	412	2.427	53.6	22.1
I	22	150	510	1.961	478	2.090	58.0	27.3
1	21	180	612	1.634	512	1.952	51.2	26.2
II	25	100	340	2.941	378	2.646	59.8	22.6
II	26	130	442	2.262	357	2.801	52.5	18.7
II	27	150	510	1.961	467	2.141	48.1	22.5
II	28	180	612	1.634	526	1.901	42.9	22.6

(1) Lay =No. of 1.70 m lengths together weighing exactly 500 grammes.

(2) Dry breaking strength divided by dry weight per metre.

Table 5.35 : Hemp Net Twine.

Quotation No.	*Analysis No.*	*Wet yarn breaking strength*	*With over hand knot, breaking strength (wet)*	*With over in % of dry yarn (wet)*	*With Overhand knot breaking strength in km. (wet)*
1	24	66.8	39.4	58.9	12.2
1	23	46.2	23.2	50.2	9.6
1	22	53.6	26.8	51.3	12.8
1	21	53.2	28.8	54.0	14.7
2	25	53.1	44.4	76.1	15.27
2	26	47.7	41.9	87.8	14.96
2	27	39.1	36.9	94.4	17.46
2	28	38.4	33.2	86.5	16.67

As a considerable quantity of sample material remained unused at the end of the tests, curiosity prompted us to carry out the same breaking strength tests, using wet twine instead, incorporating an overhand knot.

The results were surprising for we suddenly found ourselves confronted with the fact that this "fishery method" (wet, knotted breaking strength) yielded results diametrically opposed to those obtained by the textile method. Quotation II was now found to be the best.

When the data in Tables elsewhere in the chapter are compared, a striking fact emerges in connection with these hemp twines, which were of special hard twist, viz. the wet, knotted breaking strength of the "higher grade" twines deteriorates by a much greater percentage than that of the "lower grade" twines.

The wet, knotted breaking length leaves no doubt that one type of twine was far superior to the other, and that there was thus every justification for purchasing the twine which would inevitably have been rejected, had the dry evaluation method been employed.

As the hemp samples (both hard twisted) came from different manufacturers, samples (normal twist) from yet another manufacturer were tested. The results of this third series of tests confirmed our findings in the two previous tests. Here, too, it was found that:

1. twines characterised by a good dry breaking length were considerably inferior in wet, knotted breaking strength (webbing as used under fishing conditions), and that lower grade twines (low dry breaking length) possess a superior wet, knotted breaking strength.
2. as this transition is fairly rapid (between 3/36 and 3/30), it cannot be attributed to the twist or diameter of the twine, but to the intrinsic properties of the raw material itself.

A happy coincidence also made it possible to carry out this series of tests with twines made of spun yarns. As they were not the same as the net twines originally tested; the results do not correspond exactly, but the general principles, defined in 1 and 2 above, still hold good.

Table 5.36

Make	*A*	*B*
Our mark	21 blue	22 blue
Runnage (metres/kg.)	385-0	383-0
Breaking strength in kg.		
dry	45.2 (100%)	37.8 (83.5%0)
wet	45-6	35.9
wet overhand knot	38-6	34.6
Breaking length in km.	17-44 (100%)	14.74 (84-3%)
Wet overhand knot breaking length in km.	14.82 (100%)	13-32(90%)
Breaking strength in grammes per tex		
dry	17-48 (100%)	14-47 (82.8%)
wet	17-56	13.75 (78.9%)
wet overhand knot	14-82	13.25 (89.4%)

A preliminary investigation indicated that in the case of flax, made from fibres of various lengths, the results differ to a certain extent, depending on whether the textile or fishery test method was used.

The long staple fibres of this material were separated from their short counterparts and both were used to produce the same type of netting twine with precisely the same lay and twist.

Considered from the dry breaking length point of view, it appears that sample A was the best, the ratio being 100 : 84.5 on the basis of breaking length. When the wet, knotted breaking length is taken as criterion, the corresponding ratio is 100 : 90.

Here, too, there is yet another indication that the "poorer" yarn (shorter fibres) stands up better to knotting.

TESTING OF COTTON NETTING

The same characteristic deviations between the dry twine and the wet knotted breaking length are also found in the case of cotton twines.

To ascertain whether these data could be used as quality indications, the figures of the wet breaking strength were measured at the mesh, the results being expressed in the number of grams breaking strength per fibre quantity employed. An abbreviation, "tex" was used for this latter unit (1 tex = the quantity of material involved when 1,000 m. weigh exactly 1 g.).

When the quality of the various net samples is evaluated by this method, it will be found that:

1. the unit "g. tex" provides a serviceable standard for comparing the strength of different net twines.
2. webbing made from Egyptian cottons is less strong than that made from American cottons.
3. the breaking length of dry twine is not a suitable criterion for evaluating the strength of wet netting. The dry twine breaking length of sample b, for example, is very low, whereas calculated on the basis of g./tex, based on the wet mesh breaking strength, this sample is the strongest. The dry breaking lengths of twine samples b, d and f are almost identical, yet they show a difference in the number of g./tex calculated on the basis of the wet mesh breaking strength.
4. Twine sample e is stronger .than a and b; the dry breaking length is also better. On the basis of g./tex, however, calculated from the wet mesh strength, a and b are stronger than c.

TESTING OF SILK YARNS

Similar deviations in results when both the textile and fishery methods were employed are also found when silk yarns were tested to ascertain the number of grammes wet, knotted breaking strength per unit of weight The unit of weight "denier" was adopted (1 denier

= the quantity involved when 9,000m. of yarn weighs exactly 1 g.). Two different lots of material were tested, and here again it was very obvious that this "fishery test" method brought to light qualities which would not have been shown by the dry test method. As this material is intended solely for use as wet webbing, the number of g./denier calculated on the basis of the wet, knotted breaking strength, is the only correct criterion to adopt when evaluating the quality of the yarns for fishery purposes.

The breaking strength in g./denier, calculated on the basis of the dry yarn breaking strength, runs parallel with the dry breaking length, as this is, in point of fact, the same unit expressed in different terms. When the breaking strength of wet, knotted yarn, is determined, the data in respect of the various yarns will be found to differ entirely.

TESTING SYNTHETIC FIBRE WEBBING

The term "g./denier" is very frequently used to express the breaking strength of synthetic twines. This, however, invariably refers to dry twines. The current trend is more and more in favour of the wet breaking strength. The fact that. some manufacturers of synthetic fibre twines regularly indicate the wet, knotted breaking strength, is certainly a step in the right direction.

Net manufacturers treat the twines (continuous filaments) which have previously been worked up into webbing, in various ways in order to "fix" the knots. The determination of the breaking strength in g./denier, calculated on the basis of the wet, mesh breaking strength, also enables the investigator to ascertain whether this treatment has damaged the twine in the knot. This determination is far more logical than simply testing the twine prior to the knot being fixed.

Table elsewhere in the chapter summarizes certain data in respect of various samples, all expressed in terms of the wet, knotted breaking strengths of both twines and webbing.

One cannot help being amazed by the great changes which nylon 6 undergoes (I was unable to examine any webbing made from nylon 66), when made into webbing with specially treated knots.

In the case of a polyalcohol yarn, the dry breaking strength or other evaluation figures based on such a test would prove misleading as a basis for the comparison.

This fact has quite rightly been publicised by the manufacturers, and it is indeed encouraging to note that the wet, knotted breaking strength of this twine is now being advocated as a quality criterion.

TESTING OF TRAWL TWINE

A single test was carried out with trawl twine, and here, too, differences came to light in respect of the wet, knotted breaking strength, despite the fact that the manila twines were of the, same runnage. This was fully in accordance with the results of our investigations on other yarns.

The investigation had to be interrupted, however, as it proved impossible to resolve the problem of the manila fibre gradings employed.

CONCLUSIONS

1. The most widespread method in use at present for the comparison of net twines of different quality is the dry breaking strength or other criteria based on it, e.g. breaking length or tenacity (the breaking strength in denier or tex).
2. As the twine is used wet and knotted in the form of webbing, it is more logical to employ an evaluation standard based on the wet, knotted condition of the twine.
3. The wet, breaking strength of knotted twine, or the wet mesh breaking strength, is therefore the most appropriate criterion. Evaluation figures, derived from the wet, knotted, breaking length or the wet, knotted breaking strength in denier or tex, can also be used for this purpose.
4. The evaluation figures, calculated according to 1 and 3 above, do not yield parallel results. In fact, in some cases, they are diametrically opposed, as a higher dry breaking length, for instance, is attended by a lower wet, knotted breaking strength. Thus, some twines, otherwise very strong, are found to be highly sensitive to strong flexion, while less strong twines suffer to a smaller degree when wet and flexed, i.e., knotted.
5. For webbing, the wet mesh breaking strength is the only logical strength criterion, because it corresponds more accurately with the conditions under which the webbing is used, and permits a more accurate comparison to be made of the various materials used.

KNOT-FIRMNESS

In spite of all well-known test results published on the subject of general, physical and chemical properties of synthetic fibres, time and again two problems emerge which cause a certain amount of trouble in the manufacture and use of nets, twines and cordage: "lateral strength" and "knot-firmness".

Lateral strength means the resistance which the material offers to any stress differing from the normal tensile stress operating in the longitudinal direction of the fibre axis. Well-known examples are the strength in a knot and loop, alternating bending strength, etc. Since in these cases there is an additional stress on the material the normal tensile strength is, as a rule, thereby reduced. Various materials react to this additional stress in very different ways and, in this connection, the type of material (e.g. Perlon or hemp), its special properties (depending on the type), the structure of the yarn or twine and other factors often play an important part.

Knot-firmness concerns the reaction of a knot to forces which cause shifting or loosening and thereby a change in the size of mesh.

TESING METHODS

Our investigations were limited to the two synthetic fibres which are at present most important for the fishing industry, at least in Europe: polyamides (Perlon, nylon) and polyesters (Terylene, Trevira, Dacron). The material was tested in many different forms: filament yarns, staple fibre yarns, twisted twines, and braided twine.

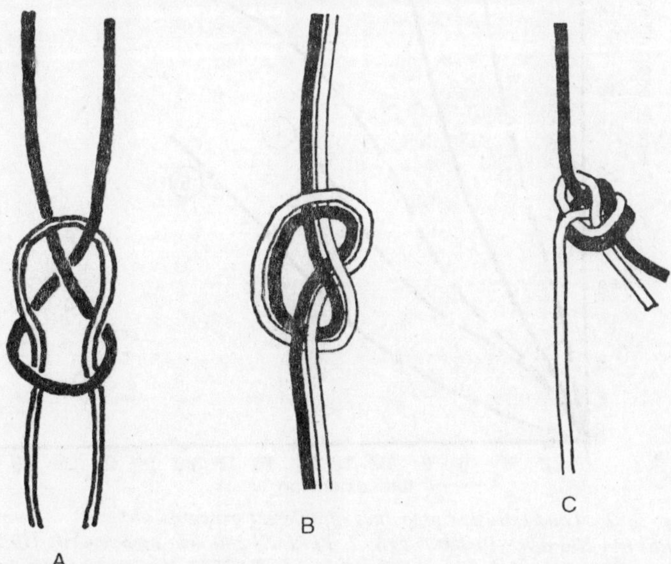

Figure 5.51 : Types of knots used in the tests. A =Fishermen's knot; B= Overhandknot; C=Double overhand knot.

The object of the investigations was to find ways and means of reducing the loss in tenacity caused by knotting and to improve the firmness of knots, particularly in very smooth materials made from continuous filament.

These efforts were successful, at least in the laboratory but the results will, of course, have to be confirmed in practice.

In our experiments, we determined the lateral strength of the various materials and structures examined by testing the strength of normal knots and customary net knots both in dry and wet conditions.

Knot-firmness is more difficult to determine by tests.

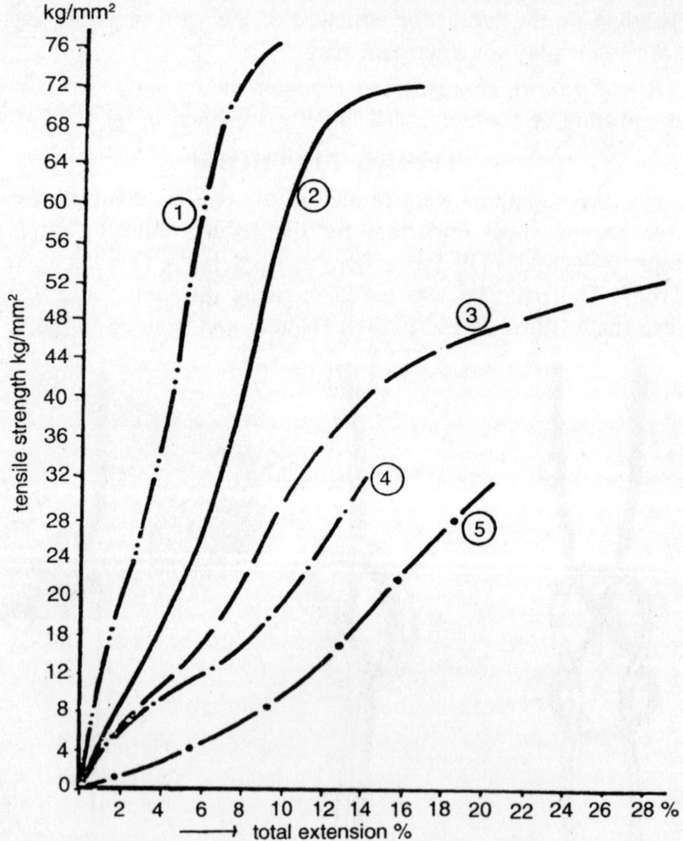

Figure 5.52 : Load-extension properties of different materials (dry).
1. TREVIRA 840 den. (10,300 m/kg.); 2. PERLON 840 den. high tenacity (10,550 m/kg.).; 3. PERLON 1,150 den. (8,080 m/kg.).; 4. TREVIRA Nm 20 (19,680 m/kg.).; 5. PERLON Nm 20 (19,500 m/kg.).

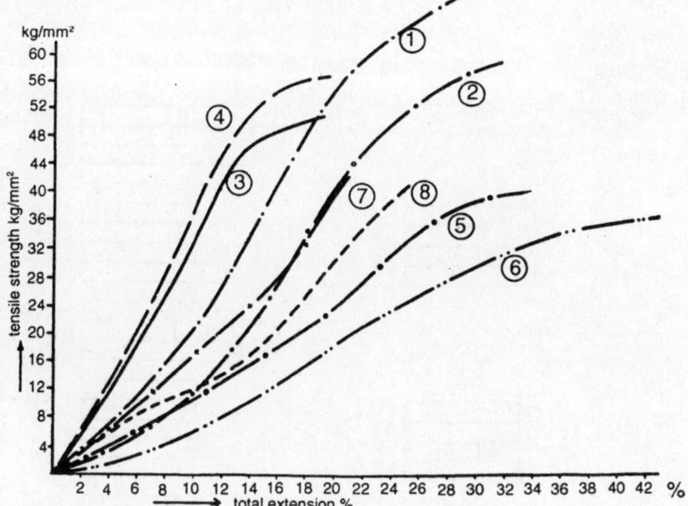

Figure 5.53 : Load-extension properties of different twines (dry and wet).
1. *Perlon 840 × 3 den. (3,160 m/kg) dry.*
2. *Perlon 840 × 3 den. (3,160 m/kg) wet.*
3. *Trevira 840 × 3 den. (3,000 m/kg) dry.*
4. *Trevira 840 × 3 den. (3,000 m/kg) wet.*
5. *Perlon Nm 20/2/3 (3,110 m/kg) dry.*
5. *Perlon Nm 20/2/3 (3,110 m/kg) wet.*
7. *Trevira Nm 20/2/3 (3,040 m/kg) dry.*
8. *Trevira Nm 20/2/3 (3,040 m/kg) wet.*

Here we also tried two ways which, although not corresponding exactly to the stress on the knot of the net in use, nevertheless allowed a comparison of the knot firmness. Since the problem of maximum resistance in a knot to displacement is obviously closely connected with maximum roughness of the surface of filaments, yarns and twines, we first tested the yarns in "loop knot" or "double overhand knot" forms in a normal tensile strength testing device. The knot was tightened by means of a constant weight. When a load was applied to the knotted yarn, the knot loosened (meaning poor firmness), or held fast while the yarn or twine broke in the knot or in some other place.

In the second method, we desisted from any tensile stress on the knotted yarns since in the net the knots will, as a rule, shift only when they are not loaded. Instead, we placed a number of short pieces of twine connected by the fishermen's knot in a rubber-lined box. A rubber roller was also put in the box. The closed box was rotated for 30 minutes at about 60 r.p.m., i.e., a total of about 1,800 revolutions. During this process, a number of knots will loosen,

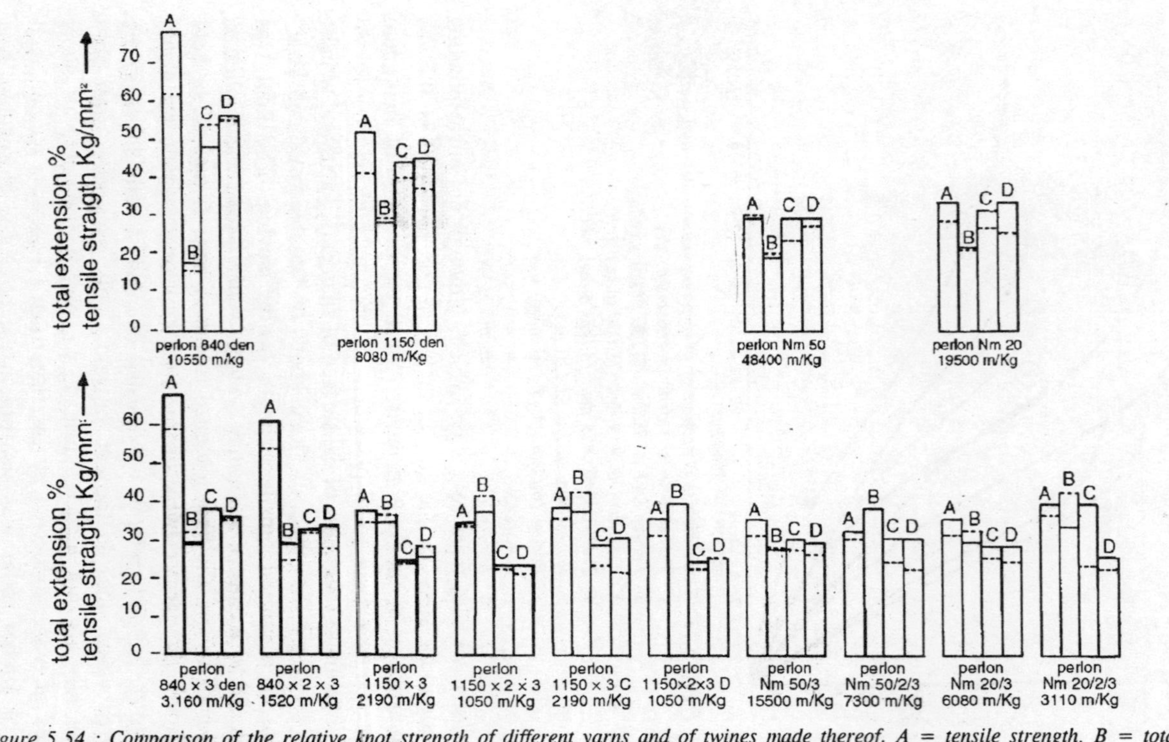

Figure 5.54 : Comparison of the relative knot strength of different yarns and of twines made thereof. A = tensile strength, B = total extension, C = kno strength with overhand knot, D = knot strength with fishermen's knot, —— = dry, ... = wet.

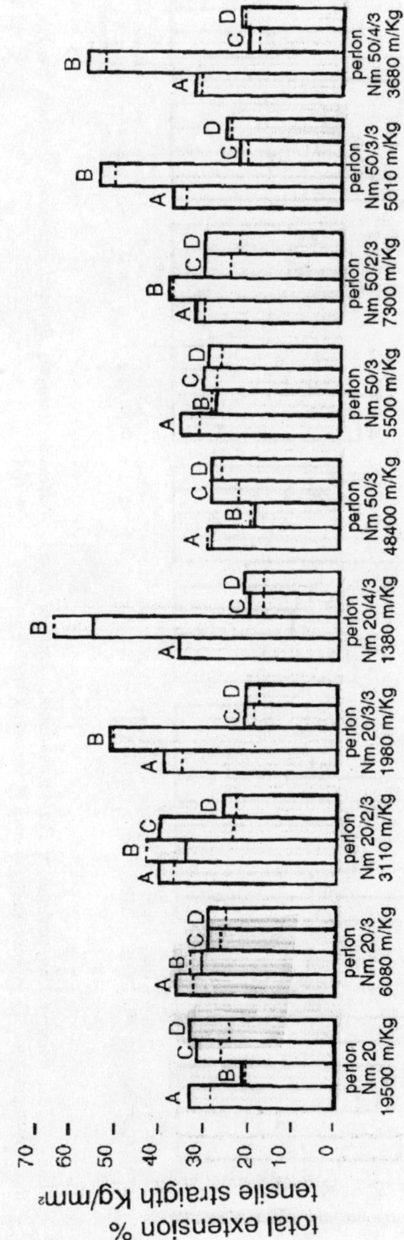

Figure 5.55 :Relative knot strength of Perlon continuous filament twines. A=tensile strength, B=total extension, C=knot strength with overhand knot, D = knot strength with fishermen's knot. —— = dry, ... = wet.

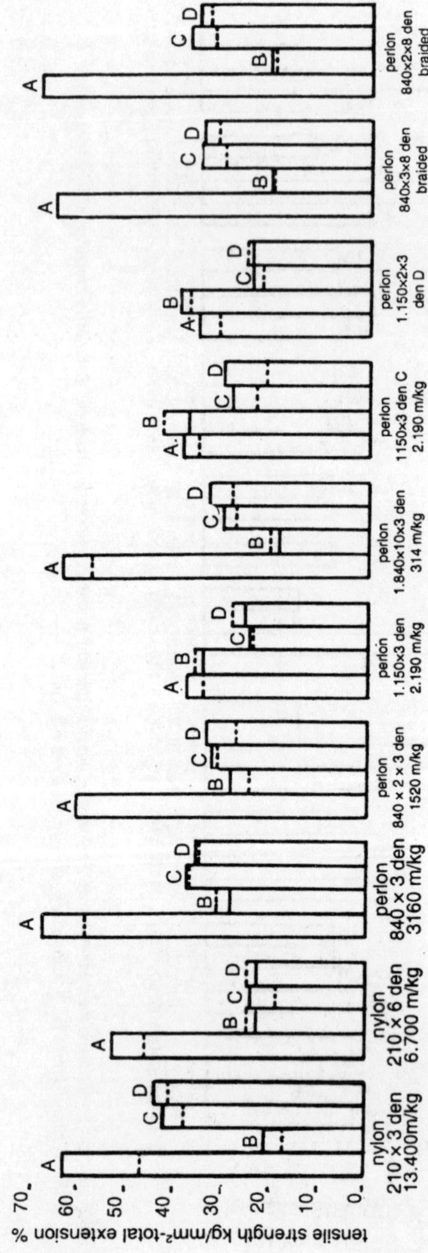

Figure 5.56 : Relative knot strength of Perlon continuous filament twines. A=tensile strength, B=total extension, C=knot strength with overhand knot, D = knot strength with fishermen's knot. - = dry, ... = wet.

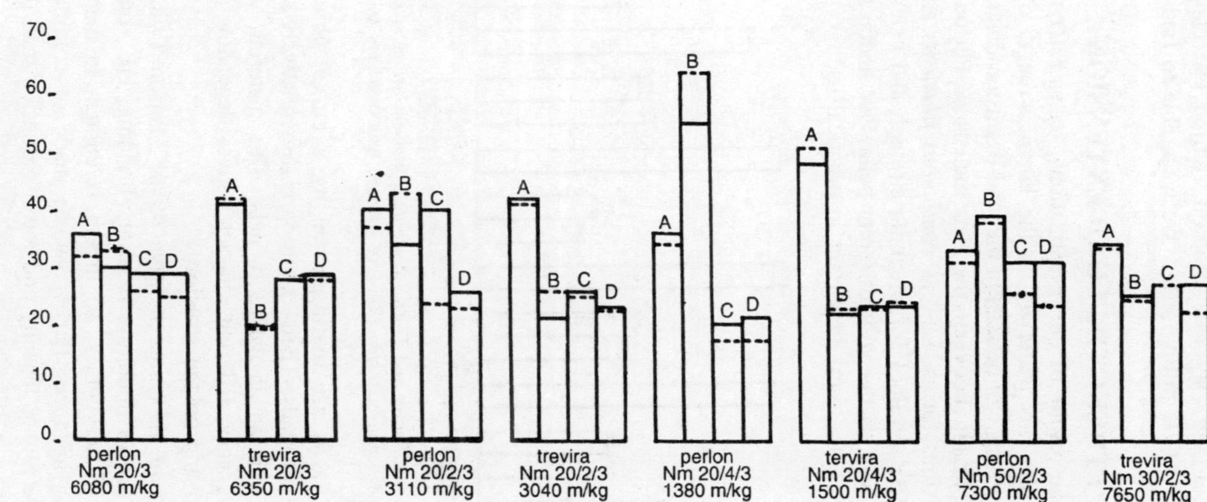

Figure 5.57 : Comparison of Perlon and Trevira staple twines in regard to their tensile strength (A), total extension (B) and knot strength with overhand knot (C) and fishermen's knot (D). ______=dry, ... =wet.

depending on material and previous treatment. This number is expressed as percentage. By this method, which corresponds approximately to the I.C.I. pilling tester, we were able to find very marked differences.

TENSILE STRENGTH, LOAD-EXTENSION

Before giving the results of the investigations, let us refer again to the influence of the raw materials on the lateral strength of the net twines. The main facts can be seen in the load-extension-diagrams. This shows a comparison between the two materials, Perlon and Trevira, in their various forms, i.e., as *continuous filaments* and as *spun yarn.* The different values for breaking strength and extension are clearly seen as well as the very different behaviour under small to medium loads, i.e., the most important in practical use.

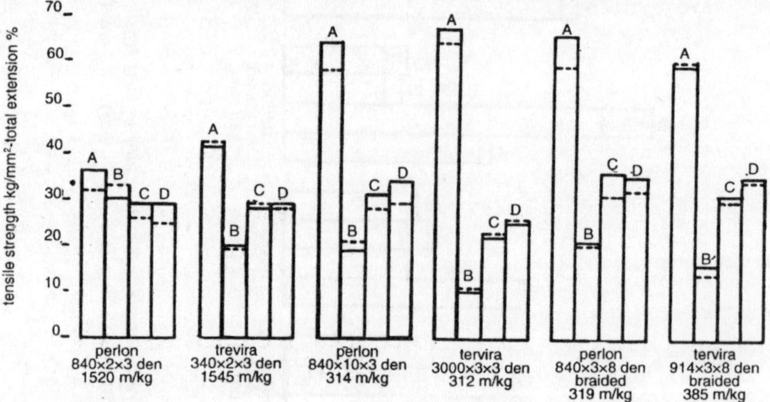

Figure 5.58 : Comparison of Perlon and Trevira continuous filament twines in regard to their tensile strength (A), total extension (B), knot strength with overhand knot (C) and fishermen's knot (D). ______ *= dry, ... = wet.*

The polyester yarns are distinguished by a curve which is relatively steep at small loads. Their extension is thus relatively small in this region, which is advantageous in nets. This material, when twisted, often shows very different characteristics, depending both on the material used and on the structure.

Dr. Must shows in his comprehensive report "Perlon Twines", the important influence of doubling, twisting, braiding, etc., on the physical properties of net twines, so that it is enough to mention this in passing. For instance, Perlon twines made of continuous filament usually have higher tenacity and lower extension than those made of spun yarn.

In net twines *wet tenacity* is of particular interest. Some comparisons between Perlon and Trevira (Terylene), the twines being of similar thickness, although some are of filament and others of spun yarns. Trevira is distinguished by a very good wet tenacity.

Trevira, in their various forms, i.e., as *continuous filaments* and as *spun yarn.* The different values for breaking strength and extension are clearly seen as well as the very different behaviour under small to medium loads, i.e., the most important in practical use.

The polyester yarns are distinguished by a curve which is relatively steep at small loads. Their extension is thus relatively small in this region, which is advantageous in nets. This material, when twisted, often shows very different characteristics, depending both on the material used and on the structure.

KNOT STRENGTH

Wet tenacity, and above all, knot strength is decisive in judging the usefulness of the material in nets and of course knot strength in a wet condition is of special interest.

It becomes obvious that high normal *tensile strength,* does not indicate equally good *knot strength.* These values for Perlon filament, staple fibre yarns and twines of different structures. The higher tensile strength of the continuous material can be clearly seen along with its higher sensitivity to knotting and plying. The initial differences in the knots still exist, though no longer in the same relation. In multiple twists, the *knot strength* of the continuous material is, under certain circumstances equal to, or below, that of twines of spur. material. Braided twines, however, usually show a better result.

This fact does not question the other advantages of twines of continuous material as far as smoothness, lower thickness, lower towing drag, etc., are concerned. Trevira gives slightly better results, especially in regard to the wet *knot strength* of spun material.

Twines made of Perlon (spun yarn as well as filament) of various structures are compared where the influence of the structure emerges clearly.

It is well known that the "critical twist" in continuous filament twines is reached much earlier than in spun material. This is also true of corresponding twist structures made of Trevira. In this case, too, the lower extension and the universally high wet strength are evident.

You see pairs of twines, each of which contains a strand made of Perlon and the other of Trevira, directly comparable or, at least,

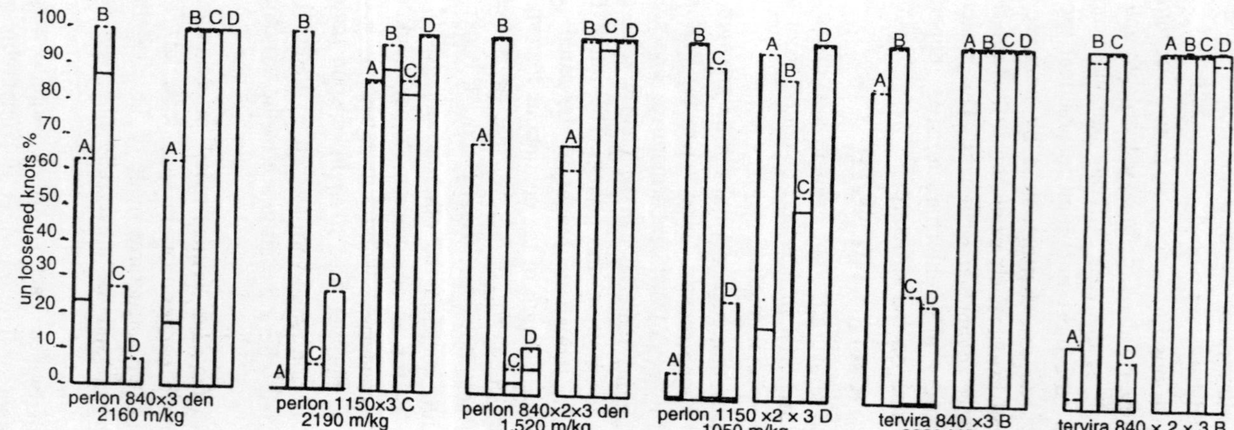

Figure 5.5.59 : Knot firmness of different Perlon and Trevira staple twines according to I double overhand knot and II fishermen's knot testing method. A = untreated, B = treatment x, C = treatment Y, D = treatment z. Results given in per cent. loosened knots. _____ = dry, ... =wet.

very similar in their total thickness (length per weight). It is tempting in this comparison to examine in detail the influence of the twist, but Dr. Must has dealt with this subject very extensively in his above mentioned report.

The purpose of this comparison-test was merely to point out the chances open to the processors in selecting *material* and structure and to warn against jumping to quick conclusions regarding the suitability of a particular material before all factors have been taken into account.

KNOT FIRMNESS

The next point concerns the problem of *knot firmness*. It is well known, and seems to be confirmed in practice, that net twines of spun yarn show sufficient firmness of knots. This is probably directly connected with the considerably rougher surface produced by the many projecting fibre ends and the position of the capillary fibres. This fact is also very clearly reflected by the values found by means of the two test methods already described. Furthermore, it was found that untreated twines, wetted for 24 hours and then dried, had a better knot firmness than unwetted twines.

Subsequently, we tried to improve the knot firmness by treating the twines with various agents. The results obtained with three different products are given below:

X means an adhesive which is almost water-insoluble and was dissolved in a solvent;

Y is a product on polyvinyl acetate basis, likewise almost water-insoluble;

Z is a product on silicate basis, likewise having a roughening effect on the surface.

The influence of these agents on the firmness of the knots is shown by both test methods so that the stability in twines of spun yarns may be considered to be practically 100 per cent. there being only slight differences between the various products.

The result of the treatment of twines of *continuous* filaments is, of course, clearer. Although the various materials and structures show different reactions to the agents, the knot firmness of the Terylene is striking.

The results of a braided twine of Perlon and of Terylene compared with Perlon monofilament are shown. The braided twines are distinguished because of their structure by a better knot firmness.

The results have been further improved by the above mentioned treatment. The position of monofilaments is somewhat different. Although knot firmness was improved, too, by the treatment, the results were not uniform. However, certain other possibilities are emerging for monofilaments, but it would be premature to report on them now.

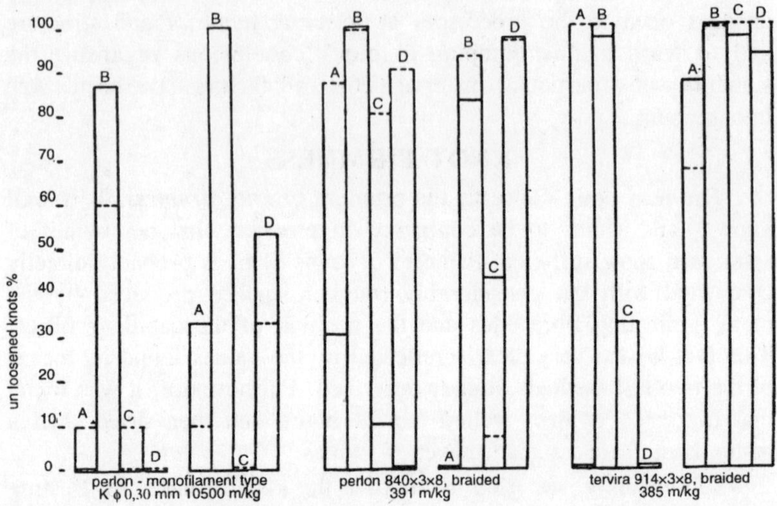

Figure : 5.60 : Knot firmness of Perlon monofilament and Perlon and Trevira braided twines according to IL double overhand knot and 11. Fishermen's knot testing method. A = untreated, B = treatment x, C = treatment y, D = treatment z. Results given in percent. unloosened knot. ——— = dry, . . . = wet.

In conclusion, it must be pointed out that the examination of this very interesting subject, has by no means been concluded. Further tests, as well as a free exchange of experience between all concerned, will have to follow if optimum results are to be achieved.

PROPERTIES OF TWINES

Dr. G. Klust (Germany) Rapporteur: This subject may be divided into two sections: Testing Methods and the Properties of Twines. Among testing methods, those of prime importance here refer to twines, nets or ropes.

Reuter, Shimozaki and Arzano list tests for net twines. These tests take into account: Construction of the twine; thickness of twine, dry and wet; weight in air and in water; length-weight unit; sinking speed; shrinkage; breaking strength, dry and wet; breaking length; breaking strength per weight unit (tenacity); knot strength; breaking extension; load-extension curve; elastic extension and permanent elongation; toughness or the ability to absorb work; abrasion resistance

; resistance to light, weathering and to water; stiffness or handiness; knot-slippage or knot firmness; resistance to bacteria, mildew and insects.

These tests were carried out under conditions as close as possible to those of actual fishing, therefore they will not always correspond with the,, methods generally used by textile industries.

Different methods often give different values and van Wijngaarden in his paper urged that there should be a certain uniformity in the measuring methods and the conditions under which they are performed; this, I believe, is necessary. This Congress will of course not be able to engage in such special problems, but it would be a considerable practical success if the papers and discussions helped promote closer co-operation between those who make such tests.

Of the testing methods specified in these papers, a few, which are not very well known at present, may be cited.

Japan Chemical Fibres Association gives a simple method of measuring the thickness of twines. Pieces of test samples are wound twenty times closely in parallel around a cylinder of about 5 cm. in diameter with a standard initial tension. Then the breadth is measured and divided by twenty. The average number is indicated in millimetres.

A method of testing the sinking speed is given in the same paper. A cylindrical glass vessel is used, having two marks at a distance of 50 cm. A piece of twine, 2 cm. long, with a knot in the middle, is put into the glass filled with water, and the time taken by the twine to sink from the top to the bottom mark is recorded.

There is no need to mention such important testing methods as those for breaking strength and extension because they are very well known and are dependent on the machines at the disposal of the examiner. In these tests too there should of course, be conformity in the methods in order to produce comparable results.

Knot slippage or knot firmness has become of great importance. We know that the smooth surface of continuous filament twines does not assure absolutely slip-proof knots. Van Wijngaarden and Stutz describe methods for determining the resistance of knots to slippage. The load-extension curves of knotted twines are drawn by a tensile tester. When the knot is slipping, the arc of the curve is interrupted by the appearance of jags or teeth. Slippage can occur suddenly or gradually. The force in kilograms or percentage of the knot strength at which the slip begins is the measure of the resistance.

Knots in smooth twines often shift without a load. To find

differences in the behaviour of twines made of different fibres, Stutz placed a number of short pieces of yarn, connected by the usual net knots, in a rubber-lined box. A rubber roller was also put into the box which was then rotated for 30 minutes. During this process a larger or smaller number of knots will loosen, depending on material and previous treatment. The number is expressed as a percentage. A similar method is described by van Wijngaarden.

Van Wijngaarden also determines the moment when the loaded knot is changing its structure so that the twine slips through the knot. The force in kg. or percentage of the knot strength at which this takes place is called the "tip-over resistance."

For measuring the stiffness of twines, van Wijngaarden winds 20 turns of the twine round a rod 4 cm. in diameter. The coil is taken off the rod and placed between one of the scales of a balance and a fixed metal plate. Weights are added to the other scale-pan until the coil is in the form of an ellipse, the short axis of which is 2.5 cm.

In this connection there is von Brandt's method, described in his book "Arbeitsmethoden der Netzforschung".

As expected, nearly all papers deal with synthetic fibres. They are now of great importance in most of the leading fishing countries and are more and more replacing natural fibres.

Needham states in his publication on the use of nylon in the fishing industry: "They bring to one of man's oldest occupations the miracle of science and, in doing so, provide easier living for the fisherman." For instance, the miracle can be noted in the rot-proof quality of the synthetic fibres. I remember the great astonishment and the scepticism of our inland fishermen when they first received gear made of PeCe. This rot-proof quality is most important in fishing geag as it ensures greater durability and eliminates the work and expense of preservation. It also enables the gear to remain in water for an unlimited time and eliminates the need to dry it.

This rot-proof quality is now taken for granted, and the developments in synthetic fibres have considerably increased the demand by fishermen or these materials.

Lonsdale outlines the desired properties of the ideal fibre for use in fishing gear, but adds that this fibre, of course, does not exist.

Teviron seems to be the cheapest of synthetic fibres, and the price of Teviron fishing net is only 30 per cent. higher than that of cotton. Its resistance to weathering is very high. The same holds true for Rhovyl and PCU.

Polyethylene fibres have the lowest density of synthetic fibres, lower than that of water, therefore ropes made of this fibre will float and, in trawling, the upper net will tend to rise of its own buoyancy.

The fibres with the greatest specific gravity are those made of polyvinylidene chloride, e.g., Krehalon. This property enables the nets to sink faster. It makes the fibre particularly suitable for setnets.

Terlene has a high tensile strength which is unaffected by wetting. The most important characteristic seems to be its very low extensibility. Gillnets of Terylene have been used successfully, principally for catching salmon and cod, but nct herring. This soft fish becomes damaged by the fine twine. Trawls made of Terylene have proved very successful.

The most important characteristic of the polyacrylonitrile fibres is a very good resistance of weathering, but the fibre does not seem to be used in fishery and none of the papers mention it as a net material.

Twines made of polyamide filament or staple fibre have a relatively low resistance to weathering, less than that of cotton. This refers especially to types of polyamides which have been delustred, and which are not suitable for use in the fishing industry. For most fishing gear this sensitivity to light is not, however, an important disadvantage. Practical experience gained by Swedish freshwater fishermen has proved that nylon nets discarded after 3 to 4 years have become unserviceable because of damage by tearing and not because of the effect of sunlight.

In Portugal, purse seine nets made of perlon staple, not dyed or prepared in any way, have been in use 1,300 fishing days and are still effective. Cotton nets frequently treated with preservatives, have a durability of 400 to 500 fishing days. The excellent physical properties of polyamide twines are of much greater importance than their resistance to light.

But there are cases where the resistance to light will be insufficient, as, for example, when part of the fishing gear remains above the water level. Colouring the net may help, or twines made of polyester, polyvinyl chloride, polyvinylonitrile or Saran may be used. Terylene polyester fibre has a better resistance to light and weathering than the polyamides and is equal to the best of the natural fibres, such as cotton, although the polyvinyl alcohol fibres are better. These fibres, such as the polyacrylonitrile (Orlon, Dralon and PAN) and especially the unchlorinated polyvinyl chloride (Rhovyl, PCU and probably the new Japanese fibre Teviron) have the highest resistance to weathering. Resistance to abrasion is very important but we do not know exactly the abrasion resistance of nets made of different types of fibres. Test methods never give absolute values, but only relative ones. They are in a high degree dependent on the method employed.

A few papers give examples for assessing the abrasion resistance. Shimozaki has chafed twined across an oil stone and a figure shows that the resistance of non-treated twines increases in the following order: Saran, Krehalon, Teviron, Cotton, Kyokurin (a mixed twine made of Amilan and Saran), Kuralon and polyamide Amilan. In tests made by Imperial Chemical Industries Ltd. twines were abrased over a hardened carbide steel bar. The abrasion resistance of Terylene twines was superior to that of cotton and flax, but less than that of nylon twines. There is a good conformity between these data and those of Must. Examples of the wet abrasion resistance of Perlon filament tissue in comparison with those of hemp and manila are to be found in the Perlon-Warenzeichenverband paper.

It seems certain that the polyamides have an especially high abrasion resistance, not equalled by any other fibre. The durability of fishing gear made of such fibres, such as trawl nets, is probably primarily due to their abrasion resistance.

What types of fishing gear ought to be made of fibre of the highest strength? First, those subjected to the heaviest stress and strain, such as trawl nets, expecially those used by large fishing boats. But it is also profitable to use strong fibres for fine gllnets. The stronger the fibre the finer the twine and the finer the twine the greater the quantity of fish caught.

The test records given in the papers show that two types of synthetic fibres are superior to others; polyamide and polyester. But here we have to distinguish between fibres of normal or ordinary tenacity and fibres of high tenacity. The factor which determines the strength of synthetic fibres is the degree to which they have been drawn out during manufacture. A high degree of drawing or pre-stretching results in a considerable increase in tensile strength, which is connected with a reduced breaking extension. But high tenacity leads to loss in lateral strength. It is important to know the decrease in strength caused by knotting and by high tenacity. Tests of knot strength are much more significant than those of the straight twine. In this connection it is interesting to note the physical properties of twines in Japanese knotless nets.

Twine and knot strengths of polyamide fibres are dependent on the degree of stretching, not on the type of fibre. In principle nylon and Perlon, for example, do not differ in strength when they are made with the same degree of drawing. (That may be seen in the paper by Arzano). Carrothers has tested twines of nylon 66 and nylon 6 or Perlon. The twine made of the latter was found to be about 40 per cent. weaker than nylon 66 of the same weight, but the two twines reacted differently to knotting. The mesh of nylon 6 or Perlon was only about 20 per cent. weaker than the mesh of nylon 66 and not 40 per cent. weaker as in the case of the twine. Carrothers has not, I think, compared twines with the same degree of tenacity, but a twine made of a highly drawn nylon with a twine made of normal Perlon. Lonsdale has found that every type of knot has a different knotting efficiency. A decrease in the angle through which the loop is formed weakens the strength. If the number of loops in a knot increases, the knot strength increases too.

As Carrothers states, tensile strength is often adopted as the only measure of quality because it can be measured relatively easily. But it is important, too, that the material is able to absorb kinetic energy. The degree of extension (elastic and permanent) is sometimes of the same or of a greater importance than tensile strength. Sta&ments of breaking extension of the twine, mostly given by manufacturers, do not provide possibilities of evaluation. In synthetic nets, the meshes of which are formed by knots, knot-strength reaches only about 50 to 60 per cent. of the breaking load. Breaking extension of twines made of different fibres may be similar, but the shape of the load-extension curves may be different. This can be of great importance to the behaviour of the twine under working conditions. Van Wijngaarden proposes to determine extension under lower load, for example, extension at 25 per cent. of the breaking load.

Net twine has a great capacity for working if total and elastic extensions are great. It will be able to absorb kinetic energy and will stand shock loads better than twine of lower extensibility. The demands on net materials differ very much between different types of gear, but for all gear it is advantageous to use fibres of high tensile strength. On the other hand, one type of gear may require a low extension while another may require a higher extension. It is always an advantage for a fibre to possess a high degree of elasticity which guarantees a good consistency of mesh size. There is no ideal fibre for fishing gear, but the various types of synthetic fibres each having different properties, provide a range of choice from which to select the best for each type of gear. So we can say with Amano "Modern fishing means the use of synthetic fibre nets".

Mr. J. E. Lonsdale (U.K.). One problem before synthetic fibre makers is to determine the exact properties required in a fishing net fibre and on this point fibre producers can learn a great deal from the fishermen and from the fisheries technologists. As synthetic fibres are man-made, they can be tailored to produce the required properties but until the manufacturers know what the fishermen want, they cannot produce the best possible fibre. Even for a material such as nylon, which has been in existence for some twenty years, there is great room for exact assessment of the properties required by the fishermen. One major complication is to differentiate between the properties required in different types of fishing and different types of nets in different parts of the world.

There has so far been no real attempt to try to rationalize these properties but once this has been done, and an attempt is made to find out exactly what the fishermen want, advances will be made in the properties of man-made fibres.

Dr. Avon Brandt (Germany) (Chairman). The polyester and polyamide fibres are used mainly in Europe and in America. The group of polyvinyl alcohol fibres is most important in Japan.

Mr. P. J. G. Carrothers (Canada). The results of my tests on Manryo were published in the Progress Report of the Fisheries Research Board of Canada, and copies may be obtained from the Vancouver Technological Station.

Manryo twine is similar to cotton in general appearance. Generally speaking, we found the properties of the new Manrya (dry twine) to be considerably better than cotton, but when knotted and wetted, the properties of both twines were similar.

So many of the tests have been on the dry twine, but wet mesh and wet knot tests are of far greater importance than a straight, dry twine test.

I would also like to say a word about the need for standardisation of test procedures. So far, these have been primarily developed from the manufacturers' point of view, but nets are used to catch fish, so the test methods are better designed from the fisherman's rather than from the manufacturers' point of view. Pertinent properties which relate to the fishing function should be measured, such as handling characteristics, the wet mesh strength, probably also toughness, etc.

Test procedures in most countries are fairly well standardised up to the stage of the dry twine, i.e., the ASTM in the States. Could not this Congress try to encourage some of these groups to consider also standard tests for netting? Many of the methods available have been developed solely for research purposes. There is a need to develop these methods further so that they can also be used for routine control purposes. It would then be possible for netting manufacturers to state how strong their netting is without fear of, say, a government laboratory or standards laboratory testing the netting by another procedure, and saying that the nets are not what they are claimed to be.

Mr. B. F. Wohnarans (Union of South Africa). About 2 years ago, the first blend of Japanese synthetic netting was put into use in South Africa. We find this is vastly superior to straightforward synthetics. Other types have been imported from U.K., Germany and Holland, but we find Marlon very much more resistant to the peculiar water and weather conditions we have in South West Africa. Firstly, nets manufactured of this material are reputed to sink very much faster; they are abrasion-resistant to a very great degree. The price factor also comes into consideration. After much experimentation our industry has concentrated on Marlon particularly in the Walvis Bay area where conditions are very bad. Kuralon is unsatisfactory, although it has been put into use in the areas further south where the water is clearer and where damage is not so noticeable.

Mr. R. S. Rack (Northern Rhodesia). A very few years ago our local fishermen were taking motor car tyres to pieces, in order to get threads for their fishing nets. Subsequently they have come to use nylon, which has superseded cotton. They are very simple people; they know cotton and nylon; it might be a blow to their vanity if I say they know very little of any other. In fact, I believe that if I were to say "this is not nylon, this is Marlon", they would say "yes, you mean Marlonnylon." There is a danger in this wide use of trade names which are being interpreted as common names. A worthless fishing twine may be sold under a high sounding name. I would therefore suggest that the trade name always be accompanied by a specification of performance. I would also suggest that the testing methods be brought under two headings: Practical forms of test which could be carried out on behalf of the fishermen to ensure that the nets sold are adequate, and more exhaustive tests which could be carried out for the satisfaction of the industry.

Could not this matter be referred to the various standard institutions who, in cooperation with users and manufacturers, might be able to provide tests of this nature? With regard to the value of synthetic fibres under tropical conditions in my part of the world we use practically all nylon. Our primary aim is to reduce the cost as much as possible. Many of our fishermen make their own nets, and therefore a nylon which is non-slipping and which could be knotted with a single knot, would be of great advantage. Abttsion is also a problem. However, the physical characteristics are not so vitally important to us. When I tell you that we lose a good many nets to crocodiles you will understand this point of view!

Mr. T. D. Iles (Nyasaland). I am concerned more with the biological side, the fisheries research in general, rather than with test of net characteristics, but I have had an opportunity on Lake Nyasa, where we are investigating a deep water gillnet fishery, in comparing cotton, flax and nylon nets. Under our local conditions, the physical characteristics of the nets such as tensile strength, do not give such a great advantage as far as the fishermen are concerned. A problem in this particular fishery concerns a fresh water crab, which causes a great amount of physical damage and, in fact, whereas a flax net will give up to 25 sets, a nylon net may only give up to 40, despite the fact that the tensile strength of the nylon is very high, far higher than is really needed for the fish that are being caught. Such problems, under these conditions at least, are far more important than physical characteristics of the net.

Mr. C. P. Halain (Belgian Congo). Nylon fishing twine —to some extent imposed upon us, owing to difficulties in finding fishing twines-was first used in the Belgian Congo. The fishermen now use nylon almost solely and cotton or linen yarn is seldom sold, but we are still looking for a better synthetic fibre.

Incidentally, some 80,000 to 100,000 tons of river and lake fish are marketed in the Congo every year.

Mr. M. K. Kramer (Israel). We have carried out three experiments with synthetic fibres. The first, with trawling, was unsuccessful, and we are now waiting for other and better fibres. Good results were obtained, however, with the fibre net in Lake Tiberias. It was found that a white fibre net doubled catches and these were tripled last year by using a red fibre net. Today there is not one gillnet in the Lake which is not of synthetic fibre.

We have now begun experiments for the change-over of sea purse-seine nets. We have begun by using fibres from various countries-three kinds from Germany, two from the U.K., and two from Japan. In 1958 we hope to experiment with Italian fibres also, and so arrive at some final result.

Dr. W. Einsele (Austria). Monofilament nets have revolutionised our fisheries. There is very little plankton in our lakes and the light penetrates very deeply. Catches were exceedingly low, especially in the summer, not because there were no fish but because we were unable to catch them. This has changed entirely since the introduction of monofil nets and this summer, for the first time, enough fish has been caught to supply the hotels, restaurants and villages in the Lake region.

Mr. E. F. Gundry (U.K.). There is a tendency, especially at the government fishery officer level, to assume that manmade fibres make the best nets to have. The biggest production of nets today is still from the orthodox vegetable fibres, and therefore the advantages of the "old-fashioned" material must not be overlooked. The great advantage of sisal or cotton over other fibres, is their lower cost, and, in an industry such as fishing, which is having economic difficulties, the lower the cost of the nets, the more attractive they are. The advantages of the man-made fibre will increase as their price decrease. Another characteristic of cctton nets is in the extensibility of the material. Herring constitutes a very big part of the fishing catch both for man and for animal feeding stuff, and there are many types of herring nets and herring gillnets. Tl man-made fibre herring gillnets are not entirely successfu , the fish becoming so deeply enmeshed, that th° catch can be extracted only with a great amount of labour and the fish often being damaged. Also in the trawler field the advantages of synthetic nets are not so apparent, because the physical conditions under which the net is operated do not permit it to last as long as their qualities might promise. The rock or the wreck on the bottom of the sea is capable of tearing and pulling loose an expensive synthetic net as easily as the one made of the less expensive natural fibre. It would be a mistake to think that there is no future for the orthodox vegetable fibre materials and those in authority and responsible for advising fishermen should inform them that good service can be obtained from the vegetable fibres as long as the nets are kept clean and maintenance is good.

Dr. J. Reuter (Holland). When a fisherman has to make his choice of material, the more figures he sees the less able he is to make his choice. This information about testing was started by scientists. Fishermen are not scientifically trained, so we should adopt test methods that are scientifically correct and at the same time be easily understood by the fishermen. Take for example the breaking strength of twine; every textile man has this as his main goal, because the stronger the twine the better it is, but he forgets that the fisherman tests his net wet by pulling on the mesh and not the straight twine. There are fibres that are strong in the dry condition but become weak when they are wet and knotted.

Messrs. G. A. Hayhurst and A. Robinson (U.K.) in a written statement: When checking twines for breaking strength and extension it is essential to avoid distorting them. If simply secured between two grips there may be a severe bend in the twine which, when subjected to the test force, will cause a stress concentration and give a most unreliable result. It is therefore recommended that quarter-circle bollards be used to lead the twine to the grip, as shown in the sketch.

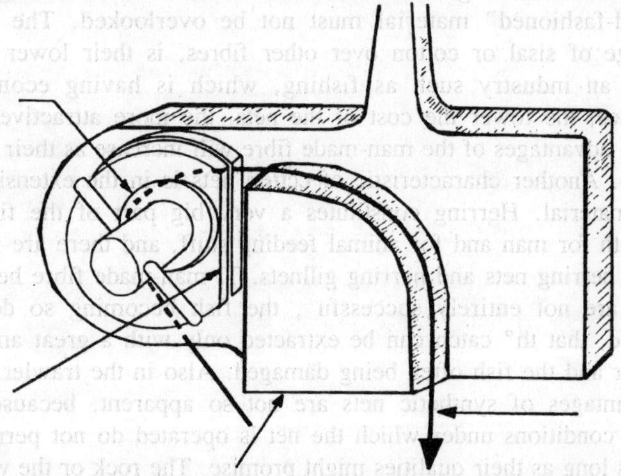

When checking extension under load, the test sample must be pre-tensioned to a slight but known extent. We always make such measurements at 1 per cent. of breaking stress.

Mr. E. A. Nilssen (Norway). Synthetic fibres were introduced about three years ago in Norway. I think that at present only synthetic fibres are used for the gillnets. Such nets bring in bigger catches, they are stronger, easier to handle, lighter, need not be dried and need no preservation. The fishermen in Norway use sisal and manila for the mounting, but they would also prefer synthetics in the mounting as they could then just leave the nets without any drying. We are now trying to find the right material for these mounting lines.

We have tried monofilament nets, but without much success. This is rather astonishing, when one thinks of Sweden, where they are using so many of these nets. As for the breaking strength, we use single-knot and double-knot nets. The double-knot nets were the best at the beginning because there was no knot slippage, but the single-knot nets have improved much in the last two years. The double-knot nets give somewhat higher strength. In Norway there is a State import monopoly for nets and raw materials; the Government also uses test machines. As most Norwegian manufacturers send the nets, the twine, etc. to this laboratory, tests are therefore uniform. The most important test is that of the knot strength. The fishermen normally pull the net in a wet and in a dry condition, we must find a twine that gives good results in both conditions. Synthetic fibre has also been tried for the big purse seines (about 180 fathoms long) in Norway. This has resulted in an increasing demand for the nets. As soon as the price of synthetic fibres comes down more or less to the level of cotton and hemp, it is possible that only synthetic fibres will be used.

Mr. R. Ocran (Ghana). In Ghana, we are trying to develop line fishing. We catch heavy fish and the breaking strength of cotton and hemp cord is not strong enough. How can synthetic fibre be applied to line fishing?

Dr. H. A. Thomas (U.K.). Some of the first trials with Courlene (a polyethylene material) were in fact with line fishing using a large number of snoods in the Irish Sea from Fleetwood, and remarkable catches were achieved. The fishermen spoke very highly of the knot strength at the point where the hooks were running out on the snood from the main line. Some purse seine nets of this yarn were then produced, in cooperation with the Fleetwood fishermen. It was found that the knot with the polyethylene yarn—Courlene □—did not have to be heat-sealed; if they were tied with the right technique, they obtained satisfactory results. We have a large number of individual reports on these fishing trials, which we shall be pleased to give in detail to anyone who is interested.

It has not been possible at this Conference to include a full-size paper dealing with materials suitable for protective clothing for fishermen. I would, however, suggest that perhaps at another Conference it might be possible to include one session on the clothing and protective equipment.

The last pages in Dr. Arzano's paper give some details of the use of man-made fibres in protective equipment. There is a development which started in Norway, namely the string vest. The object of this vest is to maintain a layer of air next to the skin to prevent clothing from becoming moist from perspiration, this also being a very good heat insulant and thereby adding to the warmth of the body. The older established man-made fibres (blend or a mixture of viscose staple and acetate staple) can be used for the string vest. To maintain the warmth of the feet it is of advantage to have an insole or interlining between the feet and the rubber boot or the gumboot. For gloves, blends of viscose staple with nylon staple and with wool seem to achieve the best results. Viscose staple and filament coated with polyvinylchloride or with linseed oil and rubber are used for such things as aprons, gumboots, overcoats and sou'westers.

Mr. A. Robinson (U.K.). We have some experience with longline fishing and particularly with snoods of nylon yarn. These have been found to be very successful indeed in various respects, not only as to durability but also for increasing the catch.

Where the speed of replacing snoods is of great importance, nylon snoods have been particularly advantageous. Very few snoods indeed have broken when the fish has been ripped from the hook, and very little time has been lost in repairs. The line itself is usually made of staple nylon in order that the snoods can be attached to it securely and do not slide on the line. The life of the staple nylon longline is very good but, of course, the price is high.

Mr. S. Springer (United States of America). In the Gulf of Mexico longline tuna fishery, we have been making use of synthetic lines-both mainlines and branch lines. Here, the advantage of the synthetic fibre is in its great strengthpossibly double that of the natural fibre lines. The nylon line that has been most successful in this fishery is a heat-stabilised line, which has less tendency to twist.

Mr. K. J. Westrop (U.K.). There is a need for standardisation of testing methods. Different methods lead to different actual figures, which do however serve usefully for purely relative testing of the fibres. For instance, the abrasing resistance described by Dr. Klust shows the superiority of the polyamide and the polyester fibres over the natural fibres, though just what these relative figures would mean when you are concerned with the actual nets, of course, is another story. As yarn producers, we can evaluate our yam as a textile material, but the fisherman himself must obviously be the man to decide what is required of a net. A practical trial, to be valid, must be made with many nets to cover the whole range of different types and the different conditions under which the nets are used.

Presumably something will come out of the Congress and these discussions as to what should be done about surveying such testing methods. I would like to add that the Imperial Chemical Industries could probably contribute to any future conference if a paper of this sort is required.

Dr. A. von Brandt (Germany). I would propose that, in addition to the Working Party on Terminology and Numbering Systems, we form a Working Party on Testing Methods, and that the following gentlemen be appointed as members: Mr. Reuter (Netherlands); Mr. Carrothers (Canada); Mr. Takayama (Japan); Mr. Percier (France). The standardisation of testing methods should be considered from the fisherman's point of view, and not from that of the textile industry.

Dr. G. Klust (Germany) Rapporteur. I should like to give a brief summary of what has been said during the discussion. Synthetic fibres have given good results from northern waters to the tropics. For instance, in the Belgian Congo, the inland fishermen almost exclusively fish with nylon nets. Several speakers stressed the extremely good results that had been obtained. with monofilament nets. Mr. Gundry defended natural fibres, especially cotton which still is being used extensively in the fishing industry, the lower price being a very important factor.

The necessity of uniformity in methods of testing net materials has been stressed. These test methods, it was pointed out, must be brought into relation with experience gained in practical fishing. A close relationship between the manufacturers of synthetic fibres and the fishing industry could give extraordinarily good results, because synthetic fibres could be produced in such a way as to meet the requirements of the consumers.

If the manufacturers know the requirements of the fishing industry they are in a better position to manufacture what is actually wanted.

PROPERTIES OF TWINES

Prof. von Brandt (Germany): The two working parties have met and started to deal with the problem of standardisation. Due to the little spare time left by the packed programme of the Congress, no definite results can be expected yet. The difficulties which such an attempt of standardisation meets have been discussed and are well known by everybody working in this field. The quality of twines for the fishery or particularly the standardisation of their denotation depends not only on' the basic material which is presently often camouflaged under a big number of trade names, and the manufacture of this material into fibres but also on the way how the fibres are processed to the twine. The final goal of the Working Party for Standardisation of the Numbering System dealing with these questions is to find out and establish a simple and unmistakable system for defining the twines used in fishery. This system is meant to enable the fishermen all over the world to know what really is offered to them and what will best suit their purpose when substituting one twine or material for another. The Working Group consists of fibre producers, twine makers and net makers. Mr. J. E. Lonsdale, U.K., has been appointed chairman. After having discussed the matter repeatedly also with many people at this Congress, the present situation is characterised as follows:

1. The presently used numbering systems are too complicated and too confusing for use in the fishery.
2. The numbering complications are caused mainly by the twine and net makers who often create own systems.
3. It is necessary for the fishing industry to use always a common numbering system at least in addition to the national or manufacturers' systems. It may in time replace those.
4. As basic units for this common system either Denier or Tex are suggested. The former is a non-decimal unit generally used in the textile industry and already introduced to a certain extent into the fishery. The latter one is a newer decimal system which is strongly recommended for international use.

The Working Party, having accepted English as working language, is undertaking to survey the numbering systems currently used throughout the world for fisheries purposes, devise methods of unification and make recommendatj#ns, In addition to the inquiries to be made to this effect, everybody who wishes to bring his view to the attention of this Working Party is requested to write to Mr. J. E. Lonsdale of the British Nylon Spinners, Ltd., Pontypool, England.

It is the intention of this Working Party, while endeavouring to bring order in the systems of counts and numbering, to take into account what has been customary in the fishing industry as to date, and the specific demands of this industry. The future count system, therefore, should at the same time indicate the main qualities important for the various fishing purposes. The following list of twine and net qualities has been set up and may be useful as a base for discussion:

(a) tensile strength
(b) knot strength
(c) knot slippage
(d) mesh size
(e) permanent elongation
(f) elastic extension
(g) softness
(h) resistance to abrasion
(i) weight
(j) water absorption
(k) resistance to rotting
(1) resistance to weathering
(m) resistance to chemicals

If applicable, these qualities should be tested in wet conditions. For nets the knot strength of the twine is of more importance than the unknotted tensile strength. The mesh size often has to be considered in regard to economic and biological problems as marketing and overfishing. The permanent elongation may be of significance for the proper mesh size. Elastic extension or elasticity may be very effective when dealing with shock loads. Softness comes in particularly with gillnets. Abrasion may be caused in different ways during handling or when towing a trawl net over the sea bed. The. weight is always considered as the weight in water not only for easy handling but also in regard to operation performance for instance in bottom trawls when a good and continuous contact with the bottom is wanted or in purse seining where the sinking speed can be of great importance. Water absorption has an obvious relation to these weight problems. Resistance to rotting is such a general problem in fisheries that no further comments are needed. Resistance to weathering is of particular importance in tropical areas with bright sunlight. The means of improving the insufficient resistance to weathering of certain synthetic materials have been discussed. They lead to the question of resistance to chemicals of which the different dying materials consist.

The proper selection of the optimal material for the fishing purpose in question is strongly affected by the mutual interference between these qualities. With a stronger material. for instance, the twine diameter can be reduced saving twine, and towing or current resistance. But, with thinner material there might be more abrasion or, in gillnetting, the fish might be damaged, or it might be inconvenient to handle. The choice, therefore, must be based on the main qualities required for the particular purpose. This working party has been limited to manufacturers of twine and webbing because they are closest to the practical fishermen.

It is the intention to submit the information compiled and the conclusions drawn in the form of a report to FAO next year. FAO will then distribute this report to all people interested in this matter. The collaboration which is hoped to be created hereby, and the comments and discussions resulting, are expected to finally result in terms and a system of numbering and defining twine and net qualities which is acceptable and most suitable for all concerned with the fishing industry.

To be comparable, the qualities listed above must, of course, not only have uniform units but must also be determined in a uniform way. The standardisation of the testing methods is the subject of the second working party having been established at the beginning of this Congress. These methods also must apply to the demands of the fishing industry rather than to the textile industry, which mainly is concerned with clothing, stockings, etc. In the short time available this working party has prepared the following short report.

"The working group was formed to find a common agreement on test methods which will be adapted to test material under similar conditions as it is used in practice. The group has appointed Prof. A. von Brandt as Chairman. The various testing methods were discussed and a preliminary division agreed upon:

(1) Test methods outside the scope of the working group, such as determination of the specific gravity of fibres, etc. which are not entirely adapted to fishing conditions but absolutely necessary for manufacture and research.

(2) Test methods which have already been accepted and standardised nationally or internationally, as for instance French elongation test etc. These test methods are also not entirely adapted to fishing conditions but absolutely necessary.

(3) Test methods of particular interest for fishery purposes on which a common agreement has still to be found-for instance, sinking speed of the net, visibility in the water, mesh knot and mesh breaking strength etc. The last group of test methods will be the main task of the working group. Prof. A. von Brandt will write to people interested in the subject who he

thinks can contribute to the unification of this subject. He will be glad to receive contributions from any participant to the present Congress. A full report will be written next year."

The working party has decided to compile the existing results of tests, discuss and evaluate them, after which a programme of work will be set up. Testing results often depend on the type of test carried out, and we must determine how we can work in agreement since so many people throughout the world are carrying out tests, there is need for some unity in this field. FAO has also emphasized to the different governments that they should help in this matter.

INDEX

V

W

X